A PLUME BOOK

THANK GOD FOR EVOLUTION

THE REVEREND MICHAEL DOWD is America's evolutionary evangelist. He has dedicated his life to proclaiming the "Great News" of a sacred view of cosmic, biological, and human evolution. During the 1980s and 1990s, Dowd pastored three United Church of Christ congregations, worked with conservative and liberal religious leaders across America on environmental, peace, and justice issues, and managed government-funded Sustainable Lifestyle Campaigns on both coasts. Since 2002, he has lived entirely on the road with his wife, Connie Barlow, an acclaimed science writer, giving sermons and presentations to numerous churches and secular organizations. Highly in demand as a speaker, Reverend Dowd is passionate about sharing the fourteen-billion-year history of everything and everyone in ways that offer both practical guidance and inspiration.

Celebrated by Nobel Laureates

"The science vs. religion debate is over. Michael Dowd masterfully unites rationality and spirituality in a worldview that celebrates the mysteries of existence and inspires each human being to achieve a higher purpose in life. A must read for all, including scientists."
—**CRAIG MELLO,** 2006 NOBEL PRIZE IN PHYSIOLOGY OR MEDICINE

"The universe took 13.7 billion years to produce this amazing book. I heartily recommend it. I am often asked how science and religion can coexist. This is a wonderful answer."
—**JOHN MATHER,** NASA SENIOR ASTROPHYSICIST, 2006 NOBEL PRIZE IN PHYSICS

"If anyone can persuade a monotheist that the science of evolution—biological, geological, or cosmological—can enrich his or her faith, I'm betting on Michael Dowd."
—**THOMAS C. SCHELLING,** 2005 NOBEL PRIZE IN ECONOMICS

"Honest students of God should welcome the revelations of science as insights, not fear them as threats. Here is a book in that spirit by an ardent believer who takes evolution to heart and celebrates it."
—**FRANK WILCZEK,** 2004 NOBEL PRIZE IN PHYSICS

"At last someone who understands that all of reality is sacred and science is our method of comprehending it."
—**LEE HARTWELL,** 2001 NOBEL PRIZE IN PHYSIOLOGY OR MEDICINE

Embraced by Religious, Scientific, and Cultural Leaders

Complete versions can be found at ThankGodforEvolution.com

"Michael's book—a sacred look at cosmic history and emergent creativity from multiple angles —will ignite a revolution in evolution for any reader. Here is a book that promises to deliver and delivers on its promises!"
—**REVEREND HOWARD CAESAR,** SENIOR MINISTER, UNITY CHURCH OF CHRISTIANITY, HOUSTON

"Dowd has given us a bridge across one of the major chasms of our times—religion and evolution. His passion for both science and religion is contagious. Reading his book, one can see that the discourse itself has just evolved to a whole new level!"
—**REVEREND MARLIN LAVANHAR,** SENIOR MINISTER, ALL SOULS UNITARIAN CHURCH, TULSA

"With the passion of a revival preacher and with grounding in mainstream evolution, Dowd has written a visionary book." —**REVEREND JIM BURKLO,** SAUSALITO PRESBYTERIAN CHURCH, BOARD OF THE CENTER FOR PROGRESSIVE CHRISTIANITY; AUTHOR OF *OPEN CHRISTIANITY*

"A gift to humanity and the Earth, especially at this critical point in human history."
—**MARY MANIN MORRISSEY,** PAST PRESIDENT, ASSOCIATION OF GLOBAL NEW THOUGHT

"Michael Dowd's new book should be made into a Hollywood film: An evangelical Christian preacher (Michael) and an atheist evolutionary naturalist (Connie Barlow) fall in love, marry each other, and give birth to their stunning new vision that promises healing for so many. If you love God, if you love the animals, if you love Jesus, if you love the flowers and Sun and Moon, here's the book that will help you gather all these loves together."
—**BRIAN SWIMME,** PROFESSOR OF COSMOLOGY, CALIFORNIA INSTITUTE OF INTEGRAL STUDIES, SAN FRANCISCO; COAUTHOR OF *THE UNIVERSE STORY*

"This book offers an enthusiastic encounter with evolutionary science in an evangelical idiom. It will change the minds and even the hearts of people who have been alienated by rhetoric that pits science against faith." —JOAN ROUGHGARDEN, PROFESSOR OF BIOLOGICAL SCIENCES, STANFORD UNIVERSITY; AUTHOR OF *EVOLUTION AND CHRISTIAN FAITH*

"Shows how the evolutionary history of the cosmos supports a deeply spiritual vision."
—JOHN B. COBB, JR., CENTER FOR PROCESS STUDIES; AUTHOR OF *CHRISTIAN FAITH, RELIGIOUS DIVERSITY*

"This is a fantastic book! A page-turner! You might as well buy ten copies because you won't be able to finish one without giving it away time and again! Be forewarned. Would I lie to you?"
—BRAD BLANTON, AUTHOR OF *RADICAL HONESTY* AND *THE TRUTHTELLERS*

"A thoughtful, timely, challenging—and readable—synthesis of science and spirituality in the spirit of Thomas Berry and Teilhard de Chardin." —JOHN F. HAUGHT, DISTINGUISHED PROFESSOR OF THEOLOGY, GEORGETOWN UNIVERSITY; AUTHOR OF *DEEPER THAN DARWIN*

"Dowd offers us an impassioned vision, forged during his encounters with countless seekers, of how the Great Story can enrich, and indeed make sense of, traditional religious orientations, while leaving plenty of room for readers not comfortable with God language. *Thank God for Evolution* is refreshing, honest, unpretentious, and deep."
—URSULA GOODENOUGH, PROFESSOR OF BIOLOGY, WASHINGTON UNIVERSITY; AUTHOR OF *THE SACRED DEPTHS OF NATURE*

"Read this book. Then read it again. Then share it with everyone you know. Dowd's comprehensive synthesis of evolutionary psychology, biology, and theology is a 21st-century operating manual for the consciously evolving human being. *Thank God for Evolution* is not only a brilliant articulation of the profound interconnections between science and spirituality; it is also a practical guidebook for living an evolutionary, radiant, scientifically informed and faith-foundationed life."
—HARRY PICKENS, PIANIST/COMPOSER, CEO, IN TUNE WITH LIFE™

An itinerant preacher who teaches evolution in the evangelical style? I was skeptical at first, but Dowd remains true to both science and the spirit of religion. He understands that what most people need to accept evolution is not more facts, but an appreciation of what evolution means for our value systems and everyday lives."
—DAVID SLOAN WILSON, DISTINGUISHED PROFESSOR, DEPARTMENTS OF BIOLOGY AND ANTHROPOLOGY, BINGHAMTON UNIVERSITY; AUTHOR OF *EVOLUTION FOR EVERYONE*

"*Thank God for Evolution* is a seminal work. It crosses the great divide between science and religion by offering a new, all-inclusive, science-based, spirit-infused way for us to move together as coevolutionary participants in the process of creation. It is a gift to all of us. I recommend it with all my heart." —BARBARA MARX HUBBARD, PRESIDENT, FOUNDATION FOR CONSCIOUS EVOLUTION; AUTHOR OF *EMERGENCE*

"Thank God for Michael Dowd and his passionate, purposeful integration of evolutionary science with revelation! This book is a tour de force and a welcome contribution to the growing literature that says that God speaks through evolution as clearly as through the events of our personal lives."
—VICKI ROBIN, COAUTHOR OF *YOUR MONEY OR YOUR LIFE*

"A voice of sanity and inspiration revealing a way not only toward a deeper Christianity but a deeper humanity." —**JOEL R. PRIMACK** AND **NANCY ELLEN ABRAMS,** AUTHORS OF *THE VIEW FROM THE CENTER OF THE UNIVERSE*

"*Thank God for Evolution* is clear, well argued, and convincing. Dowd offers an admirable and satisfying solution to the dead-end debates between theists and atheists. He gives even those of us who study evolution for a living a deep appreciation for how evolutionary theory can inspire our lives. His message is open to all faiths, including those who reject faith, and it is a message that is vital for the 21st century." —**RICHARD SOSIS,** PROFESSOR OF ANTHROPOLOGY, CONNECTICUT UNIVERSITY AND HEBREW UNIVERSITY OF JERUSALEM

"Michael Dowd is a pioneering voice for a deep spirituality that both celebrates the wisdom of science and is juicy with love, intimacy, and aliveness." —**DUANE ELGIN,** AUTHOR OF *PROMISE AHEAD* AND *VOLUNTARY SIMPLICITY*

"Michael Dowd's wonderful book inspires and provides realistic hope for the future. *Thank God for Evolution* is a guidebook for moving into a healthy future, individually, collectively, and globally." —**SENATOR LES IHARA, JR.,** MAJORITY POLICY LEADER, HAWAI'I STATE SENATE

"This is a handbook for conscious evolution. *Furry Li'l Mammal, Lizard Legacy, Monkey Mind*— these are just some of the engaging terms that the author has invented to help us understand that evolution is not something confined to museums of natural history, superseded by the invention of civilization by clever human beings, but something that is going on right now within every one of us." —**CAROLINE WEBB,** EPIC OF EVOLUTION CONTRIBUTOR

"Michael Dowd's unrestrained passionate call is as infectious as it is timely." —**ANDREW COHEN,** FOUNDER OF *WHAT IS ENLIGHTENMENT?* MAGAZINE

"No one brings evolution up closer and more personal than Michael Dowd. Convincingly making science itself a gift of God, Dowd shows us a meaningful, living universe. It's high time to know our true evolutionary potential as this book shows it and to engage in fulfilling our evolutionary destiny together!" —**ELISABET SAHTOURIS,** EVOLUTIONARY BIOLOGIST; AUTHOR OF *EARTHDANCE*

"A recovering fundamentalist reveals the profound spiritual power of our evolutionary story. A must for both sides of the debate." —**PETER RUSSELL,** AUTHOR OF *THE GLOBAL BRAIN* AND *FROM SCIENCE TO GOD*

"A stellar articulation of the emerging zeitgeist! I was only a software developer, but I am now an apostle for the Great Story. I will read this book over and over again." —**LION KIMBRO,** SOFTWARE DEVELOPER AND COMMUNITYWIKI CONTRIBUTOR

"Thank God for Michael Dowd! Finally a spiritual leader smart enough—and brave enough—to show America why biology and the Bible aren't mutually exclusive. This book will transform your life and worldview. Read it, read it to your children, take it to church and have your preacher read it from the pulpit. Dowd unlocks the secrets of the universe's most powerful pairing (science and spirituality) and helps you discover your own calling—God's purpose for your life—in the process." —**LISA EARLE MCLEOD,** AUTHOR AND SYNDICATED NEWSPAPER COLUMNIST

"A work of many brilliant and provocative insights. A timely contribution." —**DIARMUID O'MURCHU,** AUTHOR OF *QUANTUM THEOLOGY* AND *EVOLUTIONARY FAITH*

"According to Sir Francis Bacon, 'A little science estranges a man from God; a lot of science brings him back.' Now Michael Dowd has shown us how this is also true of evolution itself."

—**STEVE MCINTOSH**, AUTHOR OF *INTEGRAL CONSCIOUSNESS AND THE FUTURE OF EVOLUTION*

"Smart, provocative, and deeply personal; an inspiringly original vision of our origins!"

—**KEVIN W. MCCARTHY**, AUTHOR OF *THE ON-PURPOSE PERSON* AND *THE ON-PURPOSE BUSINESS*

"Dowd explores new terrain by revealing how evolution is relevant to ethics, community building, sexuality, spirituality, and a viable future for the Earth community."

—**HEATHER EATON**, FACULTY OF THEOLOGY, SAINT PAUL UNIVERSITY, OTTAWA

"Dowd puts more effort, intelligence, and heart into this project than anyone else I know!"

—**DAVID CHRISTIAN**, PROFESSOR OF HISTORY, SAN DIEGO STATE UNIVERSITY; AUTHOR OF *MAPS OF TIME: AN INTRODUCTION TO BIG HISTORY*

"Eloquently argues that the bitter conflict between Darwinism and religion is unnecessary. The story of evolution based on science enriches a spiritual approach to creation rather than detracting from it. This book shows how we can recover the timeless virtue of mutual respect, without anyone's having to sacrifice deeply held principles. Uplifting!"

—**PETER J. RICHERSON**, DISTINGUISHED PROFESSOR OF ENVIRONMENTAL SCIENCE, U.C. DAVIS

"Too full of unexpected and creative ideas to be missed. Michael's grand theme—that evolution is not only compatible with religious faith but is positively enriching, is exactly right, and is something that can't be said often enough."

—**PAUL WASON**, DIRECTOR OF LIFE SCIENCES, JOHN TEMPLETON FOUNDATION

"*Thank God for Evolution* sweeps us into a freshly minted universe in which both evolution and religion are transformed, enlivened, and blessed in ways they cannot be when warring against each other. They both come out shining—and profoundly relevant together, not only in the world we live in, but in the very different world that is just around the corner. Michael Dowd treats us to an alluring, prophetic glimpse of something very positive that is suddenly very possible. This is a truly original book at the near edge of a deeply hopeful future."

—**TOM ATLEE**, AUTHOR OF *THE TAO OF DEMOCRACY* AND *CO-INTELLIGENCE.ORG*

"A rare example of an easily readable book, steeped in religious language and religious interests, that celebrates evolution." —**TANER EDIS**, ASSOCIATE PROFESSOR OF PHYSICS, TRUMAN STATE UNIVERSITY; AUTHOR OF *SCIENCE AND NONBELIEF*

"Few subjects today arouse more passion than the culture war between traditional religion and evolutionary science. Dowd's timely book offers a refreshing third way and introduces us to a promising new field emerging on the edges of contemporary culture—evolutionary spirituality. Integrating breakthrough ideas from evolutionary psychology, neuroscience, physics, and evolutionary biology, Dowd makes a compelling case for why the spiritual and religious impulse is not erased but profoundly enriched by our ever-expanding knowledge of self, nature, and cosmos. With the enthusiasm of an evangelical and the open-mindedness of a scientist, he offers the reader a smorgasbord of fascinating new ideas and begins a discussion about the theological future of Christianity that is long overdue."

—**CARTER PHIPPS**, EXECUTIVE EDITOR, *WHAT IS ENLIGHTENMENT?* MAGAZINE

"Dowd transforms the sterile debate between creationism and evolutionism into the search for a deeper, more inspiring truth. A courageous and insightful work."

—**DAVID C. KORTEN**, BOARD CHAIR, *YES!* MAGAZINE; AUTHOR OF *THE GREAT TURNING*

"Writing with integrity, reverence, and passion, Michael Dowd recasts the evolutionary history of the universe as a sacred and ongoing process. With sound science and deeply spiritual theology, *Thank God for Evolution* offers an end to science-religion conflict and a new way of living as part of an integral, planetary, and evolutionary whole."

—KAREN WALSH WYMAN, DIRECTOR, NORTH AMERICAN SCIENCE AND RELIGION FOUNDATION

"Science is handing us an incredible, mind-blowing creation story. What do we do with this grand epic? How do we integrate it? What does it mean for who we are? *Thank God for Evolution* is a book we can tightly hold on to to help us through the 'shattering' and into a whole new world."

—JENNIFER MORGAN, AUTHOR OF *BORN WITH A BANG* TRILOGY OF CHILDREN'S BOOKS

"The revisioning of world religions to reflect a cosmic evolutionary perspective is part of a great evolutionary event on our planet—the emergence of conscious evolution. To read and to be moved by Michael Dowd's seminal contribution to that revisioning is to experience this event's leading edge. This book oozes evolutionary energy cover to cover."

—JOHN STEWART, AUTHOR OF *EVOLUTION'S ARROW*

"Michael Dowd describes two life-transforming conversions in his past: one to fundamentalist Christianity, the other to Darwinism. Many see these views as antithetical. Dowd shows us that religious and scientific worldviews are reconcilable. More important, he shows that our intellectual and spiritual sanity depends on how we reconcile them. *Thank God for Evolution* is a brilliant and captivating description of our sacred evolutionary epic, one that even the most loyal fundamentalist and staunchest atheist will find uplifting. It is informed by our best scientific lights, but it is also written from a personal commitment to faith. It is essential reading for anyone who thinks that a love of science is irreligious or that spirituality must remain naive."

—JOSEPH BULBULIA, PROFESSOR OF RELIGIOUS STUDIES, VICTORIA UNIVERSITY

"Every so often a book comes along with a message so fresh, so timely, and so encompassing that it seems it could actually change the world. This is just such a book—demonstrating that the scientifically revealed story of evolution is the most inspiring creation myth of all."

—CRAIG HAMILTON, NEW DIMENSIONS RADIO, AUTHOR, AND JOURNALIST

"Science and religion are the two sides to the same coin of divinity. I recommend this book to the skeptics on both sides."   —SESH VELAMOOR, TRUSTEE, FOUNDATION FOR THE FUTURE

"Michael Dowd's book, *Thank God for Evolution*, performs a crucial service for both religion and science. He shows that religion can embrace the scientific account of our origins as its own sacred story without losing any of its spiritual force. At the same time, science has nothing to fear from religion. I have nothing but praise for this book and the work that Michael and his wife, Connie Barlow, are doing as wandering preachers of the good news of the potential harmony of science and religion."   —WILLIAM IRONS, PROFESSOR OF ANTHROPOLOGY, NORTHWESTERN UNIVERSITY

"Faith seeking understanding means, these days, understanding science and its relationship with religion. Michael Dowd writes clearly and forcibly to enlighten the intelligent reader, skeptical of both creationism (and its alter ego, intelligent design) and scientism. Not 'either-or'—either religion or science—but 'both-and'; both science and religion in mutual respect. A timely, necessary book."

—REVEREND JOHN MAXWELL KERR, FORMER WARDEN, SOCIETY OF ORDAINED SCIENTISTS

"The evolutionary integrity practices, especially those in Chapter 11, are vitally important."
—PATRICIA GORDON, PROFESSOR OF ENGLISH, JOHN ABBOTT COLLEGE, MONTREAL

"This is a much-needed book, even a holy book, a scripture for a spiritual renewal available to all religions as well as people living outside organized religion. Michael writes with verve and courage. He is a true evangel, bringing a message of wholeness to everyone. His arguments for recognizing the inherent relationship of religion and science as the yin-yang of the human experience are compelling."
—BILL BRUEHL, PLAYWRIGHT

"Everyone concerned with the deepest questions of science, religion, who we are, and our place in the universe absolutely should read this book!"
—JACK SEMURA, PROFESSOR OF PHYSICS, PORTLAND STATE UNIVERSITY

"*Thank God for Evolution* is a most heartfelt love letter to the human race I have ever read. It is the ultimate guide towards the most profound psychological integration in human history—the confluence of the truly cosmic and the truly personal. Upon this reading one is invited into the timeless technicolor of the Great Story, a story that lovingly belongs to and holds all of us. It is the Alpha and the Omega brought up to date."
—CURT SPEAR, CLINICAL PSYCHOLOGIST

"Dowd does a remarkable job of integrating mainstream evolution with an authentic religious perspective."
—MICHAEL STRONG, CEO AND CHIEF VISIONARY OFFICER, FLOW, INC.

"*Thank God for Evolution* offers a powerful foundation for a new and inspiring cultural narrative. As it heals the divide between science and religion, it invites us to consider the potential for resolving other tensions in our current collective story. Imagine growing our capacity to transcend the rifts between conservative and progressive, black and white, rich and poor, and other divides by revealing a larger vision that encompasses difference. This great story, told by a gifted storyteller, offers that great possibility. What a blessing!"
—PEGGY HOLMAN, AUTHOR OF *THE CHANGE HANDBOOK*

"I'm pleased to see more and more people address the patterns of human emergence. Michael has made a significant contribution to understanding this critical aspect of human nature."
—DON BECK, NATIONAL VALUES CENTER, CEO OF SPIRAL DYNAMICS GROUP

"Something magical happened to roads in the 1920s—road maps. All of a sudden, short roads from town to town were shown as highways stretching across the nation. Road maps changed the paradigm of travel without adding a single mile of new roads. This book does the same thing for the cutting-edge sciences, especially evolution, and spirituality. It connects them all in the same way. It joins head and heart, and grounds wonder and mystery in daily experience. Approach this book as if you were 8 years old, early Christmas morning, the presents are under the tree, and you know something wonderful is imminent."
—MICHAEL PATTERSON, AUTHOR AND COMMUNITY ORGANIZER

"*Thank God for Evolution* will transform your life from a confusing, directionless struggle into an invigorating, blissful challenge filled with meaning, purpose, and rational understanding. If you miss this chance to learn about the worldview that will bring religious peace, secular cooperation, and a truly scientific age to our Earth, then you will learn about it from your children and grandchildren in the coming religious revival."
—JON CLELAND-HOST, MATERIALS SCIENTIST AND UNITARIAN UNIVERSALIST

Thank GOD for EVOLUTION

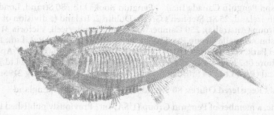

How the Marriage of Science and Religion
Will Transform Your Life and Our World

Michael Dowd

A PLUME BOOK

PLUME
Published by the Penguin Group
Penguin Group (USA) Inc., 375 Hudson Street, New York, New York 10014, U.S.A. • Penguin
Group (Canada), 90 Eglinton Avenue East, Suite 700, Toronto, Ontario, Canada M4P 2Y3
(a division of Pearson Penguin Canada Inc.) • Penguin Books Ltd., 80 Strand, London WC2R 0RL,
England • Penguin Ireland, 25 St. Stephen's Green, Dublin 2, Ireland (a division of Penguin Books
Ltd.) • Penguin Group (Australia), 250 Camberwell Road, Camberwell, Victoria 3124, Australia
(a division of Pearson Australia Group Pty. Ltd.) • Penguin Books India Pvt. Ltd., 11 Community
Centre, Panchsheel Park, New Delhi – 110 017, India • Penguin Group (NZ), 67 Apollo Drive,
Rosedale, North Shore 0632, New Zealand (a division of Pearson New Zealand Ltd.) • Penguin
Books (South Africa) (Pty.) Ltd., 24 Sturdee Avenue, Rosebank, Johannesburg 2196, South Africa

Penguin Books Ltd., Registered Offices: 80 Strand, London WC2R 0RL, England

Published by Plume, a member of Penguin Group (USA) Inc. Previously published in a Viking
edition.

First Plume Printing, May 2009
20 19 18 17 16 15 14 13 12

Copyright © Michael Dowd, 2007
All rights reserved

Pages 413–14 constitute an extension of this copyright page.

 REGISTERED TRADEMARK—MARCA REGISTRADA

ISBN 978-0-670-02045-4 (hc.)
ISBN 978-0-452-29534-6 (pbk.)

Printed in the United States of America

*I dedicate this book to the glory of God.**

*Not any "God" we may think about, speak about, believe in,
or deny, but the one true God we all know and experience.

Contents

Preface   xxi

Author's Promises   xxv

Prologue: Personal Journey   1
    The Marriage of Science and Religion   3
    Itinerant Evolutionary Evangelism   4

Introduction   7
    From "Adam and Eve" to Us—and Beyond   8
    Science and Religion Spurring Each Other to Greatness   10
    In Context   12
    Overview   13

PART I. THE HOLY TRAJECTORY OF EVOLUTION

Chapter 1. Our Big Picture Understanding of Reality   21
    Stories Within Stories   22
    What Is the Great Story?   24
    From Shape-Shifting Story to Unchanging Scripture   27
    Meaning Making   28

Chapter 2. Evolution Is *Not* Meaningless Blind Chance   31
    Interpreting Our Immense Journey   32
    The Mythopoeic Drive   34
    Time's Arrow, Time's Cycle   36
    The Role of Strife   40
    The Role of Cooperation   42
    The Role of Initiative   45

Chapter 3. Evolution and the Revival of the Human Spirit   48
    The Universe Can Be Trusted   49
    You Are Part of the Universe   56

Accept What Is and Be in Integrity   57
Grow in Evolutionary Integrity   59
Trusting the Universe Means Welcoming Challenges   61

PART II. REALITY IS SPEAKING

Chapter 4. Private and Public Revelation   65
  Beyond Belief   67
  The Birth and Maturation of Public Revelation   69
  Flat-Earth Faith Versus Evolutionary Faith   72
  Toward an Evolutionary Christianity   75
  Facts Are God's Native Tongue   77
  Religious Knowers   81

Chapter 5. The Nested Emergent Nature of Divine Creativity   84
  Thank God for the Hubble Telescope!   86
  We Are Made of Stardust   89
  The Gifts of Death   93
  Litany: "The Gifts of Death"   98

Chapter 6. Words Create Worlds   103
  Experiencing God Versus Thinking About God   105
  The Split Between Religion and Science   108
  From Clockwork "It" to Creative "Thou"   109
  Day and Night Language   113

Chapter 7. What Do We Mean by the Word "God"?   118
  No Less Than a Holy Name for the Whole   121
  Prayer in a Nestedly Creative Cosmos   123
  A Personal, Undeniable God   125
  God or the Universe: What's in a Name?   126
  Createism   128
  The Role of Humanity in an Evolving Universe   132
  Being "Faithful to God"   134

PART III. THE GOSPEL ACCORDING TO EVOLUTION

Chapter 8. Growing an Evolutionary Faith   139
  Genesis in Context   141
  Don't Throw Out the Apple   144

Chapter 9. REALizing "The Fall" and "Original Sin"   146
  Lessons from Evolutionary Brain Science   147
  Lessons from Evolutionary Psychology   154
  Resurrecting "The Fall"   158
  Your Brain's Creation Story   162
  Reclaiming "Original Sin"   166

Chapter 10. REALizing "Personal Salvation"   169
  The Challenges of Our Lizard Legacy   171
  Furry Li'l Mammal to the Rescue   174
  Thank God for Our Higher Porpoise!   176
  Discerning Your Calling Exercise   178
  Salvation Through Evolutionary Integrity   181
  The DNA of Deep Integrity   185
  Christ-like Evolutionary Integrity   191
  REALizing "Saving Faith"   195
  REALizing "the Gospel"   200

PART IV. EVOLUTIONARY SPIRITUALITY

Chapter 11. Evolutionary Integrity Practices   209
  Taming Our Monkey Mind   211
  Taming Our Lizard Legacy   214
  Growing in Trust: Nurturing Humility and Faith   215
  REALizing "Love Your Enemies"   217
  Growing in Authenticity: REALizing "Remove the Plank"   221
  Growing in Compassion/Responsibility: REALizing "Judge Not"   223
  Growing in Gratitude: REALizing "Love God and Your Neighbor"   226

Chapter 12. Evolving Our Most Intimate Relationships   231
  Touch and Tenderness   231
  Respectful Communication   233
  Playfulness and Humor   236
  Meaningful Songs and Rituals   237
  Synergy and Service   239

Chapter 13. Transformed by the Renewal of Your Mind   241
  Deep Integrity Affirmations   241
  Imagination Matters! Upgrading Your Mental Software   243

PART V. A "GOD-GLORIFYING" FUTURE

Chapter 14. Collective Sin and Salvation   251
   Wrongdoing in a Nestedly Emergent Universe   252
   Collective Sin in an Age of Information   255
   Confronting Institutional Sin   259

Chapter 15. The Wisdom of Life's Collective Intelligence   261
   On Earth As It Is in Heaven   262
   Collective Deep Integrity   266
   Conversation and Creative Emergence   267
   Co-Intelligent Social Technologies   269
   The Core Commons   270
   Cultivating Discernment Within the Whole   273
   Co-creating Our Evolutionary Spiritualities   274

Chapter 16. Knowing the Past Reveals Our Way Forward   277
   The Cosmic Century Timeline   277
   Aligning Self-Interest with the Well-being of the Whole   283
   Who and What Are We, Really? And Why Are We Here?   288
   Our Sense of Self and Our Role in the Body of Life   290
   Evolutionary Revivals   294

Chapter 17. Beyond Sustainability: An Inspiring Vision   298
   Major Challenges in the Next 250 Years   300
   Wild Cards   302
   Long-term and Short-term Positive Trends   307
   Likely Good News in the Next 250 Years   310

Chapter 18. Our Evolving Understanding of "God's Will"   318
   Responding to Critics Who Reject Religion Because of Scripture 319
   Transcending Biblical Values and Scriptural Morality   325
   REALizing "Holy Scripture" and "Divine Revelation"   327
   Public Revelation: "The Ever-Renewing Testament"   330
   REALizing Godly Morality and Ethics   333
   Wider Circles of Care, Compassion, and Commitment   334
   REALizing "Jesus as God's Way, Truth, and Life"   336

Conclusion   341

Epilogue 343
   Testimonial 344
   Vision 346

Appendix A. "Good and Bad Reasons for Believing"
     by Richard Dawkins 349

Appendix B. REALizing the Miraculous 357
   Miracles Through the Ages 358
   REALizing "the Virgin Birth" 362
   REALizing "Christ's Resurrection and Ascension into Heaven" 364

Invitation 369

Acknowledgments 371

Resources 375

Online Resources 380

Who's Who and Sources of Quotations 381

Index 391

Highlighted Stories

"The More Awesome My God Becomes"   11

Making Meaning of the Great Tsunami of 2004   22

"It's Hard to Be a Great Story Fundamentalist!"   26

"Is a Cheetah My Cousin?"   33

"He Said It Changed His Life"   35

"My Life Is Not Just My Own!"   44

What Is Conscious Evolution?   47

"To Connect Religiously with This Awesome Monument"   66

Evolution: Theory and Fact   78

Better Than a Smoking Gun   82

"That's Where Baby Stars Are Born!"   87

"Is This Purple Bead Darwin?"   88

"What a Mind Bender, Dude!"   92

"It Made That Feeling Go Away"   92

"Death—Don't Blame God!"   96

"I Learned That My Grandmother Will Die"   98

"I Am at Peace with His Death"   100

"What Does God Look Like on the Inside?"   107

"Two Gods?"   111

"I'm Here, Too"   115

"Read Me a Nighttime Book, Mommy"   117

"Finally, a God That Makes Sense!"   122

"What Does Jasmine Want?"   127

"Praise God, Brother, So Am I!"   132

Original Sin and the New Cosmology   145

The Parable of the Pickle Jar   163

"I Don't Know That Guy!"   170

"Do You Want Them to Gaze at Your Belly?"   173

"Using the Sexual Impulse to Evolve"   177

"My Life Purpose"   180

"What the Hell Are We Preaching?!"   183

STAR Clusters   187

Growing in Deep Integrity   189

"I've Never Told Anyone"   194

"Why Do We Think Differently About God?"   198

"The Joy of Watching Young and Old Alike Light Up"   203

"I Don't Merely Believe…I Know!"   204

REALizing "the Centrality of the Cross"   210

"Who Wants to Be Filled with the Holy Ghost?"   213

"Magic in Any Relationship"   225

"That's Your Cosmic Task!"   229

"Can We Have a Heart-to-Heart Talk?"   235

"Where's My Avocado?"   238

Atoning for Collective Sin Through "Pleistocene Rewilding"   258

"When I Repent…I Dwell in the Kingdom of Heaven"   263

Citizen Assemblies and Citizen Juries   268

"How Do You Measure Sustainable Progress?"   272

"We're Acting Like Cancer Cells"   291

"Such Hope!"   307

"Mixed Moral Messages"   326

Morality One-Liners   327

From Born Again Believer to Born Again Knower   361

Preface

As we observe the 200th anniversary of the birth of Charles Darwin and the 150th anniversary of his landmark book, *On the Origin of Species*, evolution has become firmly established as the central organizing principle of the biological sciences. Natural explanations for the growth of complexity through time ground all the other sciences, as well, from cosmology and chemistry to neuroscience and psychology. That *everything* within this universe has emerged through natural processes operating over vast spans of time is now well beyond dispute among scientists and the educated public. Yet even today, families and public school systems remain divided and the evolutionary worldview is still shunned by millions, perhaps billions, of religious believers around the world. Why?

One reason is surely that big changes in thought and perspective take time to be assimilated. A deeper reason is that humans do not live by truth alone. We require the sustenance of meaning—of beauty, goodness, relationship, and purpose. We require comfort in times of sorrow and suffering. We also require perspectives that encourage us to cooperate in ever-wider circles in order to solve ever-larger problems—problems that today encircle the globe.

So long as the scientific worldview is presented in ways that ignore these basic human values—values that religions excel in providing—there is little hope that the devoutly religious will appreciate science for anything more than its technological fruits. The good news is that the coming decades will see each of our religious, ethnic, and cultural stories embraced within a larger sacred context.

The scientific history of cosmos, Earth, life, and humanity is our shared sacred story—our common creation myth. It is an epic tale that reaches back billions of years and crowns each and every one of us as heir to a magnificent and proud lineage. This Great Story is open to improvement, as the revelations of science yield new insights, offer new ways of seeing, and alert us to misperceptions. It is open to change, too, whenever more helpful and inspiring interpretations of the facts become available. All this is possible, moreover, without scientists needing to fear that religious interpretations will skew or shade the truth. Nor must religious peoples join the ranks of atheists.

In public lectures that distill the contents of this book, time and again I have seen faces light up when I explain the distinction between *private revelation* and *public revelation* and when I advocate the importance of both *day language* and *night language*. Both pairs help us value the contributions of objective science without dismissing the subjective realms—artistic, emotive, and spiritual—that served our ancestors for thousands of years and still vitally serve us today. During seven years of itinerant evolutionary evangelism, I have watched young and old alike delight in the astonishing fact that we are made of stardust—that the calcium in our bones, the iron in our blood, and other atoms of our bodies were forged inside ancestor stars that lived and died before our Sun was born. I have seen, too, this naturalized and cosmic understanding of death comfort those whose grief would not otherwise be consoled.

Scaling down to the inner realm, I have witnessed tearful testimonials from those freed from years of guilt, shame, or resentment after learning our brain's creation story—that is, how the brain, with its embedded instincts, reflects an evolutionary trajectory from reptilian ancestors to early mammals, primates, and hominids. Others are grateful for the practical tools for improving lives and relationships that an evolutionary understanding of human nature affords. Still others have found that the supernatural claims that linger in the creeds and liturgies need not drive them from cherished traditions of their faith.

Sanity, health, and joy each emerge and are sustained only in right relationship with reality. *Thank God for Evolution* is thus a call to integrity, to wholeness, to sustainability—individually and collectively. In the year since its publication, events have validated and expanded the understanding of deep integrity outlined herein. From sex scandals in politics to crimes of greed on Wall Street, the underbelly of modernity and postmodernity is now vividly apparent. Thanks to discoveries in evolutionary psychology and evolutionary brain science, however, we can begin to improve institutions so that vital social structures can thrive despite human foibles. Equally, we can look to a future in which religious worldviews are free of the fundamentalism that fuels extremism.

How was the world made? Why do earthquakes, tornados, and other bad things happen? Why must we die? And why do different peoples answer these questions in different ways? The big questions that children have always asked and will continue to ask cannot be answered by the powers of human perception alone. Ancient cultures gave so-called supernatural answers to these questions, but those answers were not truly supernatural—they were prenatural. Prior to advances in technology and scientific ways of testing truth claims, factual answers were simply unavailable. It was not just difficult to understand infection before microscopes brought bacteria into focus; it was impossible. Without an evolutionary worldview, it is similarly impossible to understand ourselves, our world, and what is required for humanity to survive. For religious leaders today to rely on prenatural answers puts them at odds not only with science but with one another—dangerously so. Their resistance, however, does make sense. Until scientific discoveries are fleshed into the life-giving forms of beauty and goodness (as well as truth and utility), scriptural literalism will command power and influence.

A meaningful view of evolution is good news for individuals and families, and also for communities, nations, and our world. It is good news at these larger levels because a sacred, deep-time understanding of history and our evolutionary heritage is the very foundation needed for facing global challenges of our own making. It will encourage us to act, moreover, with compassion and inspired dedication. I offer this book and its stories of awakening toward this noble and necessary end.

Author's Promises

This book is intended for the broadest of audiences. The ideas have been evolving within and beyond me during six years of living and working entirely on the road, as a once traditionally religious and now exuberantly born-again evolutionary evangelist. From gothic cathedrals to cozy livingrooms, from gatherings of evangelical students to meetings of campus freethinkers, from university departments of religion and the social sciences to high school classrooms and homeschooling events, from rousing praise worship to quiet prayer circles, from local talk radio to National Public Radio: in all these venues and more, I have found diverse peoples hungering for the ideas you will encounter here. No matter who you are, and no matter what your beliefs or background, I promise that reading this book will expand the horizon of what you see as possible for yourself, for your relationships, and for our world.

To those of you who have rejected evolution ...

I promise that the secular version of evolution you have rejected is *not* the version of evolution presented in these pages. Indeed, if the understanding of our collective past and the vision of our common destiny outlined here do not inspire you to be more faithful in all your relationships, to find new ways to bless others and the world, and to awaken eagerly each morning to a life filled with meaning and purpose, then please continue to reject evolution!

To those who accept evolution begrudgingly (like death and taxes) ...

I promise that this book will provide you with an experience of science, and evolution specifically, that will fire your imagination, touch your heart, and lead you to a place of deep gratitude, awe, and reverence. You will also find here effective ways to talk about evolution to any friends, family, co-workers, and neighbors who are biblical literalists or young earth creationists.

To devoutly committed Christians . . .
Whether you are Roman Catholic, Orthodox, Protestant, Evangelical, Ana-
baptist, or New Thought, and whether you consider yourself conservative,
moderate, or liberal, my promise to you is that the sacred evolutionary
perspective offered here will enrich your faith and inspire you in ways that
believers in the past could only dream of.

To Jews, Muslims, Buddhists, Hindus, and other non-Christians . . .
I promise that it will be easy to apply most of what you find here to your own
life and faith. I also promise that if you explore the meaning of your tradition's
insights within an evolutionary context, as I attempt to do with Christian doctrine, you will provide an invaluable service to your religion and our world.

To agnostics, humanists, atheists, and freethinkers . . .
I promise that you will find nothing here that you cannot wholeheartedly
embrace as being grounded in a rationally sound, mainstream scientific understanding of the Universe. I also promise that the vision of "evolutionary spirituality" presented here will benefit you and your loved ones without your
needing to believe in anything otherworldly.

To those who embrace an eclectic spirituality . . .
I promise that this perspective will enrich your appreciation of the traditions
and practices that nourish you most deeply, while helping you find new excitement in each. It will also help you communicate and relate to others who hold
very different religious or philosophical worldviews.

To those who aren't really sure what they believe . . .
I promise that this holy evolutionary understanding will not only help you
make sense of the world; it will also provide a rock-solid moral and ethical
foundation for a life of passion and deep meaning in the midst of inevitable
difficulties.

To those who struggle with addiction or codependence . . .
I promise that if you say "Yes!" to the path of evolutionary integrity offered in
this book, you will gain a profound understanding of yourself and others. The

framework presented here is fully compatible with 12-step and other recovery programs. In coming to appreciate the deep roots of human instincts, you will see new possibilities for living the life of your dreams, while benefiting others, and you will experience a freedom and peace that "passes all understanding."

Finally...

To those with loved ones who have been unable to embrace science because of their religious faith, and those with loved ones who have been unable to embrace religion because of their scientific worldview, I promise that sharing this book will make a difference in your relationship. Discussing *Thank God for Evolution* with those you care about will open new doors of possibility between you and provide common ground where none existed before. This book is a perfect gift, not to convert others to your way of thinking but to converse with them deeply and heartfully about those things that matter most.

Request for Feedback

In the course of reading this book, if anything opens up for you—if you were helped or inspired in some way, or if you found something particularly meaningful—I would love to hear from you. Any and all suggestions for improvement are also welcomed. If you feel one of my promises was not kept, or was overstated, please tell me about that, too.

Thank you!

—Michael Dowd

Feedback@ThankGodforEvolution.com

Prologue:
Personal Journey

"Satan obviously has a foothold in this school!" I told my roommate twenty-five years ago at Evangel University. Moments earlier, I had stormed out of freshman biology class after the teacher held up the textbook we were going to use, and I recognized it as one that taught evolution. How else could I explain why a Bible-believing, Assemblies of God institution would teach evolution?

A little background...

I grew up Roman Catholic. As a teenager—like so many of my peers during the 1970s—I struggled with alcohol, drugs, and sexuality. In 1979, while in Berlin, Germany, and serving in the U.S. Army, I was "born again." Six months later I experienced what Pentecostals call "baptism in the Holy Spirit," evidenced by speaking in tongues. For the next three years, the people I fellowshipped with, the books I read, the television programs I watched, and the music I listened to all reflected a fundamentalist perspective strongly opposed to evolution.

I was taught that evolution was of the devil and would seduce people away from godly thinking and living. I believed Darwinism was the root of most social problems, and I was deeply concerned for my friends and family—especially those caught in the snares of a secular humanistic worldview. I even distributed anti-evolution tracts and was eager to debate anyone who thought the world was more than six thousand years old. So how was I to make sense of the fact, as I soon discovered at Evangel, that virtually all evangelical colleges and universities teach evolution?

The shift occurred in three steps. First, I came to know and trust several students and teachers before learning that they held evolutionary worldviews. Having already conversed, prayed, sung, and worshipped with each, I couldn't write any of them off as demonically possessed. The second influence was the biblical studies and philosophy courses I took at Evangel. Both the content and the professors reinforced the idea that "all truth is God's truth." The final element in my transformation was a budding friendship with a Roman Catholic hospital chaplain and former Trappist monk, Tobias Meeker. Before I discovered that Toby considered himself a "Buddhist-Christian," and that he embraced a process theology understanding of evolution, I had already assessed that he was the most "Christ-like" man I had ever met.

The past two and a half decades have been an amazing journey. After completing my undergraduate work at Evangel (double majoring in biblical studies and philosophy), I went on to earn a Master of Divinity degree at Eastern Baptist Theological Seminary. Although I learned to accept evolution at Evangel, I did so only with my mind—not my heart. That final shift happened suddenly, in February 1988. I was in Boston for the first session of a course titled "The New Catholic Mysticism," taught by cultural therapist Albert LaChance. Albert began by telling the scientific story of the Universe in a way that I had never heard it told before—as a sacred epic. Less than an hour into the evening, I began to weep. I *knew* I would spend the rest of my life sharing this perspective as great news. My evangelizing began shortly thereafter as an avocation wedged into the rest of my life. Even so, virtually everything I've preached and written since that epiphany has been in service of a religiously inspiring understanding of evolution, such that others, too, might experience our common creation story as gospel and be inspired to love and serve accordingly.

By no longer opposing evolution, but wholeheartedly embracing it as the "Great Story" of 14 billion years of divine grace and creativity, I now have a more intimate relationship with God than ever before. Throughout this book, I will be sharing how and why this is the case, and I will do so in ways that non-Christians and nonreligious people can also celebrate.

The Marriage of Science and Religion

Over the course of ten years, I pastored three United Church of Christ congregations—one in New England and two in the Midwest—before shifting careers into interfaith sustainability work and community organizing. In the spring of 2000, I attended a Pentecostal/Charismatic worship service near my childhood home of Poughkeepsie, New York. I've always loved the energy and enthusiasm of "Spirit-filled" worship. At a moment when the congregation was swept up in ecstatic praise, the woman who had invited me turned and grasped my hands. "I have a word from God for you," she declared. "Great!" I replied. She continued, "Thus sayeth the Lord, 'My son, I have called thee home to reveal thy true mission. Step out boldly with thy beloved and fear not. For I will bless thy steps and thy ministry more abundantly than thou canst imagine.'"

Several thoughts raced through my mind. The first: "I'm ready!" Then, "I wonder why God likes Elizabethan English so much?" Finally, "Whoa boy, did you hear that? God said, 'with your beloved.' You'd better get moving, dude. You don't even have a girlfriend!"

Several months later my friend's prophetic words were made flesh. I met science writer Connie Barlow at a lecture given by cosmologist Brian Swimme at Auburn Theological Seminary in New York City. Connie was the author of four books, and two of them had "evolution" in their titles (*Evolution Extended: Biological Debates on the Meaning of Life* and *The Ghosts of Evolution*). She, too, was a long-time "epic of evolution" enthusiast. What is more, her passion for sharing a sacred understanding of cosmic history was no less than mine. Seven months later I asked Connie to marry me. Three weeks after that, we were wed at the EarthSpirit Rising Conference on Ecology, Spirituality, and the Great Work, which was held in Louisville, Kentucky, in June 2001. Surely this was a marriage of science and religion. Connie was a self-described atheist, and her professional life was steeped in the sciences. My life was devoted to religion. Our union embraces both.

Three months later, the World Trade Center was attacked. We were living north of New York City, and Connie had a scheduled meeting in Tower No. 1 the very next day. The collapse of the towers forced us to reevaluate our priorities. A month later, we were watching the final installment of the PBS television special *Evolution: A Journey into Where We're From and Where We're Going*. That episode was titled "What About God?" It examined the struggle that conservative Christian college students face in trying to embrace both evolution and a pre-evolutionary interpretation of their faith. As the program ended, Connie turned to me and said, "You need to be out there speaking to those students. You need to show how an evolutionary understanding can enrich one's faith!"

Connie and I were still newlyweds. I had no idea she was prepared to follow through—personally—on her declaration. A few weeks later, after a frustrating day at work, I told her (not really serious, just sort of whining), "You know, I wish we could just travel nonstop, teaching and preaching the Great Story wherever we go." Her response was astounding. Looking me in the eyes, she said with utter conviction, "I'd love to do that!"

Itinerant Evolutionary Evangelism

Since April 2002, Connie and I have been full-time "evolutionary evangelists." We live permanently on the road, offering a spiritually nourishing view of evolution throughout North America. In the tradition of traveling preachers, we gave up our worldly possessions, left our home, and now carry everything we need in our van. We go wherever we are invited. Our goal is to inspire people of all ages and theological orientations to embrace the history of everyone and everything in personally and socially transforming ways.

We offer a view of our collective evolutionary journey that fires the imagination, touches the heart, and leaves people wanting more. We keep our distance from the polarized science versus religion conflict that festers in our society, particularly with respect to public school education. In the few hours or days that we engage with any given group, we present only the most compelling and alluring features of what many call "the epic of evolution" or "the Great Story." As with other leaders in this movement, we believe that the 14-billion-year story of cosmic, Earth, life, and cultural history can enrich any and all of humanity's cherished creation stories and religious paths.

In our first six years on the road, we have delivered Sunday sermons, evening programs, and multiday workshops in more than five hundred churches, convents, monasteries, and spiritual centers across the continent, including liberal and conservative Roman Catholic, Protestant, Evangelical, Unitarian Universalist, Unity, Religious Science, Quaker, Mennonite, and Buddhist groups. We have also presented audience-appropriate versions of this message in nearly a hundred secular settings, including colleges, high schools, grade schools, nature centers, and public libraries.

When we launched our ministry, we chose to display on our van both a Jesus fish and a Darwin fish—kissing. Many passersby flash a smile when they see it, although disapproving responses are not uncommon. A retired biology professor in Lawrence, Kansas, took one look at the decals and laughed, "Oh great! Now you piss everyone off!"

What Connie and I do on the road is serious, but it is best served by our maintaining a light-hearted approach. Our fishy pairing of what many regard as oppositional was thus a playful reminder to ourselves of who we wish to be along our shared journey.

Life on the road is far from a hardship. Connie and I have no home base in the usual sense, but North America as a whole feels like home to us. We are blessed to experience the stunning beauty of this vast continent. More, we rarely stay in public lodgings. Instead, we are invited into people's homes for a few days or perhaps a week at a time—and this, too, nurtures our souls.

Connie and I love being part of what is now a fast-growing movement that unites people across the theological and philosophical spectrum. Throughout this book, you will find a wealth of quotations from others who, like us, hold a sacred view of evolution. I will also share personal stories gleaned from our experiences on the road. These stories include evolutionary epiphanies—when people suddenly see the meaning of their lives in a larger context.

A dozen years before Connie and I met, cosmologist Brian Swimme issued a proclamation that we are now privileged to live: "We are in the midst of a revelatory experience of the Universe that must be compared in its magnitude with those of the great religious revelations. And we need only wander about telling this Great Story to ignite a transformation of humanity."

Amen!

Introduction

Many conservative Christians reject evolution. I commend them for their resistance. It compels those of us who do embrace evolution to find ever more sacred ways of communicating our conviction. Religious believers can hardly be expected to embrace evolution if the only version they've been exposed to portrays the processes at work as merely competitive and pointless, even cruel, and thus godless. Is it any wonder that many on the conservative side of the theological spectrum find such a view repulsive, and that many on the liberal side accept evolution begrudgingly?

Only when the evolutionary history of the Universe is articulated in a way that conservative religious believers feel in their bones is holy, and in a way that liberal believers are passionately proud of, will evolution be widely and wholeheartedly embraced. Fortunately, that time is now—not 2,000 years ago, not 200 years ago, and not even 20 years ago. Now is when we are awakening to the reality that God did not stop communicating truth vital to human well-being back when scripture was still recorded on animal skins and preserved for posterity in clay pots.

There is nothing for the religious to fear in this turn of events. A sacred view of science reveals that our faith traditions are more meaningful and grounded in undeniable reality than previous generations could possibly have known. When we focus our attention on the points of broad consensus, rather than where there is legitimate disagreement, conflicts that have festered for decades, even centuries, lose their grip.

Most people, in my experience, simply don't know that more than 95 percent of the scientists of the world—including scientists who are devoutly religious—agree on the general flow of natural history. Even those impressed by "intelligent design" arguments, I've discovered, are unaware that the leaders of the ID movement agree with evolutionists on the basic timeline of cosmic and biological emergence. That is, they agree that matter and life have undergone a sequence of irreversible transformations in measurable time. *Why* (and for some, exactly *how*) the living and nonliving worlds have morphed over the eons is the focus of debate. But the *fact* that our Universe has been transforming along a discernible path for billions of years—the fact that creation was not a one-time event—is of little or no dispute. It is this fact, this undeniable fact, of an emergently complex Universe that makes me want to shout from the mountaintops: "The war is over! The war is over!"

The war is over for another reason, too. Scientists have discovered that evolution is not a mechanistic, meaningless process. Admittedly, if one looks primarily at interpretations drawn by prominent scientists and natural philosophers not long dead, there is ample reason to conclude that the history of change in our Universe gives no guidance for how we should lead our lives and weave our legacies. But when we look at what prominent scientists alive today know and are discovering, we find a creation story that we can once again embrace as sacred, as holy, as ours.

From "Adam and Eve" to Us—and Beyond

> *"Both education and religion need to ground themselves within the story of the universe as we now understand this story through empirical knowledge. Within this functional cosmology, we can overcome our alienation and begin the renewal of life on a sustainable basis. This story is a numinous revelatory story that could evoke the vision and the energy required to bring not only ourselves but the entire planet into a new order of magnificence."* —THOMAS BERRY

Human consciousness emerged within a world of powerful and mysterious forces beyond our comprehension and control. As modes of communication evolved—from gestures and oral speech to writing and mathematics, to print, to science, to computers—so has our understanding of the scale and venerability of Creation, and the meaning and magnitude of humanity's divine calling. An inspiring consequence of seeing the full sweep of history is discovering that human circles of care and compassion have expanded over time. As we shall learn, this trend is in keeping with evolutionary forces. Truly, this is *Good News*.

Early on, owing to genetic guidance honed in a prelinguistic world, and then supplemented by knowledge that could be accumulated, retained, and shared only to the extent that spoken language would allow, our abilities to cooperate with one another were limited and localized. Anyone outside the tribe was suspect, and probably an enemy. As technologies of communication evolved, our ancestors entered interdependent relationships in ever-widening circles from villages, chiefdoms, and early nations to today's global markets and international organizations. Finally, the emergence of the World Wide Web has made possible collaborations no longer stifled by geographic distances and political boundaries. Throughout this evolution of human communities and networks, an inner transformation has also been taking place. At each stage our circles of care, compassion, and commitment have grown and our lists of enemies have diminished. Our next step will be to learn to organize and govern ourselves globally, and to enjoy a mutually enhancing relationship with the larger body of Life of which we are part.

Traditional religions have played crucial roles in fostering cooperation *within* each tribe, kingdom, and early nation—though not infrequently by

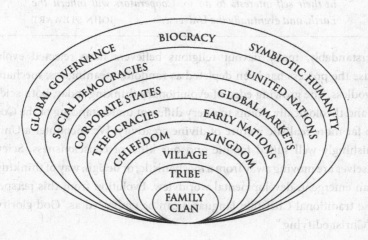

provoking suspicion and enmity of those outside the group. Now emerging is an orientation that encourages wider affinities and global-scale cooperation. For religious traditions to fulfill their potentials in our postmodern world, each will be called to harmonize its core doctrines with the evolutionary worldview. This effort will prove far more than an exercise in catching up and making do. Rather, leaders within each tradition will delight in discovering that the evolutionary outlook bolsters their core teachings. Instead of an intrusion on our faith, evolution becomes a precious blessing.

Evolutionary versions of each religion—Evolutionary Buddhism, Evolutionary Christianity, Evolutionary Islam, Evolutionary Judaism, Evolutionary Hinduism, and more—are emerging. Why is this happening? Because adherents of each tradition have discovered the same thing: *Religious insights and perspectives freed from the narrowness of their time and place of origin are more comprehensive and grounded in measurable reality than anyone could have possibly dreamed before.* Evolution does not diminish religion; it expands its meaning and value globally.

Science and Religion
Spurring Each Other to Greatness

"As evolution proceeds, living things will increasingly coordinate their actions for the benefit of the group because it will be in their self-interests to do so. Cooperators will inherit the Earth, and eventually the Universe." —JOHN STEWART

Understandably, many devout religious believers have rejected evolution because the process has been depicted as random, meaningless, mechanistic, and godless. The growing edge of evolutionary thinking today, both scientifically and theologically, points to a very different understanding of the Cosmos and a far more realistic picture of divine creativity. We encounter a Universe astonishingly well suited for life and our kind of consciousness. Scientists themselves are moving away from a mechanistic, or design, way of thinking and into an emergent, developmental worldview. Evolution from this perspective (to use traditional Christian language) can be embraced as "God glorifying" and "Christ edifying."

"The More Awesome My God Becomes"

While pastoring my first church, in rural New England, I stood under the stars one night with a parishioner, an 82-year-old farmer and amateur astronomer affectionately known as "Gramps." Gazing at the Milky Way, Gramps whispered, "You know, Reverend, the more I learn about this amazing Universe, the more awesome my God becomes!"

Two thousand years ago, it was widely believed that the world was flat and stationary, and that the Sun and stars revolved around us. The biblical writers reasonably assumed that mountains were unchanging, that stars never died, and that God placed all creatures on Earth (or spoke them into existence) in finished form. How could they have thought otherwise? The idea of a spherical Earth turning on an axis and orbiting the Sun, or of Polaris as an immense bundle of hydrogen gas fusing into helium quadrillions of miles away, or of mountains rising and eroding as crustal plates shift, or of creatures morphing over time: all these would have seemed absurd to anyone living when the Bible was written. Had anyone felt inspired to write about such things then, the early church leaders would never have considered the document authoritative. They would have thought it bizarre and dangerously misleading, and would have ensured that any such proclamations were discredited and quickly forgotten.

Many Jews, Christians, and Muslims still regard the early history of the Hebrew people, as recorded in the Torah, to be the history of humanity as a whole. We now, however, know a great deal more about what was happening in the world 3,500 years ago—two centuries before Moses was born—thanks to the worldwide, cross-cultural, self-correcting enterprise of archeological and anthropological science. Although none of this world history is mentioned in the Bible, no historian alive today would deny the following: Before Moses was born and before the story of Adam and Eve was written, southeast Asians were boating to nearby Pacific islands; Indo-European charioteers were invading India; China, under the Shang Dynasty, entered the Bronze Age; indigenous peoples occupied most of the Western Hemisphere; and the Egyptian empire's age of pyramid building had come and gone.

Each of these cultures told sacred stories about how and why everything came into being, what is important, and how to survive and thrive in the landscapes and cultures in which they lived. To interpret the early chapters of Genesis—or any of the world's creation narratives—as representing the entire history of the Universe, or to imagine them as rival rather than complementary views of a larger reality, is to trivialize these holy texts. It is also time for scientists to share their work with religionists and to understand that the traditions will not go away. The ancient religious paths are aching for coherence with the great discoveries born of the quest to understand this vast Universe, the living world, our evolved selves, and especially our innermost psyches.

In Context

"The ultimate victory for a scientific idea is to become the new commonsense."   —DAVID SLOAN WILSON

American society is rife with conflict, big and small, born of the seemingly never-ending battle between scripture-based faith and the discoveries of empirical science. A number of significant books in the realm of religion and science have been published during the past few years. Those that make the news typically fall into any of three categories:

✦ epistles countering "intelligent design" and the perennial claims that evolution is "just a theory"

✦ books that attempt to reconcile (read: make palatable) the science-based understanding of evolution with traditional religious views

✦ strident works that claim that otherworldly faith cannot be reconciled with science, and that science must triumph over supernatural religion and render it ineffectual if our species is to survive

None of those paths are offered here. Rather, my intent is to help you see what I see—*science and religion can be mutually enriching*. We are in the early stages of one of the most far-reaching transformations into which human consciousness has ever ascended. Today's conflict between science and religion is the catalyst by which both will mature in healthy ways. Neither will drive the other into extinction. Rather, both are moving in remarkable, previously unthinkable directions. As astrophysicist Joel R. Primack and cultural historian Nancy Ellen Abrams explain in their book, *The View from the Center of the Universe*, "This kind of integration of science and meaning is considered by many scientists to be a danger to science, but a science that doesn't consider its own meaning can be a danger to everyone else. Interpreting modern cosmology is—if anything is—a sacred responsibility."

This book is thus a message of realistic hope, grounded in reason and inspired by faith. Here is my vision: Within the first half of this century, virtually all of us—believers and nonbelievers alike—will come to appreciate that *evolution is a gift to religion* and that *meaning-making is a gift to science*. As the religions come to embrace the science-based history of the Cosmos, each tradition's core insights will be accessed in larger, more realistic ways than ever before. Cultures in conflict will find common ground that today seems inconceivable.

Overview

> *"The most practical belief system for a large-scale society in the long run is one that is firmly anchored in factual reality."*
> —DAVID SLOAN WILSON

Part I, "The Holy Trajectory of Evolution," delineates what I mean by, and how I will be using in this book, words such as "cosmology," "evolution," "emergence," "revelation," "God," and "the Universe." Here we shall consider what evolution is, what it is not, and why human societies require a mythic and meaningful context.

Chapter 1 examines why a people's cosmology, or "Big Picture," is so important. We cannot thrive without myth—that is, without meaningful stories that freely use poetry and metaphor to communicate what we individually and collectively experience to be true.

Chapter 2 is intended to evaporate the fears of those who reject evolution on the grounds that it is a meaningless, godless process. Here I show how mainstream science reveals that "Evolution Is *Not* Meaningless Blind Chance." Rather, biological life and human life evidence a trajectory (a holy direction). It is no coincidence, nor is it an accident, that greater complexity, cooperation, and interdependence at increasing scale are evident in the DNA and fossil records, and throughout human history as well. This does not, however, point to a designer God who planned the whole thing or who is pulling the strings. Indeed, there is compelling evidence against such a trivialized notion of the divine. (This chapter, though written for the general public, contains biological terms that may be unfamiliar to many readers. If it is too dense for your liking, skip ahead to Chapter 3, which presents many of the same concepts in simpler form.)

Chapter 3, "Evolution and the Revival of the Human Spirit," is written in sermon form and offers a passionate, contextual introduction to some of the core concepts discussed in later chapters. Here you will encounter lessons gleaned from billions of years of deep time grace. These are lessons that, while enriching traditional religious insights, can nonetheless be conveyed in ways that are also agreeable to individuals for whom religious language is off-putting.

Part II, "Reality Is Speaking," explores various modes of divine communication. The nature of human language and consciousness underpin this introduction to "the marriage of science and religion." Here we will look for overarching understandings that can be celebrated by *all* peoples, including devout religious believers of every tradition and ardent nonbelievers, too.

Chapter 4 introduces a novel distinction that will prove foundational for the rest of the book. It is the distinction between *private revelation* (divine truth sporadically received through the experience of single individuals) and *public revelation* (divine truth ongoingly obtained through the contributions made by a vast community of individuals engaged in the scientific quest). This chapter also introduces the radical idea that "facts are God's native tongue."

Chapter 5 considers "The Nested Emergent Nature of Divine Creativity." This, I contend, may be the single most restorative insight into the nature of reality gained through public revelation. Here we shall contemplate the arresting fact that absolutely everything we tangibly experience—including our own bodies—is, in truth, transformed stardust. The chapter concludes with a brief exploration of another truth that carries profound religious implications foreshadowed in mythic terms in the early Christian gospels and Book of Revelation. Death is of supreme importance in the process of divine/cosmic creativity.

Chapter 6 probes the inherently symbolic and consequential nature of human language, and why it is that "words create worlds." We shall also learn how our *day* (literal) and *night* (symbolic) experiences of reality are both important, and thus why neither an exclusively scientific nor an exclusively religious way of speaking about matters of ultimate concern would be adequate to the task.

Chapter 7 tackles the question, "What Do We Mean by the Word 'God'?" There is, of course, no one right way to express our relationship to Ultimate Reality. Nevertheless, how and where we imagine God makes a huge difference. For those of us who are religious, our images of the divine shape the largest meanings (and purposes) we attribute to our individual lives and to the collective life of our species. This chapter concludes with an invitation to *know* God and to *be faithful* to God, in a more glorious and undeniably real way than was possible before evolution was understood—and in a way that even atheists can celebrate.

Part III, "The Gospel According to Evolution," heralds the most immediately practical and personal segments of this book. *If your encounters with the scientific understanding of evolution in school, in your religious education, or via the media have not yet offered anything of value for your day-to-day living, or if the evolutionary worldview seems harsh and perhaps threatening to your faith, then you might want to dive right into this part.* Perhaps you, too, will experience the saving grace I felt when I learned how our evolutionary past is still influencing each and every one of us. From this vantage, the path to freedom becomes both obvious and achievable. By understanding our brain's creation story, new possibilities open up for overcoming long-standing personal challenges and living a life of deep integrity and unspeakable joy.

The way forward begins with this simple truth: *Your greatest difficulties (including substance addictions and other destructive habits), while your responsibility, are ultimately not your fault.* Such challenges spring from inherited proclivities that served the survival and reproductive interests of our human and pre-human ancestors. Those very same drives also give rise to some of the most precious aspects of our humanity—and what it means to be alive. They are instincts; they are not a mistake. Thankfully, we can begin to walk this path, in faith, and without becoming dour and anxious. Rather, we will learn how to harness the powers of the most ancient core of our brain—that which goes all the way back to our reptilian ancestors, and which I like to call our "Lizard Legacy." We will learn, too, why the "Furry Li'l Mammal" part of our brain is so adept at flooding us with emotions—welcome or not. Why does our "Monkey Mind" exhaust us with incessant chatter, and what can we do about it? Here, too, you will learn about the brain's most recently evolved capacity: our ability to choose a "Higher Porpoise" (higher purpose) powerful enough to override the problematic tendencies of all the older parts. Children and teens, especially, delight in these playful ways of talking about our brain.

Chapter 8 introduces a simple set of criteria for imagining how otherworldly religious concepts can be REALized—that is, made real in the world of our actual experience. Any new interpretation of a traditional understanding that could be embraced across the theological and philosophical spectrum will (a) validate the heart of earlier interpretations, (b) make sense naturally and scientifically, (c) be universally, experientially true, and (d) empower people of all ages, especially young people.

Chapters 9 and 10 explore human instincts from a "God-glorifying" evolutionary perspective. The good news is that an evolutionary appreciation of our instincts can help us navigate the troublesome issues many of us deal with related to food, safety, sex, and relationships. Here we see how and why a meaningful, deep-time view of human nature transforms lives in more comprehensive and lasting ways than classical religious or secular approaches generally can. These two chapters begin to explore traditional Christian theological concepts, such as "The Fall," "Original Sin," "Personal Salvation," "Christ-likeness," "Saving Faith," and "the Gospel," through the lens of *evolutionary psychology*. I also introduce what I call "The DNA of Deep Integrity": trust, authenticity, responsibility, and service. These virtues are central to a developmental understanding of "God's will," and they are key facets of evolutionary spirituality.

Part IV, "Evolutionary Spirituality," extends the practical emphasis by offering a solid program for personal and relational transformation grounded in evolutionary integrity. Yes, the practical *is* spiritual. Spirituality is not merely about prayer or meditation, mystical experiences, or, indeed, anything ethereal. It is about cultivating right relationships at every scale of reality, whether we are religious or not.

Chapter 11, "Evolutionary Integrity Practices," provides exercises that can bless your life and the lives of everyone with whom you are in relationship—no matter what your religion, philosophy, or beliefs. These are tools that will help you embody evolutionary spirituality in healthy and empowering ways. Each practice is crafted to support your growth in deep integrity—that is, in trust, authenticity, responsibility, and service. *This is the book's most practical and potentially life-changing chapter.* Even those who do not embrace evolution will find this chapter useful and the exercises transforming.

Chapter 12 explores the essential elements of "Evolving Our Most Intimate Relationships." Here we learn how respectful communication, touch and tenderness, playfulness and humor, meaningful songs and rituals, and service all reveal evolutionary wisdom. Attending to these, we ensure that our most meaningful relationships evolve in healthy ways.

Chapter 13 offers a smorgasbord of affirmations and visual images to support your growth in evolutionary integrity. Here you will find practical suggestions for nurturing lifegiving attitudes and habits of thought.

Part V, "A 'God-Glorifying' Future," shifts focus from the individual to the collective. How can we lovingly yet firmly confront corporate and systemic "sin" and "evil" while recognizing the vital role of chaos and breakdowns in catalyzing evolutionary creativity? How can we consciously co-evolve with the groups, communities, and institutions of which we are part? What would it mean for corporations, nation-states, and our species to be in evolutionary integrity? And what visions of realistic possibility can sustain and inspire us, and our children, well into the future?

Chapter 14, "Collective Sin and Salvation," examines the nature of group and systemic sin and suggests how we might participate in its redemption and transformation. Chapter 15 identifies ways of discerning "The Wisdom of Life's Collective Intelligence" and acting on what we discern. Here I suggest a way to REALIze another core Christian concept: "the Kingdom of Heaven." The nature of conversation is also explored here, and in a way that showcases its potential as an evolutionary force.

In Chapter 16, "Knowing the Past Reveals Our Way Forward," and Chapter 17, "Beyond Sustainability: An Inspiring Vision," I offer a hundred-year cosmic timeline and share what I and many others experience as a compelling vision of the future, grounded in an inspiring interpretation of the past. We will consider the major challenges and positive trends apparent today that will surely continue, along with potential wild cards (momentous events that may or may not take place) in the next 250 years. We also will revisit the question of who and what we humans really are in the evolutionary process, and thereby fashion a believable and empowering story of why we are here.

Chapter 18, "Our Evolving Understanding of 'God's Will'," uses our now sacralized evolutionary perspective to broaden and enrich our culture's experience of divine guidance and ethical instruction. We shall also consider fresh ways of understanding Jesus and his role in cosmic history, in the lives of Christians, and in the life of the church.

In the Epilogue, you will find my own brief testimonial, along with a vision of what a holy evolutionary future would entail. Appendix A is a previously published letter by Richard Dawkins, a renowned scientist who is no friend of otherworldly, belief-based religion. This letter, a forthright critique of the limitations of supernatural religion, is offered with loving inflection and simple analogies—for it was written to his then ten-year-old daughter.

Appendix B is my theological response to Dawkins's letter. Here I articulate an evolutionary REALizing of the miraculous stories that historically have been at the heart of the Christian tradition: the virgin birth and the resurrection and ascension of Jesus the Christ.

Some chapters begin with "prophetic inquiry" questions designed to elicit new responses to core religious concepts. Answers and interpretations beyond the ones I provide are not only possible; they are desirable in that they will encourage the collective evolution of our faith traditions. If any of my introductory responses speak to you, please make them your own—and let them live through you and your life. If a different approach for finding the sacred in our common creation story inspires you, then please share it here: ThankGodforEvolution. com. Thus might we launch our own century's contribution to a long and venerable tradition of prophetic inquiry. As more people share ideas in this way, we will generate an ever-evolving collective sense of our traditions' magnificent teachings and their relevance in this rapidly changing world.

PART I

The Holy Trajectory of Evolution

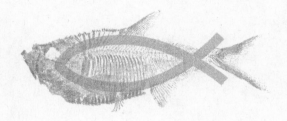

"Without a meaningful, believable story that explains the world we actually live in, people have no idea how to think about the big picture. And without a big picture, we are very small people."

—JOEL R. PRIMACK AND NANCY ELLEN ABRAMS

CHAPTER 1

Our Big Picture
Understanding of Reality

> *"All professions, all work, all activity in the human world finds
> its essential meaning in the context of a people's cosmic story."*
> —BRIAN SWIMME

Our Big Picture matters. Indeed, nothing matters more! What drives human evolution today is no longer primarily our genes. It is our sacred stories. These grand narratives furnish the context for answering life's largest questions.

For as long as humans have used words to communicate and think, we have been telling stories to answer the fundamental questions of existence:

Who are we?—the question of identity

Where did we come from?—the question of origin

Where are we going?—the question of destiny

Why are we here?—the question of purpose

What ultimately matters?—the question of meaning

How are we to live?—the question of morality/right action

What happens when we die?—the question of finality and continuity

Responses to these questions are embedded in a people's cosmology—their creation story.

Stories Within Stories

"The history of the Universe is in every one of us. Every particle in our bodies has a multibillion-year past, every cell and every bodily organ has a multimillion-year past, and many of our ways of thinking have multithousand-year pasts."
—JOEL R. PRIMACK AND NANCY ELLEN ABRAMS

Each of us is a story within stories. A child's life story is part of both her mother's story and her father's story. The story of their family is part of larger stories, too—the story of their neighborhood, their church, their town, their state, and their nation. Those stories, in turn, are contained within the story of a religious tradition, a civilization, humanity as a whole, and then beyond to the story of our planet, our star system, the Milky Way Galaxy, and, finally, to the story of the Universe itself. Each of us is thus a story within stories within stories. Each of us will have a felt relationship with the larger contexts of our existence only to the extent that we are given (or acquire on our own initiative) stories for these larger wholes that we find meaningful.

A dynamic relationship weaves into one whole each storied layer of existence. If a factory shuts down, the loss of jobs might be significant in the story of that community, and certainly in the stories of all the long-time workers (and perhaps in their children's stories, too). But a business closure would not measure as even a blip in the story of Western civilization. In contrast, if a nation suffers a severe economic downturn, goes to war, or undergoes a spiritual awakening, all the individuals and communities of that nation will have stories touched by those events. Thus the contour of any one story is affected by the *larger stories*. Indeed, when we search for the meaning of an event, we are asking: "How does this event fit into the bigger picture?" The larger the context of such interpretations, the fuller the meaning.

Making Meaning of the Great Tsunami of 2004

A poignant recent example of how humans search for larger meanings can be found in news reports of religious responses to the tsunami of December 2004, which was triggered by an earthquake of magnitude 9.0 in the ocean near Indonesia. Here is a clip from a *Washington Post* article by Bill Broadway, "Divining a Reason for Devastation" (January 8, 2005):

> In Banda Aceh, Indonesia, the hardest-hit area in the world's most populous Muslim country, imams blamed the Dec. 26 tsunami on lay Muslims who were shirking their daily prayers and following a materialistic lifestyle. Others said Allah was angry that Muslims were killing Muslims in ongoing civil strife....In Sri Lanka, which recorded the most fatalities after Indonesia, Buddhist survivors told the story of a tsunami that flooded the island kingdom 2,200 years ago when a king killed a Buddhist monk in a fit of anger. They wondered which political leader angered the sea gods this time.

Interpreting such catastrophe as a message or judgment from God or the gods was not strictly the province of religious conservatives. The same news article quoted Rabbi Michael Lerner, a well-known Jewish liberal in America, as concluding,

> The tectonic moves of the earth are part of a totally integrated moral system that has been in place since the earth began to evolve. That moral system, described by the Bible, tells us that the physical world will be unable to function in a peaceful and gentle way until the moral/spiritual dimension manifest in the behavior of God's creatures coheres with God's will: that is, is filled with justice, peace, generosity and kindness.

How could these religious responses have been otherwise? We are, after all, meaning-making creatures. Any huge event demands a huge explanation. In a pre-scientific setting, there could have been no natural explanation of the rare, horrific manifestations of what scientists now call "plate tectonics." And even today, those who do understand a scientific explanation for the tsunami may assume that a natural explanation is meaningless at best. But consider these two responses from scientists quoted in a New York Times article by William J. Broad, "Deadly and Yet Necessary, Quakes Renew the Planet" (January 11, 2005):

> "It's hard to find something uplifting about 150,000 lives being lost," said Dr. Donald J. DePaolo, a geochemist at the University of California, Berkeley. "But the type of geological process that caused the earthquake and the tsunami is an essential characteristic of the earth. As far as we know, it doesn't occur on any other planetary body and has something very directly to do with the fact that the earth is a habitable planet."
>
> "Having plate tectonics complete the cycle is absolutely essential to maintaining stable climate conditions on earth," Dr. William H. Schlesinger said. "Otherwise, all the carbon dioxide would disappear and the planet would turn into a frozen ball."

In every human society, the largest of all contexts is the story of how the world began, how everything came to be as we find it today, and where everything is going. The trajectory of our world, and our individual and collective responsibility (if any) within that trajectory, is an integral part of a culture's creation myth. The Big Picture lends meaning to every facet of individual and communal life. It is the soil out of which all our beliefs, customs, and institutions grow.

A people's cosmology crystallizes into a set of unquestioned assumptions. Like glasses with colored lenses, our cosmology colors everything we see. It determines not only the way we perceive and interpret our world, but also what we will perceive at all. Its rules and categories are generally transparent; we're not aware of them. Nevertheless, our cosmology is our reality. It underlies everything we think of as real.

> The evolutionary view of life should be as fundamental to a college degree as Psychology 101 or Western Civilization. But rather than asking students to memorize and regurgitate mountains of testable facts, we should emphasize study of the history of the discovery of evolution, its major characters and ideas, and the basic lines of evidence. This would do far more to inform citizens and prepare teachers than forcing students to remember the Latin names of taxa. We are stoning our children to utter boredom with little pebbles and missing the big picture. The drama of the story of evolution will recapture student interest.   —SEAN B. CARROLL

What Is the Great Story?

> *"The evolutionary epic is probably the best myth we will ever have."*   —EDWARD O. WILSON

The Universe, as revealed through mainstream science, does have a history. Because that history can be perceived as directional, our scientifically informed history of the Universe can become our Big Picture story. In the words of Christian theologian John Haught, "Nature is narrative to the core."

What I and others mean by the Great Story is humanity's common creation story. It is the 14–billion-year science-based tale of cosmic genesis—from the formation of galaxies and the origin of life, to the development of consciousness and culture, and onward to the emergence of ever-widening circles of care and concern. Science unquestionably provides the foundation. For this tale to be experienced as holy, however, it must don the accoutrements of myth. Bare-bones science must be embellished with metaphor and enriched by poetry, painting, song, and ceremony.

Biologists Julian Huxley and Edward O. Wilson have called this aesthetically rendered story "the epic of evolution." American conservationist Aldo Leopold called it "the odyssey of life." Anthropologist and religious naturalist Loren Eiseley referred to it as "the immense journey." Most recently, geologian Thomas Berry named it the Great Story.

The Great Story is, quite simply, *the sacred story of everyone and everything*. It springs from the grand narrative of an evolving Universe of emergent complexity and breathtaking creativity. No human story is left out. The Great Story can thus help us understand cultural, as well as natural, history in ways that honor and embrace all religious traditions and creation stories. Six core attributes of the Great Story combine to make this epic an ideal guide for humanity today:

1. A creation story not yet over. The creation of the world did not occur "once upon a time" in the distant past. Divine creation continues. Evolutionary change at all levels is ongoing, and we humans bear a responsibility for how the story will continue on Earth. We are participants in an amazing, challenging adventure!

2. A planetary perspective. Individuals from diverse cultures contribute to the Great Story, which all, in turn, can celebrate. The scientific enterprise is now global, so this story is influenced by peoples of all ethnicities, all religious traditions, and hailing from all regions of the planet. Scientists of diverse heritages are each doing their part in discerning the foundational facts of our common creation story.

3. Open to multiple interpretations. The Great Story not only tolerates a multiplicity of interpretive meanings; it welcomes them. The empirical and theoretical sciences search for material explanations of the world. When we venture into the realm of meaning, our diverse interpretations necessarily go beyond science. Multiple interpretations are encouraged by Great Story enthusiasts. Like evolution itself, this cosmological story thrives on diversity.

4. The marriage of science and religion. The Great Story seamlessly weaves together science, religion, and the needs of today's world. Because the

creation stories of classical religions and native peoples emerged well before the revelation of an evolutionary Cosmos, those venerable stories can fulfill their deep-time potential only if the ancient cosmologies are creatively reinterpreted to mesh with the fruits of today's science. In contrast, the Great Story emerges out of scientific awareness and thus evolves in step with new discoveries and with the needs and challenges of the day.

5. A metareligious perspective. The Great Story is not a new religion in competition with existing religions; rather, it offers a metareligious perspective that can deepen the profound insights of every one of Earth's spiritual traditions. The Great Story will fulfill its potential for humanity only when it is taken into and absorbed independently by each faith and worldview. Necessarily, its gifts will manifest in distinct ways in different contexts.

6. The story of the changing story. Whenever a new discovery is made and broadly verified in the sciences, our understanding of the Great Story of the Universe changes. Such change is to be welcomed—not feared.

Change is to be welcomed, not feared. Well, okay, sometimes it is inconvenient to have to change. Nevertheless, we adapt. For example, in August 2006 the community of scientists decided to change the classification of Pluto from a planet to a "dwarf planet." Three years earlier, my wife and mission partner, Connie, had written and posted on our website an evolutionary parable titled "Pluto's Identity Crisis." We've used that parable many times in our programs, calling for volunteers to read and act out the scripts of the four characters in the drama. Thankfully, a homeschool student, Bella Downey, responded to Connie's email plea for help and suggested a rewrite that beautifully incorporates the necessary changes.

"It's Hard to Be a Great Story Fundamentalist!"

Whenever I introduce the Great Story to a new audience, I do so in a lighthearted way that highlights the need for humility. "The Great Story," I say, "is the story of the changing story. Whenever the science changes or a new discovery is made, our telling of the Story has to change too. So it's really hard to be a Great Story fundamentalist!"

From Shape-Shifting Story to Unchanging Scripture

> *"If we have within us a biological need to internalize a rather permanent story of what the world is about, then not having a story, or being confronted with the changing story of science, could trigger a personal crisis. What if, however, our story becomes the story of how stories change? Even if we can't ground ourselves in an immortal story anymore, the immortality can be had in the story of how we make stories, of how we find stories through science. Rather than just celebrating the new cosmology, we could celebrate, say, this week's top science story in the journal* Nature *and the story of how that new story came to be. Keeping current means we would be celebrating the story of the changing story."* —TYLER VOLK

Imagine early humans fanning out of Africa in waves, driven perhaps by climate change, resource scarcity, overpopulation, and certainly by our inborn spirit of adventure. The only acquired knowledge or memories of the ancestral home and lifeways that would carry through the generations would be those embedded in songs, ceremonies, and orally transmitted stories.

Tens of thousands of years pass. People now inhabit all continents except Antarctica. Some have settled into fertile valleys and have domesticated favored plants and animals. Villages form. Knowledge is still transmitted orally. Then a threshold is crossed. Early forms of writing are used to record debts and other important information. Chiefdoms, kingdoms, and other large-scale organizations of cooperation and social complexity appear, like mushrooms after a rain. Finally, we witness the whir of cultural change encompassed in the last several thousand years: city-states, theocracies, and political and religious empires. The importance of writing has escalated; now even religious wisdom is preserved and passed on as *scripture*.

It is important to remind ourselves that for each of these periods—indeed, for well over 99 percent of human history—there is little evidence that any culture understood developmental time and space in a way remotely similar to how we understand it today. Nevertheless, the big cosmological questions demanded answers, and so the answers came.

Imagine parents, grandparents, and respected elders telling stories to the young about who they are, where they came from, why they are here, what really matters, and how to lead honorable, fulfilled lives. Orally transmitted stories would evolve over time as conditions changed and as generations faced new challenges. Yes, these stories would evolve—until (and if!) they were written down and declared to be the unchanging revelation of God. *When a story becomes scripture, it ceases to evolve.*

Meaning Making

"The more we learn about Earth and life processes, the more we are in awe and the deeper the urge to revere the evolutionary forces that give time a direction and the ecological forces that sustain our planetary home." —CONNIE BARLOW

We are privileged through science to know and witness the immense journey of life. Natural history is now measured in billions of years. The attraction of science is both its beauty as a heritage and its prospects for change. School textbooks, unfortunately, sometimes render science as dogmatic as any fundamentalist doctrine. In truth, science is quintessentially open to revision and discovery. Science is also open to fresh interpretation. Scientists can tell us what is and what was and, to some extent, what will be. But they cannot tell us what it all *means.* For example, Big Bang cosmology is almost universally accepted within science. Even so, it is up to each of us to choose whether we feel welcome or alien in that sort of Universe and what that means for our religious and spiritual perspectives.

Past bards of the evolutionary epic sometimes presented the story as a quest. The Jesuit mystic and paleontologist Pierre Teilhard de Chardin told the story as life "groping" toward a kind of Christic unification. Others, notably, evolutionary biologist Julian Huxley, wove a tale of enrichment emerging step by step, with no goal in sight, no lure beckoning. Philosopher Daniel Dennett's terminology is helpful for distinguishing these two worldviews. Huxley's epic is assembled like a skyscraper, by "cranes," emerging from the ground up—from the foundation to the tower—by earthly processes. Teilhard's is to some extent boosted by "skyhooks"—the pull of a higher, spiritual force. Whichever version is preferred, the tellers are challenged to offer an emotionally satisfying picture and to evoke a sense of belonging without compromising truth.

Any instructive telling of the biggest story must include an interpretive meaning, but a meaning best nuanced by regard for how any such meaning comes about. It is crucial to remember that four hundred million years ago, when an ancient lobe-finned fish set out across a tidal flat in desperate search of water, that fish had no inkling that its effort would ultimately lead to feathered flight and cathedrals. Foresight is foreign to the pre-human evolutionary process. Thanks to a big brain, however, our species has the extraordinary gift of hindsight. We can discern in the grand sweep of time a movement toward greater complexity and hence greater opportunity that was not available to a struggling fish—literally, out of water.

In hindsight, one event prepared the way for something to come. Coincidence, even misfortune, was turned into opportunity. But we should remember that at the time of each transition, the organisms involved hadn't a clue that anything grander might await their descendants. They were just looking for another tide pool. The heroes of the evolutionary epic have all been Forrest Gumps.

And isn't that marvelous! *This* version of evolution, *this* version of "In the beginning" encourages us to take on the possibility—the realistic possibility—that evolution is happening right now through us. More, what may feel like desperate fumblings might be the very stuff that launches the evolutionary epic across yet another threshold. Maybe you, maybe me, as individuals—right now—in our own little lives have made some choice that will play out powerfully over the ages. Truly, this way of seeing our place in Creation is invigorating!

I did not become a minister in order to evangelize evolution. In hindsight, however, the sequence smacks of predestination. I do occasionally think of it as one stage preparing for the next. But there is also beauty in recalling my personal story more realistically, in the way that Mary Catherine Bateson suggested in her book *Composing a Life*. We take advantage of slim opportunities, swerve ever so slightly to avoid obstacles, grasp the first hand that extends to pull us out of an abyss. Next thing we know, we are on a new life course. Theologian Gordon Kaufman has offered a striking term that reminds us of this fanciful, fluky aspect of the evolutionary epic. He calls the process underlying it all "serendipitous creativity." That puts us in partnership with the divine—yes? Not masters of our fate, but partners, groping our way forward. And here is the thing: what a difference it makes to be groping our way forward in faith—in partnership with God, or, should you prefer less traditional terminology: *trusting the Universe, trusting Reality, trusting Time.*

CHAPTER 2

Evolution Is *Not* Meaningless Blind Chance

> *"Once is an instance. Twice may be an accident. But three times or more makes a pattern."* —DIANE ACKERMAN

There is much in science at the scale of the very big (the Universe as a whole) and the very small (quantum physics) that our commonsense can't make heads or tails of. But at the scale of life and planet Earth, commonsense is quite dependable. In a million years, the ebb and flow of tides on all the sandy beaches of the world will not fashion even one instance of a multistoried sandcastle that any of us would be fooled into thinking was the work of human hands. Not in a billion years will a tornado whip together a functioning bicycle (much less a jet plane) from a heap of unassembled parts. We know this. Commonsense tells us that random, directionless processes cannot give birth to complex or sophisticated offspring.

Now here is the good news for peoples of faith: evolutionary scientists have never said otherwise. Evolution is not blind chance. Randomness yields nothing—by itself.

Each morning, when I download my email, I engage in a kind of evolutionary process. Speaking invitations I forward to my assistant; bills to my wife. Whenever I encounter spam, I hit the delete button. There is randomness, to be sure, in the order in which the emails show up on my screen. But what is far more important is my propensity to sort by function and discard anything that is not helpful.

Ever since Darwin, evolutionary scientists have been presenting biological evolution in much the same way. What Darwin called "natural selection" is nothing more than the sum of Nature's sorting processes. Random mutations that are *functional*, that help an organism survive or reproduce, will tend to be passed on to the next generation—not all the time, but often enough to serve as a shaping force. Variations that are not functional will tend to be deleted when the organisms that bear them falter in their ability to survive or reproduce. Functional mutations will be *inherited* by later generations; dysfunctional mutations will not. It is that simple.

Whatever the *sources of variation* in the genes (and especially in the regulator genes, or "genetic switches," that propel macroevolution), it is Earth's climate, topography, chemistry, and communities of life that put all novelties to the test. This is the sorting process of *natural selection*. Over eons, step by step, this natural sorting process has sculpted diversity and complexity in the stream of life. It is true: our ancestors once lived in the sea and had the personality and intelligence of a worm; our even more distant ancestors spent their entire lives within the confines of a single cell. But it is not true that out of the single cell or the worm came the human. The genetic code of ancestral worm or single cell was just the canvas. The painter was the sum total of all the forces at work on Earth (and some beyond), operating over timescales we cannot fathom.

Interpreting Our Immense Journey

"The religion that is afraid of science dishonors God."
—RALPH WALDO EMERSON

Empirical science offers these well-substantiated facts: This Universe is billions of years old. Complex atoms were forged in the cores of stars. Earth is younger than the Milky Way Galaxy. Life evolved from the simple to the complex. Dinosaurs and sabertooth cats once roamed this planet. We have "tailbones" because our ancient ancestors had functional tails. Childbirth is difficult for humans because women's anatomy is a compromise between two crucial functions: bearing large-brained offspring and running on two legs. All peoples evolved creation stories, and these creation stories differ one to the next, sourced by different environments and different life experiences. Empirical science, however, says nothing about the meaning of all these facts. That is the province of interpretation.

The perceived need to have a preordained meaning is one reason that some people cling to a literal reading of the biblical creation story. To make the leap to a fully evolutionary outlook, we would come to realize that everything in the Cosmos emerges through time. Light emerges, atoms emerge, molecules emerge, galaxies emerge, life emerges, vision emerges, flight emerges, frustration emerges, terror emerges, joy emerges, compassion emerges. So, too, has meaning emerged, and continues to emerge.

The evolutionary epic surely sparkles as an adventure tale—"the immense journey" in the words of Loren Eiseley. Awareness of this journey grounds us in what I and others like to call *deep time*, eons going back almost beyond our ability to comprehend. Here we are, by grace, we humans of today. In this deep-time grace, we are surrounded by a singing and slithering and flitting urgency of existence that is all our kin. Here we are, talking and posting messages via satellite with others of our kind around the globe—a globe that contains on distant continents the bones of our ancestors, and the shells and carapaces of cousins far removed.

"Is a Cheetah My Cousin?"

In a church classroom of elementary-age kids, Connie asked whether humans are related to monkeys. One boy declared, "Fish, too, and even microbes!" A young girl then asked, "Is a cheetah my cousin?" "Yes!" replied Connie. Beside herself with joy, the girl responded, "Cheetah is my favorite animal!"

We are, as Edward O. Wilson puts it, "life become conscious of itself," or as Julian Huxley expressed it, "evolution become conscious of itself." If we choose to celebrate the deep-time grace of this pulsating stream of life, if we choose to envision ourselves as life waking to an awareness of the breadth and depth of its own existence, then what science gives us in the way of spiritual fare is extraordinary!

The Mythopoeic Drive

"The predisposition to religious belief is the most complex and powerful force in the human mind and in all probability an ineradicable part of our nature. It is one of the universals of social behavior, taking recognizable form in every society from hunter-gather bands to socialist republics. It goes back at least to the bone altars and funerary rites of Neanderthal man."

—EDWARD O. WILSON

The meaning drawn out of science by each interpreter is a constructed, but not arbitrary, product of the human imagination. To find meaning in the epic of evolution is no less legitimate than to have an aesthetic response to a landscape. Others may have a different response, but to be fully human is to have a response of some sort.

"In the beginning was the Big Bang."
"In the beginning was the Great Radiance."
"In the beginning God created the heavens and the earth."
"In the beginning was the Word, and the Word was with God."

So far as can be known, we are the only animals blessed and burdened with a mythopoeic drive. We simply cannot *not* make events mean something. For example, each of us tends to recall the events of our own life in ways that render the whole into something meaningful, a coherent pattern that explains how we became who we are today and how we got to where we are. We exclude (or forget) the seemingly extraneous events that don't move the story forward, that don't contribute to the pattern that connects the dots into a coherent picture. Similarly, beginning in early childhood, we yearn to learn the story that came before and that can make sense of our birth. This is the story of our parents, the story of our people, the story of how everything we see (or learn about on television and the Internet) came to be. Historian of myths Mircea Eliade goes so far as to suggest that we are not *Homo sapiens*, but rather *Homo religiosus*.

The mythopoeic drive is a fine example of what the late paleontologist Stephen Jay Gould called an *exaptation* (distinct from an *adaptation*). That is, it is unlikely that the forces of nature directly selected for an urge to ask, and answer, the Big Picture questions. Indeed, perhaps the contrary: all these pesky questions might do us harm if we are obsessing with them rather than getting on with the practical business of making tools, finding food, securing someone to copulate with, caring for the children, and maintaining vigilance against hostile interlopers. But something else surely would have been advantageous—something from which the mythopoeic drive inexorably would emerge.

Consider this scenario: When our ancestors came down out of the trees, stood upright, and developed a relationship with rocks sufficient to defend themselves from formidable predators, they were able to embark on a new way of life that would secure them a great deal more protein than had been available to our lineage before. They could now safely head out onto the treeless savanna and scavenge the carcasses of large beasts. Even more, when their relationship with rocks had advanced to the point where they could themselves become formidable predators and engage in the hunt, another threshold was passed and the lifeway changed yet again. Finally, when their ability to think and communicate passed a later threshold, this would have been the result: A hunting party could spot a set of hoof prints in the sand and from that collectively make and consider interpretations that would be superior to what any individual was capable of making alone. Past events would be remembered as stories, as causes linked with consequences, that would help the group correctly interpret what they saw before them now. How long ago had the animal passed? Was it in good health? Where was it going? What stories did our elders tell us about their own hunts in conditions such as these? And thus, will it be worth our effort to attempt to overtake this animal?

Making sense of the past in order to interpret the present in ways that best secure a desirable future are capacities that were selected for and amplified in our evolutionary past. They continue to serve us today. They are the ground from which our urge to gather stories into a Great Story has emerged.

"He Said It Changed His Life"

The morning after I presented my two-hour digital slide program, "Thank God for Evolution!" at a church in Texas, I downloaded my email and found this, to my delight: "I brought my nine-year-old son to your program last night. He said it changed his life and made him feel better about many things. He talked about it all the way home and then wished he didn't have to go to bed so he could stay up and talk about it some more. Thanks for putting poetry and magic to the things I have been telling him."

Time's Arrow, Time's Cycle

"A mission of art, science, and religion alike is to teach us to
see the beauty in everything that's true, not just in what also
happens to be pretty."   —TIMOTHY FERRIS

The evolutionary epic is first and foremost a celebration of the arrow of time.
The history of the Universe is a glorious parade of emergent novelty upon novelty. The evolutionary epic is also a tale of enrichment. Unicellular life, acted
upon by all the forces of Earth—and the forces that impinge upon Earth—
gives rise to multicellular life; life in the sea creeps onto land; prostrate land
plants explore the vertical and forests are born; insects and pterosaurs take to
the air, followed by birds and even gliding mammals that coexisted with the
great dinosaurs. Throughout, animals take on every color of the rainbow to
attract a mate or to broadcast toxicity; later, flowers do the same to woo pollinators. Teeth and nematocysts, alarm calls and seductive songs, all these innovations yield an escalating torrent of creativity.

The arrow of evolution buds offshoots, too, that curve back on themselves
to become cycles. There is repetition in the epic as well as invention. There is
pattern to be discerned, pattern to be interpreted. Today, well-watered lands
with moderate climates support rich forests of angiosperm trees (like oaks
and hickories). When the long-neck dinosaurs reigned, the gymnosperms
(ancestors of spruce and pine) had their heyday. Before them, the niche of
towering photosynthesizer was filled by trees whose only living descendants
are horsetails and evergreen club mosses that hug the forest floor. Turning
to the ocean realm, we learn that reefs were built first by calcareous algae,
then by sponges, then by rudist clams. At the end of the Cretaceous period
(65 million years ago), the rudists gave way and the corals claimed the reef-
building niche, which they have carried on through today. One might look
upon that sequence as algae to sponge to mollusk to coelenterate. But one
might also interpret it as reef-builder, reef-builder, reef-builder, and reef-
builder—all variations on the same theme. There is a cycle of attainment and
replacement, attainment and replacement. Similarly, at the end of the Cretaceous, when Earth was struck by an intruder from space, the dinosaurs were

not to return. But various mammalian lineages, including that of the modern rhinoceros, evolved into the form and function of tank-like herbivore, echoing Triceratops. Tall brontotheres and, later, camels and giraffes stretched their necks into the trees as the brontosaurs had done tens of millions of years earlier. One group may have laid eggs and the other suckled young, but in form and ecological function the players past and present are remarkably the same.

The arrow of evolution moves not just from sea to land but cycles back to the sea, in the form of eelgrass, whale, walrus, sea otter, manatee—each foray independent of the other and each body form bearing conspicuous signs of past lives on land. Earlier, of course, there were reptiles that returned to the sea and who also had to surface periodically for air. These were the monstrous mosasaurs, the dolphin-like ichthyosaurs, and the long-necked plesiosaurs.

From land to air and back to land is the shared path of the distinct lineages that led to all the flightless birds: the ostrich of Africa, the rhea of South America, the emu of Australia, the kiwi of New Zealand, the extinct elephant bird of Madagascar. The circuitous course of each is apparent in the skeletal architecture, which is surely no product of engineer-like intelligent design, but clearly *is* the product of intelligent innovation and adaptation.

Stephen Jay Gould popularized an understanding of evolution that focused on the role of randomness and chance. "Rewind the tape" of evolution, he would say, and imagine the whole process unfolding from the start once again: everything would be different. At one level, this interpretation is indisputably true: the species would surely be different. There would be no white oak, no gray whale, no emu. But at another level, the level that matters most to me and surely to many others, the central issue is whether there would be eyes to see, whether there would be trees reaching into the sky, whether there would be creatures scampering on land, flying through the air, and perhaps even swimming in the sea but needing to surface for air. We wonder, too, whether there would be a form like us who would come to know and celebrate the 14-billion-year story of the Universe.

I am convinced that the best answer is an unqualified Yes! We can have confidence in this conclusion for one compelling reason. These forms and life-ways have independently evolved, time and again, during the actual 3.8 billion year epic story of life on Earth. This propensity, this drive, for life to evolve in the same ways in unrelated lineages is known as *convergent* or *parallel* evolution. Birds and bats and insects and pterosaurs have wings not because a common ancestor had wings but because wings independently evolved multiple times in very distinctive lineages. Their *dis*-similarities give the lie to their imagined common origins. Their similarities give us confidence that developmental constraints, functional demands, and niche opportunities in Earth's environments can be counted on to produce and reproduce the same core patterns. A shared inheritance of simple photoreceptor cells—cells sensitive to changes in light and dark—was ramped up independently in snails, squids, insects, spiders, and vertebrates, yielding complex, yet distinctly different, ways of seeing shapes, textures, and color. Surely, the Universe was determined to see itself!

An awe of the power and performance of convergent evolution profoundly shaped the worldviews of great biologists of the past—notably, Charles Darwin and Julian Huxley. The classic examples of convergent evolution are the striking similarities between Australia's pouched marsupial mammals and the look-alike placental mammals found elsewhere in the world. Among plants, the succulent cactuses and yuccas of the New World are remarkably similar to the succulent euphorbs and aloes of the Old World. Both of these convergent sets were known in Darwin's day and were the subject of wonder and speculation. Although convergent evolution had faded as a topic for serious study and for popular writing after about the middle of the twentieth century, in the 1990s a renaissance began. Esteemed voices within biology began to alter the professional and popular understanding of "evolution" to include, once again, a kind of developmental trajectory. Richard Dawkins, Edward O. Wilson, John Maynard Smith, Simon Conway Morris, Mark McMenamin, David Sloan Wilson, and Sean Carroll are among the biologists who readily point out the patterns in evolutionary history, including instances of evolutionary convergence. They find compelling evidence that many very distinct structures, functions, physiological processes, senses, and behaviors have evolved independently in a number of unrelated organisms. Here are a few quotations, drawn from this illustrious list of scientists.

Simon Conway Morris:
   It is now widely thought that the history of life is little more than a contingent muddle punctuated by disastrous mass extinctions that in spelling the

doom of one group so open the doors of opportunity to some other mob of lucky-chancers. The innumerable accidents of history and the endless concatenation of whirling circumstances make any attempt to find a pattern to the evolutionary process a ludicrous exercise. Rerun the tape of the history of life, as S. J. Gould would have us believe, and the end result will be an utterly different biosphere.... Yet what we know of evolution suggests the exact reverse: convergence is ubiquitous and the constraints of life make the emergence of the various biological properties very probable, if not inevitable.

When you examine the tapestry of evolution you see the same patterns emerging over and over again. Gould's idea of rerunning the tape of life is not hypothetical; it's happening all around us. And the result is well known to biologists—evolutionary convergence. When convergence is the rule, you can rerun the tape of life as often as you like and the outcome will be much the same. Convergence means that life is not only predictable at a basic level; *it also has a direction.*

Richard Dawkins:

It seems that life, at least as we know it on this planet, is almost indecently eager to evolve eyes. We can confidently predict that a statistical sample of reruns [of evolutionary life on Earth] would culminate in eyes. And not just eyes, but compound eyes like those of an insect, a prawn, or a trilobite, and camera eyes like ours or a squid's, with color vision and mechanisms for fine-tuning the focus and the aperture. Also very probably parabolic reflector eyes like those of a limpet, and pinhole eyes like those of *Nautilus,* the latter-day ammonite-like mollusk in its floating coiled shell.... And if there is life on other planets around the Universe, it is a good bet that there will also be eyes, based on the same range of optical principles as we know on this planet. There are only so many ways to make an eye, and life as we know it may well have found them all....

Like any zoologist, I can search my mental database of the animal kingdom and come up with an estimated answer to questions of the form: "How many times has X evolved independently?" It would make a good research project, to do the counts more systematically. Presumably some Xs will come up with a "many times" answer, as with eyes, or "several times," as with echolocation. Others "only once" or even "never," although I have to say it is surprisingly difficult to find examples of these. And the difference could be interesting. I suspect that we'd find certain potential evolutionary pathways which life is "eager" to go down. Other pathways have more "resistance."

Sean B. Carroll:

The DNA record also reveals that evolution can and does repeat itself. Similar or identical adaptations have occurred in the same way in species as different as butterflies and humans. This is powerful evidence that, confronted with the same challenges or opportunities, the same solution can arise at entirely different times and places in life's history. This repetition overthrows the notion that if we rewound and replayed the history of life, all of the outcomes would be different.

The Role of Strife

> *"Thus, from the war of nature, from famine and death, the most exalted object which we are capable of conceiving, namely, the production of the higher animals, directly follows. There is grandeur in this view of life, with its several powers, having been originally breathed by the Creator into a few forms or into one; and that, whilst this planet has gone cycling on according to the fixed law of gravity, from so simple a beginning endless forms most beautiful and most wonderful have been, and are being, evolved."* —CHARLES DARWIN

The history of life can be read in many ways. Some biologists detect a direction to that story, an inspiring meaning, and perhaps even moral lessons for humanity. In contrast, others see in the pattern of speciations and extinctions only a meander, with our own self-reflective powers and technological prowess an accidental triumph, but not a culmination and by no means a goal. Factual knowledge of the history of life and the Universe may not therefore satisfy our mythopoeic drive. We want to learn what impels that history, and we want to establish a relationship to those powerful forces.

A look at the processes underlying evolution opens a new level of questioning, with its own philosophical gloss. The "tools" of evolution and the metaphors chosen to represent them reveal a diversity of worldviews. For example, Charles Darwin used the example of *wedges* to illumine his theory of natural selection. Imagine a tree stump into which wedges are being hammered, one after another. Eventually, as the population of wedges grows, for any additional wedge to enter, the very act of hammering in the new will expel one of its neighbors.

Some thirty years ago, a new idea entered biology that reflected the Cold War ethos of a nuclear arms race between the United States and the then Soviet Union. It was called *evolutionary arms race*, and it means simply this: if a predator becomes more proficient in capturing prey, then the prey species will be pressured to evolve better means for avoidance or defense, which in turn drives the predator to improve yet again. Arms race escalations in the natural world are by no means limited to predator/prey associations. Richard Dawkins urges us to see the beauty of a forest in a whole new way:

> Why, for instance are trees in forests so tall? The short answer is that all the other trees are tall, so no one tree can afford not to be. It would be overshadowed if it did. This is essentially the truth, but it offends the economically minded human. It seems so pointless, so wasteful. When all the trees are the full height of the canopy, all are approximately equally exposed to the sun, and none could afford to be any shorter. But if only they were all shorter, if only there could be some sort of trade-union agreement to lower the recognized height of the canopy in forests, all the trees would benefit. They would be competing with each other in the canopy for exactly the same sunlight, but they would all have "paid" much smaller growing costs to get into the canopy. The total economy of the forest would benefit, and so would every individual tree.

Thank God for evolutionary arms races! Were all forests a meter high, there would be little room for birds and an inordinate diversity of flying beetles. There would be no pathways through which humbled humans would stroll beneath a wondrous canopy of green. Indeed, there would have been no impetus for the evolution of large, arboreal creatures with grasping hands—including our own primate ancestors.

The Role of Cooperation

"Ecological communities are not simply gladiatorial fields dominated by deadly competition; they are networks of complex interactions, of interdependent self-interests that require mutual adjustment and accommodation with respect to both the other co-inhabitants and the dynamics of the local ecosystem. The necessity for competition is one half of a duality, the other half of which includes many opportunities for mutually beneficial co-operation."   —PETER A. CORNING

"Survival of the fittest" is the leitmotif of evolutionary interpretations in which competitive strife is seen as the primary tool of evolution—interpretations for which wedges and arms races are useful metaphors. While few biologists today would deny that competition and strife play vital roles in evolutionary emergence, some conclude that cooperative processes are equally or even more important. From this perspective, evolution is driven by "survival of those that fit best" into the web of ecological relationships. The role that mutual support has played in evolution has been powerfully explicated by biologists Lynn Margulis and Mark McMenamin. When they view the forest, they see an entirely different face of natural selection at work:

> Symbiosis has shaped the features of many organisms. The great evergreen forests that spread across the northern latitudes would wither and die without the threads of symbiotic fungi that extract nutrients from rocks and soil and convey them to the tree roots. Termites would be no threat to houses, except that their guts contain myriad bacteria and other, larger creatures capable of digesting the cellulose in wood. The giant tube worms that live near hot springs on the ocean floor lack mouths; they take nourishment from symbiotic bacteria that live in their tissues.

Symbioses—beneficial partnerships—not only shape the ecologies of our world. The cooperative aspect of the living world is responsible for many of the significant innovations in the evolutionary journey of life. Most notable was the merger of different kinds of bacteria to form the first eukaryotic cells (cells with a nucleus). This was the innovation that opened the door for the evolution of all lineages of multicellular beings: plants, animals, and fungi. Scientists exploring beneficial partnerships do not envision evolutionary change through symbiosis as inimical to the "red in tooth and claw" school of thought. In fact, they suspect that the paths to win-win solutions have been fraught with conflict—that the oxygen-using mitochondria in our cells entered our ancient, single-celled ancestors as invasive parasites; that the hapless predecessors of photosynthetic structures in plant cells were ingested as food but resisted digestive juices.

Systems scientist Peter Corning uses the term *synergy* to apply to any and all associations with mutually beneficial outcomes, no matter what the initial circumstances. Let us focus on the universality of the outcome, he urges—not the details of the different paths for getting there. Synergism is, of course, widespread in the human realm; indeed, there would be no civilization without it. For our remote human ancestors, cooperation existed only within small family groups. Cooperative organizations expanded to produce multifamily bands, then tribes, then agricultural villages, cities, and empires, then nation-states, and now some forms of economic and social cooperation that span the globe.

This trajectory can be seen as progressive, in the sense that an increase in complexity is a progression. The same holds for the multibillion-year journey of life. To view biological history as progressive doesn't mean that every lineage is becoming more complex or that more complex organisms are somehow better or more important than simpler ones. It just means that as evolution proceeds, more complex organisms and systems tend to show up and that each stage of evolution transcends and includes (incorporates and builds upon) earlier stages. So long as there are synergies to be had of association, and so long as those synergies outweigh the costs and burdens of cooperation, evolution can be expected to bring them forth—in one lineage or another, and quite possibly in more than a few. Often such increased complexity enables the handling of greater flows of energy, matter, and information and can, in human systems, produce an increase in options and freedom of choice (which, of course, brings with it greater responsibility—and complexity).

Evolutionary history teaches us a vital lesson in *how* complexity takes shape. Emergence of more inclusive structures and collectives seems to happen only when the activities of the parts of a system do not diminish the workings of the larger whole. What makes complex systems work is that the "interests" of the whole and the parts all come into alignment. It must genuinely be in the interest of the parts to cooperate in the service of the whole—that is, where the parts benefit by doing so or are disadvantaged in some way by failing to do so. I shall return to this vitally important point in Chapter 16.

> ### "My Life Is Not Just My Own!"
>
> Throughout our travels, Connie and I have been videotaping "stories of awakening" (religious testimonials) from those who have come to embrace the epic of evolution as their cherished creation story. Those interviewed represent a wide range of spiritual and philosophical worldviews. On one occasion, the climax of the interview came when the subject explained the core of the shift that had happened to him. He spoke about how he had begun to see the evolutionary process at work not only "out there," but within him, too. He saw the larger whole of which he was part. "I get it!" he declared. "My life is not just my own!"

As we have seen, cooperation and competition are each fundamental to the evolutionary epic. Looking up in a forest, one can witness the results of a savage "arms race" of competition for sunlight that long ago made trees into towers, driving the redwoods and the pines and the beeches into a frenzy of skyward longing. Today, they rise as high as the pull of water transpiration and the thrashing of storms will allow. Below ground, in contrast, lies a kingdom of goodwill, visible to the human eye only when mushrooms push up through the duff to deliver spores to the wind. Below ground pulses a symbiotic partnership of fungal threads delicately probing a network of roots. Soil minerals, essential to life, are delivered by fungi to the vertical giants. In turn, sun-ripened sugars of the green canopy stream downward, to be shared with the underworld.

Strife co-created tooth of predator with hoof of prey: wolf and caribou, lion and zebra, *T. rex* and Triceratops. It was strife that suggested armor to the armadillo, quills to the porcupine, shell to snail, carapace to terrapin. It was strife that gave keen eyesight to coyote and eagle, night vision to owl, and wings to archaeopteryx. It was strife, in turn, that gave the rabbit its ears, octopus its ink, chameleon its camouflage. Equally, it was strife born of a challenging climate that cloaked rodent and bear in fur and coaxed them to consider hibernation. It was strife that compelled plants to shed their leaves in seasons of frost or drought. And it was strife in its ultimacy—death—that made room on a finite planet for the sheer excess essential for experimentation, novelty, and hence enrichment.

Meanwhile, the synergy born of mutual aid paired petal with pollinator. Symbiosis brought algae and fungi together into the hardiest beings of all, lichens. Symbiosis made possible the evolved partnership of ant and the aphid it milks for honey, the cooperative agreement struck by plant-eating mammals and the plant-digesting microbes in their rumen or gut. Herds, flocks, swarms, schools, hives sing the benefits that come with sociality. Successful partnerships shaped the ecosystems of the world and the biosphere as a whole.

In summary, cooperative synergy and competitive strife are the yin and yang of biological evolution. It is perhaps human nature that drives us to enthrone one or the other as primary. Like hoops within hoops, what appears as competition at one level may be driven by or result in cooperation at the next level up or down.

The Role of Initiative

> **Agent Smith:** *"Why, Mr. Anderson? Why do you do it? Why get up? Why keep fighting? Do you believe you're fighting for something? For more than just your survival? Can you tell me what it is? Do you even know? . . . You can't win. It's pointless to keep fighting. Why do you persist?!"*
> **Neo:** *"Because I choose to."*
>
> —*THE MATRIX REVOLUTIONS*

In one sense, evolution impelled by strife and synergy is something that happens *to* a lineage; in another, evolution is something that a lineage works upon itself. There is agency as well as passivity in the epic of evolution. Antelopes evolved fleetness because at some crucial juncture an ancestor chose to flee. In contrast, rhinoceroses evolved body mass and deadly facial weapons because an ancestor chose to stand firm and face the enemy. It was only because an ant took the initiative to prod the abdomen of an aphid that the ant–aphid symbiosis could evolve. And what is the peacock's tail if not the mark of an extravagant willfulness—the sexual preferences of a long line of pea hens?

Stephen Jay Gould celebrated the role that initiative plays in evolution: "Organisms are not billiard balls, struck in deterministic fashion by the cue of natural selection and rolling to optimal positions on life's table. They influence their own destiny in interesting, complex, and comprehensible ways." Gould's interpretation of evolutionary history puts ordered entities in charge of their own evolution, at least to some degree. No longer passive and pummeled into shape by a ruthless and fickle environment, living and quasi-living systems are seen as agents of their own fates.

"Self-organization" is an umbrella term that unites physicists, chemists, biologists, and computer scientists who are searching for what underlies the growth of order and complexity in a Universe in which powerful forces of entropy and dissolution are also at play. Complexity theorist Stuart Kauffman has written: "We have come to think of selection as essentially the only source of order in the biological world. If 'only' is an overstatement, then surely it is accurate to state that selection is viewed as the overwhelming source of order in the biological world." Kauffman continues,

> It is not that Darwin is wrong, but that he got hold of only part of the truth....It is this single-force view which I believe to be inadequate, for it fails to notice, fails to stress, fails to incorporate the possibility that simple and complex systems exhibit order spontaneously. That spontaneous order exists, however, is hardly mysterious. The nonbiological world is replete with examples, and no one would doubt that similar sources of order are available to living things.... Much of the order we see in organisms may be the direct result not of natural selection but of the natural order selection was privileged to act on.

Self-organization in the abstract, its role in assembling the building blocks of life, may seem far removed from the willfulness we humans know to be at the base of our own lives, our struggles to persist and grow, our quest for selfhood and community. Is there a role for something beyond automatic self-organization in the evolutionary process? Is there perhaps a role for initiative beyond that of an ancestral prey species choosing to flee or to stand its ground, beyond that of an ant prodding the abdomen of an aphid, beyond that of a bird assessing the color and length of each suitor's tail?

Yes! There is a role for conscious evolution. And that is where the human enters the picture. Indeed, the possibility that evolution is open to conscious intent within the realm of human society is the very reason I am writing this book.

What Is Conscious Evolution?

The website of Barbara Marx Hubbard (evolve.org) includes this description of *conscious evolution:*

"Due to the increased power given us through science and technology, we are learning how nature works: the gene, the atom, the brain. We are affecting our own evolution by everything we do. With these new powers we can destroy our life support systems—or we can move toward a hope-filled future of immeasurable possibilities.

"We are the generation of choice, and we do not have much time to choose! Conscious Evolution is the worldview that has arisen precisely at this moment in history to deal with the new human condition. It is a vision and a direction to help us navigate through this transitional period to the next stage of human evolution. As Einstein admonished, humankind cannot solve its problems from the same place of consciousness in which we created them. A new place of consciousness is required.

"In simple terms Conscious Evolution takes place when we intend to grow in consciousness and use our increasing awareness to guide our actions and achieve a positive future. Bela H. Banathy, author of *Guided Evolution of Society,* offers this additional understanding of Conscious Evolution:

✦ It is a process by which we can individually and collectively take responsibility for our future.

✦ It is a process of giving direction to the evolution of human systems by purposeful action.

✦ And most importantly, Conscious Evolution enables us, if we take responsibility for it, to use our creative power to guide our own lives and the evolution of the systems and the communities in which we live and work. It is a process by which individuals and groups, families, organizations, and societies can envision and create images of what should be, and bring those images to life by design."

CHAPTER 3

Evolution and the Revival
of the Human Spirit

I wrote the following sermon in 2002 just before my wife and I launched our itinerant ministry as evolutionary evangelists. I always speak extemporaneously, so this text served only as a template for many of the actual talks I delivered at ecumenical and secular gatherings during our first several years on the road. It reflects only a fraction of the material you will encounter in this book—a small portion of what I see as Great News in a holy view of cosmic history—and it doesn't address any expressly Christian topic (which I will attend to in later chapters). Nevertheless, it serves to introduce two foundational themes. First, a meaningful interpretation of what mainstream science teaches about our vast evolutionary past can enrich virtually any religious or philosophical worldview. Second, a sacred rendering of our evolutionary journey offers enormous practical benefits for leading joyful, on-purpose lives and for recovering from life's inevitable calamities.

A quotation from the great philosopher and father of American pragmatism, William James, addresses the practical difference it makes whether we view the Universe as benign or indifferent. James writes, "From a pragmatic point of view, the difference between living against a background of foreignness [an indifferent Universe] and one of intimacy [a benign Universe] means the difference between a general habit of wariness and one of trust."

✦ ✦ ✦

Have you ever wondered what it would be like to feel such passion for life, gratitude, and a sense of purpose that you could hardly wait to jump out of bed each morning?

What I'm about to share with you I call the gospel of evolution. I call it the gospel of evolution because "gospel" means "good news" and this message is indeed good news. It's the good news of how you can be free of judgment and guilt, how you can access the guiding wisdom of the Universe on a daily basis, how you can have inner peace in times of accelerating change, and how you can find realistic hope when you look into the future.

For your effort, if you pay close attention to what I'm about to share with you and apply this message to your own life, beginning today, I guarantee that this season, and indeed this year, will be one of your best ever, no matter what life throws your way.

The Universe Can Be Trusted

To begin, let us all rejoice that there now exists a vast, worldwide consensus within the scientific community about the nature and history of the Cosmos—from the very, very small (subatomic realm), to the very, very large (the Universe as a whole). We are astonished by the picture of the Universe that scientists paint, by the grandeur and magic of our common creation story, the epic of evolution.

Consider that a hundred years ago we had no idea that atoms were created in the bellies of stars, nor that our Sun was a third or fourth generation star. We also didn't know that stars were organized into galaxies. Even two decades ago, before the Hubble Space telescope, we could not have conceived that galaxies numbered a quarter of a trillion! Your grandparents and great grandparents did not know that the continents slowly slid around the Earth on vast tectonic plates, nor that genetic information is stored within the architecture of a double helix molecule. As I look around this room, I would guess that almost none of us learned in school that the dinosaurs were annihilated by an intruder from space. This knowledge is just too new.

Yet how many of us live our lives with an awareness of the *story* of the Universe, a story of 14 billion years of the comings and goings of stars and planets and life? The Universe is just a place, right? A place where *our* stories unfold, right?

No! There *is* a story, a great story. In fact, it could be called *The* Great Story because it's the story that embraces and includes every other story that has ever been told or ever will be told. Even if extraterrestrials prove to exist, it is their story too because it's the *Universe* story, and everything that exists is, by definition, part of the Universe.

Consider that no matter what happens in the Universe, the story can always be counted on to move in the same five-fold direction: the direction of greater diversity, greater complexity, greater awareness, greater speed of change, *and* greater intimacy with itself.

Imagine a zygote: a fertilized egg in the womb. When your father's sperm and your mother's egg came together, you got your start as one undifferentiated living cell. That cell then doubled and doubled and doubled again. Some cells became eye cells. Others became ear cells, kidney cells, bone cells, and so on. Importantly, after a few months, when your fetus self in utero was able to distinguish light from dark, it would have been silly to think of that transformation as, "Well, the eye cells can see now," or "The eye is seeing now." Rather, what we say is, "The *child* is now able to see." So, too, with hearing, and with all the other senses. When you began to distinguish your mother's and father's voice, would you have expected your parents to think, "Isn't it wonderful that her ear can now hear us?" or "Well, his ear cells seem to be hearing now." No. The excitement is, "Our baby can hear!"

The Universe, first and foremost, is a *Uni*-verse. It is a singularity, a holy whole. It started out as undifferentiated energy and has been expanding and becoming more complex, more aware—and thus more intimate with itself—throughout the last 14 billion years. When the first eye was fashioned, when early creatures could distinguish light from dark, shapes and movement, it wasn't just those creatures that were seeing. It was the *Universe* that was learning to see. It was the *Universe* becoming more aware of itself and experiencing

itself in a more intimate way. When the first ear developed, when living creatures could distinguish different sounds, it was the *Universe* learning to hear. It
was the *Universe* becoming more aware and more intimate with itself through
the complexity we call hearing. So, keeping the analogy of a zygote in mind, let
us now look at the five-fold direction in which the Universe is headed.

Since the beginning of time, the Universe as a whole could be counted on—it
could be trusted (though not without interruption or occasional setbacks)—
to move in these five directions:

✦ greater diversity—more variety and novelty over time.

✦ greater complexity—wholes becoming parts of larger wholes, from atoms
  to molecules to cells to organisms to multicellular organisms, all the way
  up to democracies, satellite telecommunications, and the Internet.

✦ greater awareness—creatures appearing over time with a larger sphere
  of potential consciousness. For example, a turtle has more awareness
  than an amoeba. A horse has more awareness than a turtle. A human
  has more awareness than a horse. Humans living today are *collectively*
  aware of more than what humans living a few hundred years ago
  could have possibly known, before the advent of orbiting telescopes
  and electron microscopes.

✦ greater speed of change—creative breakthroughs happening more
  quickly than in prior times, because each evolutionary advance
  encourages further breakthroughs. Atoms, molecules, life, photosynthesis, multicellularity, vertebrates, mammals, primates, humans,
  agriculture, industry, computers, the Internet. Whew! Each advance
  happens faster and opens the way for the next breakthrough to
  happen even faster. Don't expect things to slow down. Greater speed
  of change is intrinsic to this evolving Universe. That doesn't mean, of
  course, that we can't lead peaceful, centered lives. It *is* possible to have
  peace of mind in the midst of enormous and fast-paced change.

✦ greater intimacy with itself—when the first eye developed, and the
  first ear, it was, literally, the Universe learning to see and hear *itself*,

and thus becoming more intimate with itself. Similarly, the onset of sexual reproduction expressed a new level of intimacy, as did the inception of predation some billion years ago when creatures started eating other creatures. Mammals are the living forms through which the Universe begins to experience its own depths of feeling in ways that reptiles and earlier forms of life could not. It is now through the human that the Universe awakens to its wholeness and to the wonder of existence.

The Universe can be counted on over time—it can be trusted, deeply trusted—to move in the direction of more diversity, more complexity, more awareness, more transformation and growth, faster and faster, *and* more intimacy.

There are two other things the Universe can be counted on doing. First, we can depend on the Universe to feistily hold on to its learnings, its creative breakthroughs, its evolutionary advances. To use relational language, it is fiercely loyal. Since the Universe learned to create complex atoms from simple hydrogen, it has always been able to do so. Once it learned to eat sunlight, or form multicellular organisms, or live on land, it never forgot how to do these things either. In fact, there is no evidence that the Universe has ever lost a major evolutionary breakthrough. Sure, there are plenty of what scientists call "evolutionary dead ends." But as far as we can tell, none of these contributed uniquely to the creative trends of the Universe we just discussed. The Universe has a stubborn habit of preserving its greatest achievements, especially those that contribute to greater diversity, complexity, awareness, intimacy, and speed of change.

Finally, the Universe can be counted on—deeply trusted—to provide every creature and every age with all sorts of problems and breakdowns, stresses and difficulties, and occasionally even full-scale cataclysms to deal with. What we've recently discovered, and what your parents and grandparents never knew, is that problems and breakdowns are normal, natural, even healthy for an evolving, maturing Cosmos. Indeed, they seem to be essential for creativity. As it turns out, every evolutionary advance and every creative breakthrough in the history of the Universe, as best as we can tell, was preceded by some difficulty, often of great severity.

Is this process beginning to sound familiar, perhaps like our own lives?

Too often we confuse what feels bad *to* us with what is bad *for* us. Yet the two are not the same. Who among us has not experienced something we labeled "bad" in the moment—a problem, disaster, whatever—that set in motion events that eventually offered up "good"? When we can let go of our judgments and resentments and come to trust Wisdom beyond our understanding, we discover that *life seems bent on taking what manifests in the moment as bad news and transforming it into good news and further evolutionary development.* Let me give a few examples from Earth's history.

One of the most violent events to regularly occur in the Universe is a supernova explosion. A star is a huge ball of hydrogen gas, compressed by gravity, fusing into helium, which releases an enormous amount of heat and light. In large stars, at least eight times as large as our Sun, when the hydrogen fuel is all used up and even the helium has mostly fused into carbon, a marvelous sequence of cosmic alchemy begins: carbon is fused into neon; neon into oxygen; oxygen into silicon; silicon into calcium, magnesium, and eventually on up to iron. Because iron fusion doesn't produce energy (it requires energy), the star's iron core implodes under excruciating heat and pressure. Heavier elements such as cobalt, nickel, copper, tin, gold, and uranium are formed when the star rebounds in an explosion so spectacular and breathtaking that its brightness briefly outshines its entire galaxy. The explosion seeds the galaxy with a rich assortment of elemental stardust, essential for planets and for life. Except for hydrogen, every atom of your body and everything around you was forged in the womb of an ancestral star. Remember the 1960s song "Woodstock"? "We are stardust, we are golden…" That is literally true. We are stardust evolved to the place that it can now think about itself and tell its own story. We do not merely believe this; we know it.

A supernova explosion is about as violent as the Universe gets. A star is obliterated. Boom! And yet without these explosions, the basic stuff of planets—and life—would not exist. The story would be diminished, and there would be no beings to learn the story, to tell the story, to delight in the stupendous story of Creation! Can you begin to see why Edward O. Wilson, professor of biology at Harvard University and one of the most respected scientists in the world, refers to the epic of evolution as perhaps the greatest religious story we will ever have?

A supernova explosion is goodness. Here is another example of catastrophe catalyzing creativity:

Our planet's first pollution crisis was bad news for virtually everything alive at the time. The early Earth's atmosphere was not at all like it is today. It contained almost no free oxygen. One day, some clever bacteria figured out a way to extract hydrogen from water and passed this skill on to their descendants, who did the same. These bacteria spewed their waste directly into the air, with no concern for the health of the environment. Over time Earth's atmosphere became so polluted with a deadly, toxic poison—oxygen—that life suffered horribly. Bacteria were dying all over the place. But, as always, the Universe pulled another rabbit out of its creative hat. Because it was precisely this bad news that forced different kinds of bacteria to cooperate in ways that they had never done before, which eventuated in cells with a nucleus that could breathe oxygen, then multicellular organisms, then communities of multicellular organisms. So! No oxygen pollution crisis equals no cooperation, no community—nothing more exciting than anaerobic bacteria.

Another example calls up our memory of the dinosaurs. Dinosaurs ruled the continents for more than 150 million years (*much* longer than us). During this time, which scientists call the Mesozoic era, mammals were small scruffy creatures who stayed in burrows and mostly came out just at night, because, as you can well imagine, they were terrified of big, ugly, carnivorous dinosaurs. Then one spring day, a terrible catastrophe struck. An asteroid 10 miles across, traveling at a speed of 50,000 miles per hour, crashed into our planet just off the Yucatán peninsula of what is today Mexico, punching out a crater 100 miles wide. Imagine all the nuclear weapons that our species has ever created being launched and arriving at the same destination at exactly the same moment … and then multiply that by a thousand. That's right. This event, 65 million years ago, was a thousand times more powerful than all our nuclear weapons combined.

The meteor impact that wiped out the dinosaurs turned the sky into a cauldron of sulfuric acid. It also triggered a magnitude 12 earthquake, which is a million times more powerful than a magnitude 6 earthquake. This, in turn, unleashed at least six mega tsunamis, several of which were more than 300 feet high. The impact ignited a global firestorm that incinerated perhaps a quarter of the living biomass, releasing so much carbon dioxide that the average global temperature (after first plunging into cold, owing to the cloud of dust obscuring the Sun) later rose by 20 degrees Fahrenheit and stayed that way for a million years. Whether taken out by the firestorm, the acid rain, the tsunamis, or the extreme fluctuations in temperature, three out of every four species alive at the time went extinct. The biggest creatures were hit the hardest, and thus each and every species of what we loosely call "dinosaurs" went extinct, along with all the pterosaurs of the air and mosasaurs and plesiosaurs and ichthyosaurs of the sea.

All in all, it was not one of Earth's better days. But thankfully, from our perspective, it was precisely this catastrophe that allowed those mammals who survived in their burrows to flourish and diversify, culminating in all the amazing mammals of the world today, including ourselves. So: no catastrophe, no whales or dolphins, no dogs or cats, no giraffes or elephants, no lions and tigers and bears (oh, my!), and, of course, no me, no you.

So the next time a comet crashes into your psyche or your life feels like sulfuric acid is raining on your head, or the next time a magnitude 12 earthquake rocks your world and you feel like you want to hide away in a dark hole for several months, just remember: in a few million years, things will be fine. Seriously though, the next time you're greeted by a 300-foot tsunami at home or at the office, just remember that you are part of an amazing, creative Universe that turns chaos and catastrophes into new growth and opportunities as regularly as day follows night. This is *very* good news.

Let's look at one last example of goodness wrapped in a strange package. It is a case study a whole lot closer to us in time than the dinosaur extinction. One of the most traumatic periods in human history began in the 14th century when Asia and Europe were ravaged by the bubonic plague, also known as the Black Death. Reaching Europe in 1347, this plague was so devastating that within just a few years it killed off some 25 million people—one-third of the population of Europe. Almost half the inhabitants of Florence died within three months—and no one knew why. Even when the worst was over, smaller outbreaks continued, not just for years but for centuries. Survivors lived in perpetual fear of the plague's return, and the disease didn't disappear from Europe until the 1600s.

As you can imagine, medieval society never recovered from the effects of the plague. So many people had died that there were serious labor shortages all over Europe, which led workers to demand higher wages. But landlords refused those demands, so by the end of the 1300s, peasant revolts broke out in England, France, Belgium, and Italy. The disease also took its toll on the religious establishment. People throughout Christendom had prayed devoutly for deliverance from the plague. Why hadn't those prayers been answered? A new period of political turmoil and philosophical questioning ensued. So...bad news, right? Of course! Yet many scholars and historians conclude that not only did the plague lead to major political and religious reforms but, apparently, it might have been the primary impetus that launched modern scientific inquiry, which eventually gave us the Great Story, the epic of evolution. How so? Because many educated people during this time refused to believe that all the anguish and suffering and loss were simply the result of God's wrath. So there began a fierce drive to figure out how the world works, which eventuated in, among other things, the discovery that the plague was spread by fleas on rats. So...no plague, quite possibly no awareness of much of what we now know to be true about the nature of the Universe and our role in it.

What becomes obvious from these examples is that, to speak metaphorically, the Universe seems resolutely determined to take bad news and turn it into new creativity. That is, on the other side of Good Friday is Easter Sunday. And so the first insight, the first affirmation of faith and confidence in the gospel of evolution, is that *the Universe can be trusted*. Specifically, it can be trusted to move in the direction of greater diversity, complexity, awareness, intimacy, and speed of change. It can be trusted to preserve its breakthroughs. And it can be trusted to provide a wealth of problems and breakdowns that fuel the creative process.

Now, all this is well and good for stars and planets and life. But what about you and me? What about our own little stories within the Great Story? Can we trust that the wisdom of the Universe is at work in our own era, in our own little dramas?

You Are Part of the Universe

A Native American elder, Black Elk, said this: "The first peace, which is the most important, is that which comes within the souls of men and women when they realize their relationship, their oneness, with the Universe and all its powers."

Thanks to science, we now can see that a relationship with the Cosmos awaits us. We begin to catch a glimpse of the awesome role of the human in the Universe process—and of our destiny as *Homo sapiens*, or wise humans. For with us, the Universe brought into existence a creature by which it can begin to know—to reflect on—its vastness, its beauty, its amazing journey. You see, we didn't come *into* this world, we grew out from it. Black Elk knew this, at least intuitively. We humans are not separate creatures *on* Earth, *in* a Universe. We are a mode of being *of* Earth, an expression of the Universe.

A human being looking through a telescope is literally the Universe looking at itself and saying, "Wow!" A student looking through a microscope is Planet Earth learning in consciousness, with awareness, how it has functioned unconsciously and instinctually for billions of years.

We are a means by which Nature can appreciate its beauty and feel its splendor. We also, of course, contribute in our own ways to the chaos and breakdowns that will, in turn, catalyze further creativity. As we have already seen, if the Universe can be trusted to do anything, it surely can be trusted to do that!

What would it be like to spend the next day or week or month playing with this mindset? What would life be like if you viewed your problems as blessings in disguise—knowing you are part of the Universe, a mode of expression of the Universe? If you got that far, then it's just a stroll through the woods to come to embrace the third point in this gospel, this good news, of evolution: Accept what is and be in integrity.

Accept What Is and Be in Integrity

Accepting what is and being in integrity is another way of doing what most other animals and species do naturally, without effort, and without hesitation or foreboding. Other animals don't waste time grumbling about events that come their way. As far as we can tell, they don't fret about all the things that could go wrong. They don't tell themselves that this or that should or shouldn't have happened. They just accept whatever is real, whatever happens to them and to the world, and then they make the best of it.

For us humans to attain a similar freedom from judgment and regret we must begin by trusting the Universe, fostering faith in God, *making life right*. These are three ways of saying essentially the same thing. Each approach nurtures a peaceful heart and mind, which in turn allows for clear communication between us and the larger Creative Reality in which we all live and move and have our being. It's trusting that the same wisdom and intelligence that has brought the Universe along for 14 billion years is still at work. And it's trusting that all the difficult, painful, or discouraging experiences in our own lives, and in the world as a whole, are nevertheless part of the creative process, and can be embraced by the arms of faith.

What this means for me is that now when something painful or traumatic happens, or when something frustrating occurs, the first thing I do is stop and really feel my feelings. Then I act *as if* both my feelings and the triggering event are gifts and blessings in disguise—that the Universe is conspiring on my behalf. Whether this is true or not, I cannot know. It really doesn't matter. This way of perceiving is transformational and empowering. "The Universe can be trusted" is a very useful belief. When I act *as if* all things work together for the good of those who love Reality and are called to serve a higher purpose, I love my life! What more could I want?

"Accepting what is," "trusting the Universe," "making life right," "celebrating Reality" are all inclusive ways of saying what many religious people—and people in recovery—mean when they speak of "having faith in God" or "trusting their Higher Power." It really is the ticket to emotional, psychological, and spiritual freedom, and to a peace that transcends understanding. Accepting life on its own terms, just as it is—with all the challenges—is to remove oneself from the judgment throne of deciding which things are good and which things are bad. It's letting the Universe/God play that role.

To have a powerful relationship with your own intuition and instincts—and thus to have a clear channel of communication with the creating, sustaining Life Force of the Universe (whatever you may choose to call It/Him/Her)—one must cultivate humility in this sense: *Stop assuming that you know best how things are supposed to go in the world.* Rather, try on an attitude of gratitude—not just for what is easy to be grateful for, but also for those challenges and difficulties in life for which you cannot yet detect a silver lining.

Having faith and being in integrity means trusting that each and every one of us is doing the best we can, given what we've got to work with at the time. It's trusting that, from the perspective of the Universe, everything may be "right on schedule." This is a powerful way to live, and it need not diminish the urgency to act and to be of service in the world. A trusting attitude will actually strengthen the urge to be in action, because we know that the Universe works through us, too—through our own deeds. For example, I trust that our Western consumer culture is not a cosmic mistake, but I am also doing all that I can to help it recover from its addictive patterns and thus to mature beyond its present self-destructive and Earth-destructive habits. Looking within, I trust that my shadow—my prideful, arrogant, selfish, seductive side—serves a purpose, but I am also committed to being a humble, thoughtful, compassionate, and faithful man. We can trust that those who oppress others are less evil than they are ignorant, and at the same time we can do everything within our power to ensure that freedom and justice prevail. Thus, trusting the Universe also means trusting that the anguish and anger that we sometimes feel over what is happening to the oppressed and to our world, and the yearnings we have for a more just and sustainable society, are part of the Universe too, and are meant to propel us into action.

It does take effort to remind ourselves, especially when troubles abound, that "the Universe can be trusted" and that "I am part of the Universe." But I promise, it won't take long for this outlook to become second nature. The mind will almost always take the most empowering route available to it, when given the chance. But don't take my word for it. Play with it yourself. See how it works for you.

Grow in Evolutionary Integrity

To live life fully and love the life you live you must be committed to deep, or evolutionary, integrity. If you want to experience real joy, true peace, and lasting fulfillment, there's really no other way. To grow in evolutionary integrity means getting right with Reality (God/the Universe) by growing in humility, authenticity, responsibility, and service to the Whole.

Why humility? Because what's undeniably so is that the Universe is primary and you are derivative. Said another way, you are not the center of the Universe and your ego does not run the show. We were not thrust *into* the Universe, we were born out of it. You couldn't exist without it and the Universe would do just fine without you. And that's the truth! Humility and its twin sister, trust, are thus essential because only when you're coming from a place of humility are you in touch with Reality as it actually is.

Why authenticity? Because only by being authentic are you aligned with Reality. Honesty, transparency, and authenticity enable the feedback necessary for individuals and groups to evolve in healthy ways. We may sometimes be tempted to lie or present ourselves in false ways because of the promise of a cheap thrill. Deception never, however, provides lasting joy.

Why responsibility? Because what's really *real* is that there is only one person responsible for the quality of your life, and that person is you. As Jack Canfield advises in his book *The Success Principles*,

> If you want to be successful, you have to take 100 percent responsibility for everything that you experience in your life. This includes the level of your achievements, the results you produce, the quality of your relationships, the state of your health and physical fitness, your income, your debts, your feelings—everything! This is not easy. In fact, most of us have been conditioned to blame something outside of ourselves for the parts of our life we don't like. We blame our parents, our bosses, our friends, the media, our co-workers, our clients, our spouse, the weather, the economy, our astrological chart, our lack of money—anyone or anything we can pin the blame on. We never want to look at where the real problem is—ourselves.

Only by taking full responsibility for our lives, and the wake we've left, can we know heaven on Earth. Righteous indignation may feel good in the moment but blame never yields true happiness.

Why service to the Whole? Because it is everything to us: our source and sustenance, our Alpha and Omega, beginning and end. Whatever we may choose to call the Whole, and whatever metaphors or analogies we use to describe it, the undeniable fact is that Ultimate Reality is creator of all things, knows all things, reveals all things, is present everywhere, transcends and includes all things, expresses all forms of power, holds everything together, suffers all things, and transforms all things.

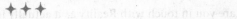

The good news here is that while it is possible to *feel* alienated from the Universe (when we are out of integrity, judging events negatively, or casting blame), the fact is that it is impossible to ever *be* alienated—no matter what. You *are* part of the Universe. Achieving enlightenment, freedom, salvation, empowerment is as easy (and as challenging) as developing a habit of trusting what's real and growing in humility, authenticity, responsibility, and service to the Whole—that is, growing in evolutionary integrity.

Trusting the Universe Means Welcoming Challenges

Your shadow side—your mistakes, your shortcomings, your sins, the places you've been out of integrity—all these are part of the Universe as well, and can be trusted just like everything else. Your sins and shortcomings, your vices and transgressions, are a necessary part of the creative process of the Cosmos. They're as essential as comets and earthquakes and plagues. Isn't that a comforting thought!

Now, this is really good news. All my life I thought my problems and difficulties were evidence that either *I* was fundamentally flawed, or someone else was to blame. Now I see my problems and difficulties, and our world's challenges as well, as gifts for my and our evolution, evidence that we're all alive and growing, and evidence that our species is maturing. So we can forgive ourselves, and others, for all the unloving, stupid, selfish, arrogant, ugly, petty, nasty things we've all said and done because the Universe/God has already forgiven us. In fact, I'm not sure forgiveness is quite the right word. Let's just say our sins and failings can be used as compost for new growth.

You see, as Christians claim—ultimately, we *are* "saved by grace through faith." Of course, this doesn't mean that Jews and Buddhists and Muslims and Taoists and Confucians are wrong. Each religious tradition on the planet, and every philosophical belief system, has unique gifts and limitations. Different religions are like different flowers. Each one has its own special fragrance and beauty. The Great Story embraces them all for the simple reason that it's the *Universe* story, which, of course, includes the story of how every religion came into existence at a particular time and in a particular place, how each supported group-level cooperation, and how each has evolved.

Here is an affirmation that when you commit it to memory and rehearse it throughout the day, will add tremendous value to your life: "I trust that everything is perfect for my growth and learning." Here's another: "I trust the

Universe. There is nothing I can't forgive or find a way to appreciate or have compassion for."

Finally, I invite you to consider that this message is not only good news for us as individuals; it's also good news for us as a species. When we look around and see what's happening to others and to our world, and when we realize how much work still needs to be done, it is all too easy to give up in despair. And yet, what if the Universe really *can* be trusted? What if the enormous challenges ahead are precisely what we need to compel us to make healthy changes? What if these challenges are exactly what we need to move us beyond our disputes? Is it possible that all really *is* right on schedule? I, for one, choose to trust that this is so. In the words of cultural historian Thomas Berry,

> The basic mood of the future might well be one of confidence in the continuing revelation that takes place in and through the Earth. If the dynamics of the Universe from the beginning shaped the course of the heavens, lighted the Sun, and formed the Earth, if this same dynamism brought forth the continents and seas and atmosphere, if it awakened life in the primordial cell and then brought into being the unnumbered variety of living beings, and finally brought us into being and guided us safely through the turbulent centuries, there is reason to believe that this same guiding process is precisely what has awakened in us our present understanding of ourselves and our relation to this stupendous process. Sensitized to such guidance from the very structure and functioning of the Universe, we can have confidence in the future that awaits the human venture.

PART II

Reality
Is Speaking

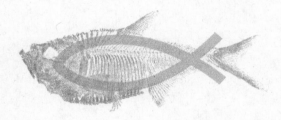

"I like to think of our challenges as weaving. We are all weavers.
The weaver constructs the warp, anchoring it to the loom, and
then, by working the weft in and through the warp, creates
patterns and the entire tapestry. The Epic of Evolution, in
scientific form, is the warp on which all present and future
meaning for our lives must be woven. There is no single correct
way in which the weaving will take shape, no single authorized
manner in which the Epic must appear in our worldviews. All
of the various weavers of meaning will find something common
in the warp. In the cultural crises that face us all, each will
learn from how others move within the loom's constraints and
possibilities. We humans are the cultural religious animals of
evolution on our planet. We are here to weave the spiritualities
that are lifegiving for our phase of the Epic of Evolution and
for the next generation."

—PHILIP HEFNER

Reality Is Speaking

CHAPTER 4

Private and Public Revelation

"Science is the structuring and organization of knowledge such that it can be tested, archived, restored, communicated accurately, built upon, and extended—and such that new varieties can easily appear. Most importantly, science is how knowledge is structured so that it can be structured further."

—KEVIN KELLY

We are at a turning point in human history, and it has everything to do with embracing a holy view of deep time. *Deep time* lends a perspective that extends for billions of years—not just hundreds or thousands of years—into the past and the future, during which the Universe, Earth, life, and humanity evolved and, importantly, are still evolving. Catalyzing this transformation is the modern method by which we collectively access and expand our understanding of the nature of reality. New truths no longer spring fully formed from the traditional founts of knowledge. Rather, they are hatched and challenged in the public arena of science. This is the realm of *public revelation*.

In contrast, *private revelation* entails claims about reality that arise primarily from personal experiences—some of which are compelling. Alas, private revelations enshrined for centuries in sacred texts cannot be empirically verified today. Such claims cannot be proven because they are one-person, one-time occurrences, obscured by the passage of time. Accordingly, private revelations must be either believed or not believed. When private revelations reside at the core of religious understandings, people are left with no choice but to believe them or not.

For example: Is it true that the entire Universe was created in six literal days, as suggested in the first chapter of Genesis? Today millions of people believe so—and millions do not. The result personally: families sundered by theological differences. The result collectively: intense conflict over the teaching of science in America's public schools.

"To Connect Religiously with This Awesome Monument"

When Connie and I offer programs in secular or religiously liberal contexts, audience members not uncommonly assume that these self-proclaimed "evolutionary evangelists" disapprove of young-Earth-creationist interpretations of natural history. Yes—and no. During our visit to the Grand Canyon, we were pleased to find that the park's interpretive center was selling a controversial book. *Grand Canyon: A Different View*, by veteran Colorado River guide Tom Vail, is a beautiful picture book that interprets all features of the Grand Canyon as persuasive evidence of Noah's Flood, and thus of a young Earth in which evolution plays no role.

I explain to my audiences, "I'm thrilled that *all* people who visit the Grand Canyon can be given a way to connect religiously with this awesome monument of Earth history!" Similarly, when Connie visited the famous Paluxy River site of dinosaur tracks at a state park in Texas, and then visited the nearby Museum of Creation Science, she surprised her host by announcing that she was happy to see a busload of Christian Academy kids taking a tour of that museum. "I'd support just about anything that helps kids feel a historical connection to the marvels of this place!"

Connie's and my chosen work, our mission, is to do all that we can to make the evolutionary story more appealing (religiously and emotionally, as well as intellectually) to our largely Christian culture. But until we, and others, have a big impact, we're happy to see even the old flat-earth cosmological interpretations used in ways that connect an otherwise "amythic" and unstoried culture to the grandeur of this continent's natural heritage. I share this perspective first voiced by Connie: "I'd rather have kids learn about evolution in meaningful ways in church than meaninglessly in school."

Is it in fact the case that devout Jews and Christians will burn forever in hell because they do not embrace as the Word of God the teachings of the prophet Mohammed, as recorded in the Qur'an? Hundreds of millions of Muslims believe this is so. And hundreds of millions of non-Muslims (as well as many liberal Muslims) believe not. The result personally: good people who come to harbor judgment and resentment against other good people, as well as the heartache and estrangement that happens when those "others" are kin. The result collectively: communities and nations are divided and at war.

Is it historically true that God intentionally, purposefully drowned billions of animals and tens of millions of human beings in Noah's flood and instructed Moses to kill millions of men, women, and innocent children, as the Bible literally reads? (e.g., Genesis 6–9; Exodus 32:27–28; Deuteronomy 2:34, 3:4–5, 7:1–2; Joshua 11:12–15) Countless people believe that these stories reveal God's unchanging moral character. Countless others believe they do not. The personal result: millions who leave their religious traditions, unable to worship such a God. The collective result: warring nations, each convinced that God is on their side.

These are the conundrums that worldviews based on private revelation, embedded in unchangeable scripture, inescapably promote. And they are by their very nature unresolvable. That is, short of worldwide conversion to one belief system or worldwide expulsion of all such belief systems, the future of humanity will continue to be compromised by adversities born of conflicting beliefs—especially in a world in which weapons of mass destruction now come in small packages.

Is there, perhaps, another way?

Beyond Belief

"The new naturalism does a superior job of telling everybody's story. It is more durable than metaphysical perspectives precisely because it rejects claims to finality, inviting upon itself empirical scrutiny and falsification. It gives us what is to date the most reliable and satisfying account of where we came from, what our nature is, and how we should live." —LOYAL RUE

Private revelations, as subjective claims for which no evidence for or against would be universally compelling, can only be believed or not believed. Private revelations, thus, cannot be *known*. In contrast, the arena of public revelation offers opportunities for us to learn ever more about the nature of reality—and to recognize and revise mistaken notions. People of all philosophical and religious backgrounds can therein come to agree on the same basic understandings, regardless of differences in how those shared understandings will be interpreted. All religions and worldviews already contribute some of their most inquisitive, capable, and devoted citizens to the now-planetary effort of public revelation—otherwise known as the scientific endeavor.

The mindset that welcomes public revelation is marked not only by openness and curiosity. Such a mindset is grounded in a trust so solid that nothing that might be revealed would shake its foundation. What is this trust? Here is mine:

> *I have faith in the God-given ability of myself and others—individually and especially collaboratively—to interpret any new discovery made in any of the sciences in lifegiving ways that serve the whole.*

Thanks to the scientific method, assisted by the wonders of modern technologies (themselves a gift of the scientific enterprise), public revelation emerges when claims about the nature of reality are based on measurable data and can be tested and modified in light of evidence and concerted attempts to disprove such claims. This process typically results in understandings so distinct from belief and so removed from cultural contexts that they can be regarded, for all practical purposes, as factual. From this perspective, the history of humanity is a fascinating and universally relevant story of how Reality has progressively revealed itself to human beings, which is tied to how we acquire, share, store, and reconsider knowledge.

One of the most fruitful ways biologists have come to think about the major transitions in life's several-billion-year history turns on how learned information (knowledge) is stored and shared in increasingly sophisticated ways that give rise to greater awareness, freedom, complexity, and cooperation. Crucially, "knowledge" in this sense is something that humans have no exclusive claim to; lineages of bacteria have been on the knowledge-sharing path for eons. One way to think about the major transitions in evolution, as John Maynard Smith and Eörs Szathmáry proposed, is to look at how information is organized within and between various forms of life and how it carries forward from one generation to the next. In this view, improvements in how information is transmitted into the future are what account for advances in biological and cultural complexity.

For example, by making possible an ordered recombination of traits from each partner, rather than just the random diversity of gene-by-gene mutations, the invention of sexual reproduction was a significant leap in evolvability. The emergence of multicellular entities—plants, animals, and fungi—and then, later, cooperative groups of beings, such as insect and primate societies, are other examples of advances in knowledge organization and transmittal. A monkey differs from a jellyfish in many ways, but none more important than the additional layers by which information flows through the monkey from one generation to the next. For example, instruction of young by social mammals is a huge evolutionary leap beyond the genetic transmission of behaviors. We see the same dynamic at work culturally. Driving social complexity—from tribes and villages to chiefdoms and kingdoms to early nations to global markets— are the ways information is stored and shared at each level.

The Birth and Maturation of Public Revelation

"What we have is a new story of the Universe we can see and imagine, set within an ultimate mystery, a myth like all the preceding ones of whatever culture or religion. Except that this one is woven by all of humankind from far wider strands of experience and imagination, fed by electronic circuitries and seen through electronic eyes, and shepherded by a common scientific method. It is our ultimate personal history, tracing our individual existence beyond our families, beyond our species, beyond life, beyond the earliest galaxies, and finally beyond even the primordial fireball to the time when there was no time, the place where there was no space, and there was only mystery." —ERIC CARLSON

Scientific knowledge has grown through time in much the same way as life evolved from simple to complex. Major leaps in both realms spring from innovations in the way that information is organized, stored, and carried into the next generations. Indeed, our own Information Age is the latest innovation of the immense journey of life on Earth. Cultural innovation is the newest expression of biological innovation. Kevin Kelly, founder of *Wired* magazine, posted an essay on his website (kk.org/thetechnium) that tracks the evolution of information technologies in the human phase of the epic of evolution. Kelly's "Major Transitions in Technology" takes the reader on a quick tour of the achievements in information technologies, which in turn have made the realm of public revelation not only possible but ever expanding and ever more reliable.

Kelly begins this tour at the language threshold, when primate forms of communication gave way in our own lineage to symbolic form. "No transition has affected our species, or the world at large, more than the creation of language," he writes. Storytelling provided our ancestors (and still provides us) with an emotionally compelling and memorable way to learn and retain important bits of information. More, the lessons one learns in even a brief life could be passed on to future generations even if the discoverer dies childless. Symbolic language allows information to leap ahead of purely genetic ways of carrying into the future the successes of the past.

The transition from orality to literacy yet again expanded the human capacity to share, retain, and pass forward information. Now, Kelly writes, "ideas could be remembered more accurately, and just as importantly, their organization could be examined and analyzed." With the advent of writing came the possibility of calendars, inventories of harvests, records of sales and debts, noteworthy events, and, of course, scripture. With the invention of the printing press, the capacity and the accuracy of information storage and transfer ramped up yet again. Kelly writes, "As printing became ubiquitous so did symbolic manipulation. Libraries, catalogs, cross-referencing, dictionaries, concordances, and publishing of observations all blossomed, producing a new level of organization of bits present everywhere—to the extent that we don't even notice that printing covers our visual landscape."

Enter science as a networked enterprise, no longer merely the province of geniuses working in isolation: "The scientific method followed printing as a more refined way to deal with the exploding amount of information humans were generating," Kelly writes. "Via scholarly correspondence and later journals, science offered a method of extracting reliable information, testing it, and then linking it to a growing body of other tested, interlinked facts." The new knowledge made it possible for humanity to create and improve tools that would further expand scientific access to new facts and understandings.

Meanwhile, because information was both ordered and retrievable, innovators from all realms of life put the new knowledge to use. Kelly writes, "When the scientific method was applied to craft, we invented mass production of interchangeable parts, the assembly line, efficiency, and specialization. All these forms of information organization launched the incredible rise in standards of living we take for granted." Today another transition in the way information is stored, retrieved, and carried forward is under way. Kelly concludes,

We are in the midst of a movement where we embed information into all matter around us. We inject order into everything we manufacture by designing it, but now we are also adding microscopic chips that can perform feats of computation and communication. Even the smallest disposable item will share a sliver of our collective mind. This all-pervasive flow of information, expanded to include manufactured objects as well as humans, and distributed around the globe in one large web, is the greatest (but not final) ordering of information. And it marks the most recent major stage of technology.

In another web-posted essay, Kevin Kelly provides a chronology of how the scientific enterprise steadily improved the scope and reliability of factual knowledge gained about the Universe at all nested levels. Posted in 2004 on his website, The Technium, Kelly summarizes the "Evolution of the Scientific Method" via this list:

280 BCE	Libraries with Index
1000 CE	Collaborative Encyclopedia
1410	Cross-referenced Encyclopedia
1550	Distinction of the Fact
1590	Controlled Experiments
1609	Scopes and Laboratories

1650	Societies of Experts
1665	Necessary Repeatability
1687	Hypothesis/Prediction
1752	Peer Review Referee
1780	Journal Network
1920	Falsifiable Testability
1926	Randomized Design
1937	Controlled Placebo
1946	Computer Simulations
1950	Double Blind Refinement
1974	Meta-analysis

In reflecting on this chronology, Kelly observes:

> Today, the organization of knowledge within science is extremely layered, richly convoluted, and present at many levels. In research we have double-blind clinical trials and tests for the validity of simulations, for example. The scientific method today bears little resemblance to the earliest attempts at science 400 years ago, before the advent of experiment, report, peer review, and other inventions.

Flat-Earth Faith Versus Evolutionary Faith

"It's all a question of story. We are in trouble just now because we are in between stories. The Old Story—the account of how the world came to be and how we fit into it—sustained us for a long time. It shaped our emotional attitudes, provided us with life purpose, energized action, consecrated suffering, integrated knowledge, guided education. We awoke in the morning and knew where we were. We could answer the questions of our children. We could identify crime, punish transgressors. Everything was taken care of because the story was there. But now it is no longer functioning properly, and we have not yet learned the New Story."

—THOMAS BERRY

A distinction must be made at this point between *flat-earth faith* and *evolutionary faith*, as I shall use these terms throughout the rest of this book. What I mean by flat-earth faith is *not* people believing the world is flat. Rather, it refers to any perspective in which the metaphors and theology still in use came into being at a time when peoples really did believe the world was flat—that is, when there was no reliable way for humans to comprehend the world around them by means of science-based public revelation. Religious traditions that are scripturally based, and whose texts have not changed substantially since the time of Copernicus, Galileo, Newton, Darwin, Einstein, Hubble, Crick, Dawkins, and Hawking become, necessarily, flat-earth faiths when interpreted literally.

The term "flat-earth faith" applies to Eastern religions as well as Western. The Vedas, Upanishads, Bhagavad Gita, and Smriti of Hinduism are flat-earth texts. They were written before anyone knew empirically that Earth resided in the heavens no less than does Venus, that stars are suns very far away, that for billions of years single cells were the only forms of life, that polar bears evolved after *Homo habilis*. If these sacred Hindu texts are interpreted literally today, then that form of Hinduism is, by definition, a flat-earth faith.

"Flat-earth faith" also applies to literal interpretations of Buddhism's Tripitaka. It applies to literal forms of Taoism, grounded in the Tao Te Ching, and to literal forms of Confucianism, grounded in the Analects. It applies to Judaism if the Torah or any other component of the Tanakh is read literally. It applies to Christianity if the Bible is mistaken for a science text. It applies to Islam if the cosmology embedded in the Qur'an is regarded as inerrant.

"Flat-earth" also characterizes literal understandings of *commentaries* on sacred scriptures that also predate (or ignore) the evolving cosmological perspective that is the fruit of modern science. Examples would include allegiance to literal interpretations of the Talmud and various midrashes in the Jewish tradition. Perhaps the most prominent flat-earth commentary in the Christian tradition is the Nicene Creed ("We believe in one God, the Father, the Almighty, maker of heaven and earth, of all that is, seen and unseen ... ").

Because the Nicene Creed is an *interpretation* of the core message of the early Christian scriptures, its flat-earth metaphors require modern Christians to speak as if they accept a literal translation of the Bible, although in their minds they may be making the requisite translations. Some liberal Christians bristle at the prospect of having to recite "We believe" for something that they do not in fact believe. Thus in many ways, today's continuing use of flat-earth commentaries to interpret flat-earth scriptures is the most problematic of all. Other Christian commentaries that often contain flat-earth components include papal encyclicals and thousands of sermons delivered every Sunday from pulpits throughout the world.

A flat-earth commentary from the Islamic world (which, when taken literally by young Muslim men can bring untold suffering to innocents) is one section of the Hadiths. Here, Islamic martyrs are promised not only eternal life in heaven but special dispensations that young men living in a sexually conservative culture find especially attractive.

Overall, "flat-earth" applies to any perspective whose understanding of Creation and whose guidelines for human behavior do not square with public revelation. It applies to any worldview whose cosmology assumes that divine creation was something that happened only in the past, or that salvation is something that happens only in the future. "Flat-earth" refers, as well, to any system of ethics that does not enthusiastically teach that six billion human beings are all members of one family, to be treated as kin, with all their irksome foibles.

As I shall describe in Chapter 9, "flat earth" is also an apt description of any psychological perspective that presumes humans are inherently (and only) good. Such understandings haven't kept pace with the discoveries of evolutionary brain science and evolutionary psychology. To ignore that we all have built-in capacities for ill as well as good is to fail to appreciate the complexity of our instincts.

Flat-earth perspectives are also those that elevate humans at the expense of Earth's millions of other species. Traditions that fail to teach the basic principles of ecological living in this technological and populous era threaten planetary, as well as our own, well-being. Without such instruction, it is all too easy for humans to live in ways that harm the Earth community and future generations.

Finally, and most important (because this is the underlying cause of all other forms of flat-earth religiosity), "flat earth" applies to any spiritual perspective that does not, in effect, thank God for evolution.

Toward an Evolutionary Christianity

"How important it is that we learn the sacred story of our evolutionary Universe, just as we have learned our cultural and religious stories. Each day we will do what humans do best: Be amazed! Be filled with reverence! Contemplate! Fall in Love! Be entranced by the wonder of the Universe, the uniqueness of each being, the beauty of Creation, its new revelation each day, and the Divine Presence with all!"
—MARY SOUTHARD

A holy understanding of evolution will usher the world's religions into their greatness in the 21st century. Thomas Berry cautions that we will not be able to move into an ecologically sustainable future on the resources of the existing religious traditions—and we can't get there without them. Nothing is more important for the future of the human species than the emergence of evolutionary forms of every religious tradition. Here is how I envision deep-time celebration of my own faith:

> *Evolutionary Christianity is an integral formulation of the Christian faith that honors biblical and traditional expressions, conservative and liberal, while enthusiastically embracing a deep-time worldview. Evolutionary Christianity interprets the entire history of the Universe in God-glorifying, Christ-edifying, scripture-honoring ways.*

Too many spiritual options on the modern menu are still stuck (and thus undermined at their core) by one or more flat-earth assumptions. The most limiting assumption of all is that private revelation is still where it's at—and the more ancient, the better. Public revelation is still widely regarded suspiciously, and as nothing other than secular science. Even so, religious fundamentalists do have one thing very right: reconciling ancient texts with modern understandings is no solution. What the human soul craves is not concordance at the expense of enthusiasm. It is not right to command a flat-earth faith to jump into a lukewarm bath drawn by science or to step into a cold shower of reason. Thus, to my mind, mere reconciliation of the old with the new is a cop-out. The translation must be bold; it must resonate with the core truths of the tradition. Apostles of evolution must strive for an Evolutionary Christianity, an Evolutionary Hinduism, an Evolutionary Islam, an Evolutionary Judaism, and more.

Let us admit to ourselves and to one another that the glory of our religious traditions is compromised by flat-earth cosmologies. Let us open to the possibility that our human community as a whole, with divine guidance, can turn this crisis into a creative and healthy transformation. Crucially, let us cultivate faith in the possibility that any and all of our cherished religious and spiritual traditions can mature into vibrant evolutionary forms, forms that enrich rather than diminish what we cherish most from the past.

To begin to craft evolutionary forms of our traditions, we must engage in the task of translating the symbols, metaphors, doctrines, and theology of each. Such translations will come from *within* the traditions. You will not see me offering any such translations for Islam, Buddhism, or Hinduism in this book. Only my own faith heritage, Christianity, will be given the rudiments of an evolutionary gloss. But I do hope that the course I chart toward an evolutionary form of Christianity will encourage the adventurous in other traditions to undertake similar projects—and for other Christians to join me in this particular effort.

My intent for Evolutionary Christianity is not only to reinvigorate scripture by updating the cosmology. Translation is also required for meshing the ancient texts ethically with the ways in which our understandings of human nature have evolved, and with the ways in which human societies have evolved as well. An evolutionary form of Christianity will thereby flourish in its refreshed ability to offer sound guidance for living in a postmodern and highly interconnected world.

The core teachings of Christianity will remain foundational. The marvels of public revelation will not unseat them. Jesus as "the way, the truth, and the life" will still be central in an evolutionary form of Christianity, just as the backbone of our common ancestor who swam in the sea more than 400 million years ago is still within us, providing vital support. Moreover, Jesus as "the way, the truth, and the life" will be universalized. The doctrine will be presented in ways that make sense to those in other traditions who are doing the work of evolving their heritages.

Of necessity, this evolutionary effort will also mean that some of the teachings will be translated almost beyond recognition, just as our skin is so unlike that of our scaly reptilian ancestors. Then, too, some passages will have so little utility that they will disappear, just as the primate tail was lost within our lineage of apes. Some of the ancient teachings may, of course, be problematic no matter what the translation. Biological evolution will likely never be able to fix the problem of the tube in our throat that does double-duty—putting us at risk of asphyxiation each time we swallow a morsel of food. And some of our religious inheritance will carry into the future in forms that, although reduced, will put us at risk of life-threatening illness. The sickness might originate and fester in just a tiny subset of the body, but with little warning it may burst in ways devastating to the whole in the same way as the rudimentary appendix of the human intestine can provoke life-threatening sepsis.

Virtually every religion on the planet today has already begun what will surely be a decades-long process of moving from flat-earth understandings of its core insights to evolutionary interpretations of those same doctrines. But this effort is still in its infancy. Any who choose this path now have an opportunity to make a profound and lasting difference. We are the fishes climbing out onto land for the first time, trying to make the most of our new situation. This is truly a moment of grace.

Facts Are God's Native Tongue

"Is evolution a theory, a system, or a hypothesis? It is much more: it is a general condition to which all theories, all hypotheses, and all systems must bow and satisfy henceforth if they are to be thinkable and true. Evolution is a light illuminating all facts, a curve that all lines must follow."

—PIERRE TEILHARD DE CHARDIN

Flat-earth religion has its scripture, its words revealed by God. These words of God represent the standard against which believers reconcile their thinking. Believers conform to these words, they submit. Evolutionary religion's alternative to reliance on ancient scriptures is *empirical data*. In a way, the data are our scriptures—and to these we submit.

What about theory? Scientific theories represent a kind of sacred commentary, a midrash, on the data. Theories are well informed, empirically based products of our reason and imagination. The data themselves are products of observation.

Theories do not command the same knowledge status as do observable data. Theories must submit to the data. To elevate a theory to the level of data would, to my mind, be a kind of idolatry. So there's a humility that we, as honest seekers of truth, must cultivate. But an evolutionary trajectory, as well, can move theory into fact.

For example, it was once *just a theory* that the Earth orbited the Sun. This was the theory that Copernicus proposed, based on his new interpretation of the observed data that astronomers and mathematicians of his day and earlier had found repetitive enough to be regarded as fact (science's form of "truth"). To propose a Sun-centric theory was a radical move in his day, and for several generations to come. Five hundred years later, that the Earth orbits the Sun is now itself a fact, as humans and our machines have actually left our planet and have seen it from the outside. As more and more reliable data (facts) are marshaled in support of a theory, and as attempts to disprove the theory fail, a theory can itself become a fact—as Earth's movement relative to the Sun has so become.

Similarly, less than two hundred years ago, when Darwin proposed that the complexity and diversity of life on Earth were not the result of supernatural and instantaneous creation, that proposal was *just a theory*—an outlandish, scandalous theory at that. Today, that life on Earth came into being over a vast span of time, and that complex forms emerged from simpler forms is *fact*. You may or may not feel comfortable calling this biological history of life on Earth *evolution*. To be sure, our scientific understanding of the *causes* of biological change are still very much in the discovery phase (witness the fascinating new science of evolutionary development, or "evo-devo"). Nevertheless, Darwin's *theory* that life emerged over a long history, and by means internal to natural Earth processes, has now become *fact*.

Evolution: Theory and Fact

"Let me try to make crystal clear what is established beyond reasonable doubt, and what needs further study, about evolution. Evolution as a process that has always gone on in the history of the earth can be doubted only by those who are ignorant of the evidence or are resistant to evidence, owing to emotional blocks or to plain bigotry. By contrast, the mechanisms that bring evolution about certainly need study and clarification. There are no alternatives to evolution as history that can withstand critical examination. Yet we are constantly learning new and important facts about evolutionary mechanisms." —THEODOSIUS DOBZHANSKY

"In the American vernacular, 'theory' often means 'imperfect fact'—part of a hierarchy of confidence running downhill from fact to theory to hypothesis to guess. Thus creationists can and do argue: evolution is 'only' a theory, and intense debate now rages about many aspects of the theory. If evolution is less than a fact, and scientists can't even make up their minds about the theory, then what confidence can we have in it?...

"Well, evolution *is* a theory. It is also a fact. And facts and theories are different things, not rungs in a hierarchy of increasing certainty. Facts are the world's data. Theories are structures of ideas that explain and interpret facts. Facts do not go away when scientists debate rival theories for explaining them. Einstein's theory of gravitation replaced Newton's, but apples did not suspend themselves in midair pending the outcome. And human beings evolved from apelike ancestors whether they did so by Darwin's proposed mechanism or by some other means yet to be discovered....

"Evolutionists have been clear about this distinction between fact and theory from the very beginning, if only because we have always acknowledged how far we are from completely understanding the mechanisms (theory) by which evolution (fact) occurred. Darwin continually emphasized the difference between his two great and separate accomplishments: establishing the fact of evolution, and proposing a theory—natural selection—to explain the mechanism of evolution." —STEPHEN JAY GOULD

"Today, nearly all biologists acknowledge that evolution is a fact. The term *theory* is no longer appropriate except when referring to the various models that attempt to explain *how* life evolves...it is important to understand that the current questions about how life evolves in no way implies any disagreement over the fact of evolution." —NEIL A. CAMPBELL

"Biology without evolution is like physics without gravity."

—SEAN B. CARROLL

Nearly a half century ago, Thomas Kuhn wrote a now-famous book, *The Structure of Scientific Revolutions*. He asked, Does science progress additively or does it lurch haphazardly by replacement of one paradigm, or understanding, with another? On the one hand, our scientific understanding has grown like a great edifice that we keep adding to. On the other hand, there have been episodes in which scientific ideas once widely held have been scrapped and replaced with very different ideas. The Copernican Revolution, the Newtonian Revolution, the Darwinian Revolution, the Einsteinian Revolution, the Information Revolution are examples of grand new theories replacing (or including and transcending) the old. From both points of view, however, facts are foundational. Thus, facts are God's native tongue.

Facts are God's native tongue!

> If there are scriptures beyond the holy texts of Earth's various religious traditions...
>
> If God didn't stop communicating knowledge crucial for humans centuries ago...
>
> If it is possible for new understandings to arise in ways more widely available and testable than what can be channeled through the hearts and minds of lone individuals...
>
> Then surely this is it: God communicates to us by publicly revealing new facts.

The discovery of facts through science is one very powerful way to encounter God directly. It is through the now-global community of scientists, working together, challenging one another's findings, and assisted by the miracles of technology, that "God's Word" is still being revealed. It is through this ever-expectant, yet ever-ready-to-be-humbled, stance of universal inquiry that God's Word is discerned as more wondrous and more this-world relevant than could have possibly been comprehended in any time past. (I'll revisit this subject in the final chapter.)

"Religion will not regain its old power until it can face change in the same spirit as does science."

—ALFRED NORTH WHITEHEAD

A message the United Church of Christ has been promoting in recent years is "God is still speaking," (with emphasis on the comma). I like this, and offer the title of this section and an exclamation point as a fitting completion: "God is still speaking, and facts are God's native tongue!"

Religious Knowers

> *"Only after we had absorbed Darwin and recalculated the age of the Universe, after the vision of static forms of life had been replaced by a vision of fluid processes flexing across vast tracts of time, only then could we dare to guess the immensity of the symphony we are part of."* —CHRISTOPHER BACHE

The fruit of public revelation is not only factual knowledge. It is factual knowledge ever open to wider truth and revision as each generation continues to learn and discover, thus building on the best of the past. The honing and improvement of factual information by processes that transcend the minds and lifespans of individuals is a more-than-human accomplishment. Surely, this too is revelation—revelation of a distinctively modern cast. Factual knowledge, in turn, offers more than mere facts. Facts become the springboard for meaningful interpretations. Meaningful interpretations then become the foundation of religious responses.

In the distant past, peoples around the world used the best factual evidence they and their forebears had acquired to answer the big picture questions— questions that a people must answer if they are to function as individuals and cohere as a group: *Where did we come from? What is our relationship to other humans, to other life forms, and to the power (or powers) of the Universe that so forcefully impinge on our lives? How are we to live in accordance with those powers? What is our role as a people? To what greater purpose do our individual, brief lives contribute?*

Nothing has changed in that regard—save for the volume and trustworthiness of facts generated and the updatable relevance of interpretations based on such facts. And that, in turn, changes everything—or, at least, it could. Yet today, relations between the realms of science and religion are strained by the sad fact that age-old private revelations have for centuries been held hostage by a pervasive idolatry of both the written word and tradition. Private revelations have thus been denied the lifegiving breath of ongoing public revelations.

Scientists are telling us today that it is no longer a question of believing or disbelieving that the Universe has been evolving for billions of years and that we're part of the process. The issue has moved beyond belief. Well over 95 percent of the scientists of the world now agree on the fact of evolution. When the DNA evidence of life's shared heritage becomes more widely known, surely the few remaining holdouts will be won over. As with Galileo's heresy, this new understanding has come about not through private revelation, but public revelation.

Better Than a Smoking Gun

In the introduction to his 2006 book, *The Making of the Fittest: DNA and the Ultimate Forensic Record of Evolution*, Sean B. Carroll writes,

"Genomics allows us to peer deeply into the evolutionary process. For more than a century after Darwin, natural selection was observable only at the level of whole organisms such as finches or moths, as differences in their survival or reproduction. Now, we can see *how* the fittest are *made*. DNA contains an entirely new and different kind of information than what Darwin could have imagined or hoped for, but which decisively confirms his picture of evolution. We can now identify specific changes in DNA that have enabled species to adapt to changing environments and to evolve new lifestyles.

"The new DNA evidence has a very important role beyond illuminating the process of evolution. It could be decisive in the ongoing struggle over the teaching of evolution in schools and acceptance of evolution in society at large. It is beyond ironic to ask juries to rely on human genetic variation and DNA evidence in determining the life and liberty of suspects, but to neglect or to undermine the teaching of the basic principles upon which such evidence, and all biology, is founded. The anti-evolution movement has relied on entirely false ideas about genetics, as well as the evolutionary process. The body of new evidence...clinches the case for biological evolution as the basis for life's diversity, beyond any reasonable doubt."

Those who base their religious orientation, as I do, on public revelation are more aptly categorized as *religious knowers* than as *religious believers*. This core distinction will be revisited throughout this book, wherever I contrast a knowledge-based, evolutionary form of religion with belief-based, flat-earth religion. But first, let us review what the fruits of public revelation can teach us about the Universe—and about the nested emergent nature of divine creativity.

CHAPTER 5

The Nested Emergent Nature
of Divine Creativity

"The epic of evolution is the sprawling interdisciplinary narrative of evolutionary events that brought our Universe from its ultimate origin to its present state of astonishing diversity and organization. Matter was distilled out of radiant energy, segregated into galaxies, collapsed into stars, fused into atoms, swirled into planets, spliced into molecules, captured into cells, mutated into species, compromised into ecosystems, provoked into thought, and cajoled into cultures. All of this (and much more) is what matter has done as systems upon systems of organization have emerged over 14 billion years of creative natural history."

—LOYAL RUE

A fundamental truth born of public revelation, made possible by technological advance, reworked over generations, and now widely held as accurate is this: *The whole of reality is creative in a nested emergent sense, and we are part of the process.* Like nesting dolls, smaller realities are contained within larger ones—from the infinitely small to the infinitely vast—and every one of them is divinely creative. That is, each scale of reality is blessed with an ability to bring forth novelty through natural processes of emergence gloriously specific to its unique station within the nested whole. "Divinely creative" expresses, too, that each nested level has the power to bring into being something that never existed before. This is a primary characteristic of God: "Creator."

Generations of scientists have progressively discovered the wondrous fact of nested creativity—a fact that Moses, the Apostle Paul, Mohammed, and Thomas Aquinas may have glimpsed but could not possibly have grasped, and could not have rendered into teachings accessible to their contemporaries. But now any among us can take in this revelation, and this truth can be made accessible through metaphors familiar to our time and place.

Following Arthur Koestler and Ken Wilber, I use the term *holon* to refer to the nested nature of the Universe. A holon is a whole that is also part of a larger whole and is itself composed of smaller wholes. Everything is part of something bigger and is made of smaller components nested within it. Each of those whole/parts, or holons, is creative. So the Universe is made of creative holons.

We are holons, too. Within, we find organs, tissues, molecules, atoms, subatomic particles. Without, we form families, societies, planets, solar systems, galaxies. Every holon is creative in ways distinct from the powers that operate at both larger and smaller scales. *Nested creativity thus is the source of emergence, of continuing creation by the collective and within each of its parts.* At the human scale, we find ourselves smack in the middle of this creative enterprise. Ultimate Reality, or God, is the One and Only Whole (Holy One) that is not part of some larger, more comprehensive reality.

Because we are a subset of the whole and cannot step outside to examine it, we shall never grasp the full nature of Ultimate Reality. Joel R. Primack and Nancy Ellen Abrams use the ancient metaphor of an uroboros (a snake eating its tail) to help us sense our necessarily restricted, yet exalted place, in the vast nestedness of the Cosmos:

I stand here on the Cosmic Uroboros, midway between the largest and smallest things in the Universe. I can trace my lineage back fourteen billion years through generations of stars. My atoms were created in stars, blown out in stellar winds or massive explosions, and soared for millions of years through space to become part of a newly forming solar system—my solar system. And back before those creator stars, there was a time when the particles that at this very moment make up my body and brain were mixing in an amorphous cloud of dark matter and quarks. Intimately woven into me are billions of bits of information that had to be encoded and tested and preserved to create me. Billions of years of cosmic evolution have produced me.

We now know that stars create almost all the atoms in the periodic table of elements. Atoms in community give rise to molecules. Molecules assemble into living cells. Out of cells emerge multicellular plants, animals, and fungi. Ants, termites, crows, prairie dogs, and human beings generate societies. Societies spawn cultures and technologies. Cultures yield artistic and religious expression. And the creativity of all of it, at every level, is possible only because of the Ultimate Whole of Reality, which I enthusiastically call "God."

The importance of nested emergence cannot be overemphasized. Creation is a self-organizing, nested, emergent process of divine creativity—creative wholes that are part of larger creative wholes within still larger wholes. Each level, or holon, has its own "intelligence," capabilities that its constituent holons do not have full access to. As philosopher Ken Wilber suggests in *A Brief History of Everything*, this "holarchical" (or nestedly hierarchical) view of reality will be the first principle, the solid foundation, for 21st century science, philosophy, and theology.

As we will examine in Chapter 6, one of the greatest ironies in the history of Western thought is this: by examining Reality as though it were a machine, humanity has discovered that Reality is not a machine. The *mechanistic paradigm* opened our collective sensibilities to the presence of a decidedly non-mechanistic Universe that now requires our use of non-mechanistic metaphors (e.g., "nested creativity," "self-organizing Universe," "holarchical Cosmos") in order for that Universe to faithfully be made accessible to human understanding. And all this can be accomplished "to the glory of God"—that is, in ways that serve the Whole.

Thank God for the Hubble Telescope!

"Most people tend to identify themselves with fairly narrow categories—a nationality, a race, a religion—which leads not only to conflicts but also to a stunting of imagination and potential. The wider our sense of identity, the more likely we will be able to experience our genuine connection to the universe. If

> *a lost child who knew nothing of her background and had been raised by an indifferent family suddenly discovered that she was the direct descendant of an illustrious house traceable back many centuries, her sense of identity would expand momentously even before anything else changed.... The discovery of our own cosmic genealogies may have a similar expansive effect. We humans are luminous, stardust beings."*
>
> —JOEL R. PRIMACK AND NANCY ELLEN ABRAMS

From a holy evolutionary perspective, God is no longer envisioned as a supreme landlord residing off the planet and outside the Universe. This ancient view of the divine is now much too small to embrace the vast, intricate, and nuanced realities that have been revealed by science in the past few hundred years. My grandparents were born in a time when our best scientists thought that the entire Universe was the Milky Way Galaxy. Astronomers of that era surmised that our Sun resided near the center of the galaxy—and thus at the center of the Universe. I was in grade school when Earth finally had the technological know-how to launch a piece of itself toward the moon in a way that it could see itself, as if in a mirror, for the first time. My youngest daughter learned to read when the Hubble Space Telescope had expanded our understanding of the Universe to encompass more than two hundred billion galaxies, each averaging a hundred billion stars. Four generations, and such a vastly different reality!

"That's Where Baby Stars Are Born!"

Connie was setting up her posters of Hubble space photographs in the church classroom where she would be teaching a children's program. An eight-year-old girl and her younger sister arrived early and immediately stepped up to the posters. The older girl pointed to one of the most widely recognized and beloved of all Hubble photos—three golden pillars of gases rising against a background of green. She asked, "Is that the Eagle Nebula?" "Yes!" Connie replied. The girl continued, "That's where baby stars are born!"

God did not stop revealing truth vital to human well-being thousands, or even hundreds, of years ago. This is a crucial realization. Scientific discoveries are themselves opportunities for religious feelings to blossom. For me and for many others, the view of Earth from space, along with the entire and growing archive of Hubble (and other space telescope) photographs, are among the most spiritually resonant images of divine beauty that humans have ever produced. One need not be a religious liberal to concur. On a drive from Colorado Springs to Durango, Connie and I passed a rural Christian academy that displayed an extraordinary sign. There, with the words "God's Creation," was a stunning color image of the Eagle Nebula, the Pillars of Creation.

For many of my programs, I display a half dozen or more posters of Hubble Telescope photographs. After one such program delivered in a Unitarian Universalist church, I received a thank-you letter from a woman who included a poignant testimonial of her awakening to a sacred relationship with the Universe. She wrote of her struggles to believe in God, her longing to enter into relationship with the divine. "Then one day I saw the pictures sent back by the Hubble Telescope." She continued, "And I knew that, finally, I had seen the face of God."

"Is This Purple Bead Darwin?"

When guest teaching in elementary schools, Connie always wears her double-loop necklace of Great Story beads. She begins, "I have the entire fourteen-billion-year Story of the Universe right here, represented in beads!" She then slips off the necklace and displays the beads as one giant loop. "Pick a bead, any bead," she announces, "and I will tell you its story."

At a public Montessori school in Denver, children lingered after class, eager to continue the game of bead pointing and bead telling. Eighteen months later, Connie returned to that school to teach the sixty five-million-year story of North America. After class a sixth grader approached Connie and asked if she could examine the necklace. "Is this purple bead Darwin?" the girl asked. "Yes!" Connie exclaimed. "How did you know that?"

"I remember from the last time you were here. I was in fourth grade then, and I asked you about that bead."

We Are Made of Stardust

"We are the local embodiment of a Cosmos grown to self awareness. We have begun to contemplate our origins—starstuff pondering the stars!"
                                                              —CARL SAGAN

The epic of evolution begins not with the origin of the first cell on Earth, but with the origin of the entire Universe, the Big Bang—or, as Connie and I prefer, the "Great Radiance." During the past several decades the discoveries of astronomers, the musings of theoretical physicists, and the speculations of cosmologists have been breathtaking. Estimates of the number of galaxies in the Universe and of stars in our own Milky Way consistently challenge our minds to work at unaccustomed scales. We must think in terms of billions, even hundreds of billions. Black holes and pulsars, photons and quarks: these denizens of the Cosmos are dazzling in their strangeness. For the Universe story to become *our* story, however, amazement is not enough. We need to feel relationship. We must make a connection, sprout an umbilical cord to the Cosmos. What out there can offer us such relationship?

Simply this: ancestral stars are part of our genealogy. We can now know and feel a familial bond to the heavens. Every atom in our bodies, other than hydrogen, was forged in the fiery belly of a star who lived and died and recycled itself back to the galaxy before our own star, the Sun, was born.

A woman in her late twenties told us that Carl Sagan's *Cosmos* aired when she was seven years old and that it changed her life. "How?" we asked, as this was not a show for children. Her response: "I knew that I was related to *everything*!" Indeed! We, and everything else, are made of stardust—or "starstuff," as Sagan enthused.

We are made of stardust! We can trust this fact in the same way that we know that Earth is a planet that revolves around the Sun. How can we be so sure?

The formation of chemical elements is not something that happened merely in the distant past. Rather, it is ongoing. Thanks to spectrometers and other instruments that discern chemical signatures in the spectra of light emitted by stars, scientists can sample the material composition of heavenly bodies they can never hope to visit. They can witness the genesis of starlight born of "stellar nucleosynthesis" (stars making new atomic nuclei)—a form of divine creativity happening right now in stars throughout our galaxy and in all the distant galaxies, too. Indeed, by isolating streams of photons arriving at Earth from any star, scientists can study the absorption spectra to see which chemical elements in the star's own body are absorbing distinctive wavelengths of light, and thus blocking photons of that wavelength. Such measurements of starlight confirm the predictions first made in 1957 by scientists who used their knowledge of mathematics, nuclear chemistry, and thermodynamics to calculate the steps by which different sizes and ages of stars use nothing more than immense gravity to fuse hydrogen atoms birthed in the Great Radiance (Big Bang) into progressively more complex atoms.

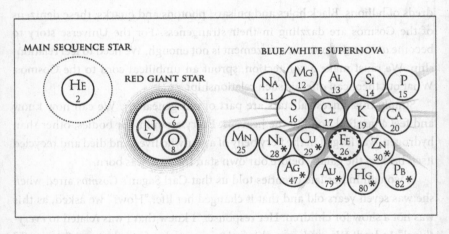

A New Periodic Table of Elements. In the beginning, condensing out of the Great Radiance (Big Bang) was hydrogen and helium. Hydrogen nuclei nearly 14 billion years old are, right now, combining in the core of our own star, the Sun, into the second most complex element: helium (two protons in the nucleus). All Main Sequence stars, regardless of size, are at this stage: fusing hydrogen into helium. Five billion years from now, our Sun will run out of hydrogen fuel and begin to fuse helium into carbon, thus entering its Red Giant phase. Stars at least eight times more massive than our Sun burn so brightly in their initial stage of hydrogen fusion that, instead of yellow, they are a brilliant shade of blue. When they run out of hydrogen fuel (very quickly, as it turns out), they become either a Super Red Giant or a Blue Giant. At last, when one of these biggest stars has fused elements all the way up to iron (Fe, 26 protons), it collapses on itself, then rebounds in a dazzling Supernova explosion in which all the heavier elements (including precious metals, like gold and silver) are created and propelled back into the galaxy at large.

Our middle-aged Sun is an average yellow star. For five billion years it has been generating light and heat by fusing hydrogen nuclei into helium. Some mass is converted into energy during this process, as in Einstein's most famous equation: $E=mc^2$. When our Sun becomes an elder (in another five billion years), it will undergo its biggest transformation since birth. Having used up almost all the hydrogen in its core, it will begin to fuse helium into carbon. Its whole demeanor will change. Its core will heat up, and the outer layers will expand beyond the orbits of Mercury, Venus, and finally Earth. Those cool outer layers will make it visible as a very large reddish star: a Red Giant.

The ancients could not have known these facts, nor could Moses, Jesus, Mohammed, Confucius, or the Buddha. Indeed, had any such prophet been capable of discerning this understanding through some sort of private revelation, his followers surely would have balked. So far beyond reasonable belief, such proclamation would not have been faithfully transmitted for later installation as sacred scripture. Thank God for public revelation! And thank God for stellar nucleosynthesis! Without the latter, there would be no complex atoms and thus no rocky planets or life in this Universe. Without the former, we would have no awareness that stars are suns very far away, that all such suns are mortal, too, and that stars that died long ago are to be counted among our very own ancestors.

As my wife, Connie, likes to tell children, almost all of whom know and love the 1990 Disney movie, *The Lion King*, "Puumba the warthog and Mufassa the Lion King are both right: Stars are 'balls of burning gas billions of miles away' and stars are our ancestors!" She then teaches them a new verse to a song that, for the two centuries of its existence, has been faithfully transmitted from one generation to the next:

> Twinkle, twinkle little star
> Now I know just what you are
> Making atoms in your core
> Helium and many more
> Twinkle, twinkle little star
> Now I know just what you are

"What a Mind Bender, Dude!"

I was invited to teach the story of chemical evolution to high school students at a Native American boarding school near Gallup, New Mexico. After presenting the basic science, I asked rhetorically, "Do you get this? I mean, do you *get* this? We are stardust now evolved to the place that the stardust can think about itself!" I paused and looked intently at the teens. After seven or eight seconds, like popcorn, one face after another began to light up with amazement. One particularly vocal boy (the kingpin of the class, I discovered later) exclaimed, "Wow, what a mind bender, dude!" From that point on, I had the rapt attention of virtually everyone.

Southwestern Colorado was on our itinerary in the summer of 2005 and again in the summer of 2006. On our return, a young mother said to Connie, "I have to tell you something amazing. Last time you were here you told a story to the kids about how their bodies are made of stardust. My youngest son was three and a half then, so it is surprising that he remembered anything at all. But obviously he did, because not long ago I was telling him about something that happened to our family in the past, and he asked, 'Was I born yet?' 'No,' I said. 'Was I in your belly yet?' Again, no. 'Oh!' he replied. 'I must have still been stardust!'

"It Made That Feeling Go Away"

A Unitarian Universalist minister told Connie, "I first learned that we are made of stardust when I read Brian Swimme's book, *The Universe Is a Green Dragon*. It meant so much to me that I've been teaching it ever since. One year, I taught the story of stardust at a summer camp. That evening I came upon the teens outdoors, all lying on the grass, feet touching in a star formation, saying things like, 'I think I'm from *that* star!' and 'I wonder if the light from *that* star left before I was born?' The next day, during a private counseling session, a boy told me that he had come to camp thinking that he might commit suicide here, but that hearing the stardust story and where the atoms of his own body came from made that feeling go away. I couldn't help but think that maybe, just maybe, his learning of his deep origins was akin to his finding his way back to God."

Those of us who have embraced the scientific story of our roots know ourselves to be reworked stardust with a multibillion-year pedigree. We know these facts deeply, and for us the story is as compelling as any tale that has ever come alive in the flames of a fire at the mouth of a cave or in the vaulting echoes of a cathedral. For us, the history of life and of the Universe as given by science becomes more than a sequence of strange and arresting events. It becomes our personal and shared story, our creation myth. As Joel R. Primack and Nancy Ellen Abrams remark,

> There's a joke among cosmologists that romantics are made of stardust, but cynics are made of the nuclear waste of worn-out stars. Sure enough, the complex atoms coming out of supernovas can be seen either way, but these atoms introduce into matter the possibility of complexity, and complexity allows the possibility of life and intelligence. To call them nuclear waste is like calling consumer goods the waste products of factories. A cosmology can be a source of tremendous inspirational and even healing power, or it can transform a people into slaves or automatons and squash their universe into obsession with the next meal or with trivial entertainment. The choice of what attitude the twenty-first century will adopt toward the new Universe may be the greatest opportunity of our time. The choice between existential and meaningful is still open.

For more on the science and meaning of stardust, and for playful ways to experience this aspect of the Great Story, see this page on our companion website: ThankGodforEvolution.com/thegreatstory.

The Gifts of Death

> *"Life spirals laboriously upward to higher and even higher levels, paying for every step. Death was the price of the multicellular condition; pain the price of nervous integration; anxiety the price of consciousness."*
>
> —LUDWIG VON BERTALANFFY

A core function of any religion is to provide a meaningful and comforting context for anticipating one's own death. That context must also help us to fully, yet safely, feel and express our grief when loved ones die. In Western culture we are reticent to talk about death because of unexamined assumptions that hail from another era. These flat-earth assumptions are not up to the challenges encountered today—especially the ethical conundrums arising from our arsenal of technologies that forestall death, even at great material and emotional cost.

Perhaps there is no more alluring portal for discovering the benefits of evolutionary spirituality than death understood in an inspiring new way. Thanks to the sciences of astronomy, astrophysics, chemistry, geology, paleontology, evolutionary biology, cell biology, embryology, ecology, geography, and even math, we can now not only accept but celebrate that

✦ Death is natural and generative at every level of reality.

✦ Death is no less sacred than life.

During the past 500 years, scientists and explorers made important discoveries that revealed how death, more often than not, is a cosmic blessing. For example, not until 1796 did biologists have to face up to the fact that God (or Nature) might have created some species (i.e., mastodons and mammoths) that later went extinct. The news was unsettling because no one with a pre-evolutionary mindset could understand why "anything so imperfect" would have been created in the first place. A satisfying answer became available in 1859, when Charles Darwin explained that species death was necessary for life to become increasingly complex during a vast span of time.

Similarly, in the science of embryology, the study of fetal growth (notably, in chicken eggs) indicated that without the death of particular cells during development, animals would look something like spheres. Much later, we learned that "programmed cell death," even in our elder years, is crucial for the health of the body—and for prevention of cancer. During the past century, when the science of ecology came to the fore, we learned that it was death—specifically, an understanding of who ate whom—that gave "the web of life" its very structure. As well, we had to acknowledge that most of the animals hatched or born were destined to become food well before they had a chance to mature and reproduce.

In 1930 Hans Bethe made an astonishing discovery in the field of astrophysics. He calculated that in about five billion years our Sun would use up its store of hydrogen fuel and begin its slide toward death. A quarter century later, new calculations helped us appreciate that the inevitable death of our own star was not a demerit in the grand scheme of Creation. As already discussed, a team of scientists discovered in 1957 this astonishing fact: *without the death of stars, there would be no planets and no life.* More recently we learned that our very own galaxy, the Milky Way, has been enlarging for billions of years by consuming smaller galaxies—and that in maybe two or three billion years it will merge with another huge galaxy, Andromeda, toward which it is now heading.

The science of geology got its start a little more than two centuries ago. That's when we began to learn that mountains are mortal, and thank God that they are! Without the wasting away of mountains over millions of years, there would be no small particles to form continental soils. Had the ice sheets not plowed their way across northern latitudes, destroying all life in their path, the prairie states would have no rich blanket of windblown silt—and America would be missing its vast and productive farmlands. In the late 1960s we learned that even continents are torn asunder (as when the ancient supercontinent Pangaea began to break up during the Age of Dinosaurs). Even so, such destruction is what created the Atlantic Ocean and, later, carved out a chunk of Antarctica and sent it northward into the tropics. This emigrant land mass, whose northward momentum is still pushing up the Himalayas, we now call India. Similarly, had Madagascar not been severed from continental Africa and launched eastward, Earth would have lost one entire family of primates. The fossil record indicates that whenever and wherever monkeys arrived, the resident lemurs vanished. Continental break-up thus dedicated one large island to the task of providing lemurs with the isolation they required.

Turning to geography and math, we can reflect on the first time that any civilization managed to circumnavigate the globe. In 1522, the expedition that Ferdinand Magellan had launched three years earlier arrived back in Spain— only one of the original five sailships completing the round-the-world voyage. That Earth was round was no news to educated Westerners in 1522. Nevertheless, Magellan's expedition was undeniable proof, and it still symbolizes the end of flat-earth thinking—at least for geographers. How did the voyage help us naturalize our understanding of death? It took awhile, but in 1798 Thomas

Malthus published his "Essay on the Principle of Population," which Charles Darwin cited as having played a crucial role in his own evolutionary thinking. Malthus demonstrated how, given the exponential rate at which all species reproduce, it is only the reality of death that prevents any animal (including humans) from overpopulating the range and running out of food. Europeans, of course, kept the Malthusian specter at a distance for several more centuries— by killing one another, by dying from diseases of overcrowding, and by expanding their "range" onto other continents, where they could vanquish the natives and declare ownership of new habitat.

Today, of course, our commonsense understanding of ecology has most of us convinced that even God could not have designed a functional world in which birth is allowed but death is banished. This is not, however, the way some religious believers interpret the sacred scriptures of the West. Romans 5:12 reads, "Wherefore, as by one man sin entered into the world, and death by sin; and so death passed upon all men." (For more on the science of death as natural and generative, see: ThankGodforEvolution.com/thegreatstory.)

"Death—Don't Blame God!"

Below is the transcript of an audio podcast posted February 23, 2007 on the Answers in Genesis *website. In the podcast's introduction, Ken Ham is presented as "a popular speaker at Answers in Genesis seminars."*

QUESTION: Ken, why is it that when a loved one dies, I can hear even committed Christians cry out and say, "Why did God do this?!"

KEN HAM: I believe that because much of the Church has been taught to believe in millions of years of Earth's history, Christians just don't understand what death is all about. You see, most people have been indoctrinated with the idea that the fossil record took millions of years to be laid down. For example, dinosaur bones are supposedly millions of years old. Now, when Christians believe this—and, sadly, the majority probably do—then they've accepted that there were millions of years of death, suffering, bloodshed, violence before God made the first man. Therefore, they believe that God used death, suffering, and violence over millions of years as part of His creating. So when a loved one dies, it's logical for these people to blame God for death. But Genesis makes it plain that death entered the world *after* Adam sinned. It was a Judgment because of sin. The Bible describes death as an enemy. When a loved one dies, we should fall on our knees before our Holy God and recognize that death is in the world because *we* sinned.

Whether or not "committed Christians" truly believe that Earth is several thousand rather than several billion years old, and whether or not a majority of biblical literalists consciously subscribe to the notion that death came into the world as a punishment for Adam's disobedience, it is a truism that American culture has a deep aversion to even talking about death. Something is *wrong* with the presence of death in this world. Our institutions reflect this reticence—most notably, the medical profession. And that is why the hospice movement has been such a godsend for many in America, including my wife when her mother was dying.

An evolutionary understanding of death in no way diminishes the grief we suffer when a loved one dies. That is not its purpose. What this perspective does offer is a solid and trustworthy "cosmic container" in which grief can fully manifest, while protecting the bereaved from the risk of falling into an abyss of anger and despair. More, if we acknowledge that *there is something profoundly right with death*, with the fact that we grow old and that we must die, it will be easier to clean up unfinished business before it is too late. Meaningful conversations with family and friends will ensue—including expressions of gratitude, apologies, and forgiveness.

Crucially, we can adopt a new and expanded view of death as natural and generative, while respecting diverse poetic imaginings (what I refer to as "night language" in the next chapter) about what happens to consciousness, or soul, or spirit after death, and while honoring the teachings of each religious faith. By viewing the cycle of life through deep-time eyes, we can come to regard death as not only natural and right but as that which makes possible virtually all that we hold most dear. Many people have reported to us over the last few years that this awareness has helped them deal with the pain of losing someone. Several also said they were especially grateful that they could embrace the positive role of death in the Universe without needing to give up cherished notions of what will happen to them or their loved ones on the other side.

"I Learned That My Grandmother Will Die"

Connie loves to be invited into religious education classes and private schools to teach elementary-age children her "We are made of stardust" program. One reason that she chooses to teach this particular episode of the epic of evolution is that it provides an opportunity to talk about death as a natural and creative part of the Cosmos: Without the death of ancestral stars, there could have been no planets or life.

At some point in the program, she usually asks the children, "Do any of you have a grandparent who has already become an ancestor?" Instead of hesitancy, the children readily raise their hands. At a church in New Jersey, a boy proudly proclaimed, "My grandma became an ancestor on January 26, 2004!"

Connie usually concludes these stardust lessons by seating the kids in a circle on the floor. After a song and ceremonial application of fairy-dust glitter (stardust) on one another's foreheads, she will ask the children, "Did any of you learn something here that you didn't know before and that you think you will remember for the rest of your life?" At a church in Mississippi an 8-year-old girl responded, "I learned that my grandmother will die."

What a wondrous and celebratory way to introduce our children to the reality of death—ideally, before the death of a loved one (including a pet) confronts a child with the reality of death in an emotionally charged way.

Here is a litany that Connie wrote to accompany our presentations on death:

Litany: "The Gifts of Death"

Without the death of stars, there would be no planets and no life.
Without the death of creatures, there would be no evolution.

Without the death of elders, there would be no room for children.
Without the death of fetal cells, we would all be spheres.

Without the death of neurons, wisdom and creativity would not blossom.
Without the death of cells in woody plants, there would be no trees.

Without the death of forests by Ice Age advance, there would be no
   northern lakes.
Without the death of mountains, there would be no sand or soil.

Without the death of plants and animals, there would be no food.
Without the death of old ways of thinking, there would be no room
   for the new.

Without death, there would be no ancestors.
Without death, time would not be precious.

What, then, are the gifts of death?

The gifts of death are Mars and Mercury, Saturn and Earth.
The gifts of death are the atoms of stardust within our bodies.

The gifts of death are the splendors of shape and form and color.
The gifts of death are diversity, the immense journey of life.

The gifts of death are woodlands and soils, ponds and lakes.
The gifts of death are food: the sustenance of life.

The gifts of death are seeing, hearing, feeling—deeply feeling.
The gifts of death are wisdom, creativity, and the flow of cultural change.

The gifts of death are the urgency to act, the desire to fully be and become.
The gifts of death are joy and sorrow, laughter and tears.

*The gifts of death are lives that are fully and exuberantly lived, and then
   graciously and gratefully given up, for now and forevermore.*

Amen.

"I Am at Peace with His Death"

In the summer of 2004, Connie and I were jointly presenting an evening workshop at a Unitarian Universalist church in Ohio. Connie talked about the creation of atoms inside stars, and the importance of those stars dying and giving back to the galaxy what they had created during their lives. A woman sent us an email afterwards: "During Connie's talk about stardust, I knew why I had come. My elderly father, who took his own life in May, always told me I was made of 'starstuff.' After hearing you, I am at peace with his death. His spirit is with the Goddess, but even stars die, and his substance will continue on as new life. Thanks so much!"

When I think about death as a cosmic blessing, I am struck by how this public revelation mirrors the core message of the early Christian scriptures: on the other side of Good Friday is Easter Sunday. The profound similarity between ancient and evolutionary insights occurs to me as more than coincidence. I am convinced that a straightforward reading of the death and resurrection of Jesus the Christ, as portrayed in the Christian gospels, is a magnificent celebration of a universal truth that could not possibly have been revealed in a way that all people could embrace until telescopes, microscopes, and computers became available. The truth is this: *Death is a doorway to greater Life.*

This view of death does not, of course, lessen the pain and grief we experience when a loved one dies. Whenever we are in relationship with someone or something—a person, an animal, a tree, a special place—and that relationship is severed (for whatever reason), sadness and grief are natural and healthy responses. But knowing that death never has the final word, that it virtually always contains the seeds of new life, can provide a safe harbor for experiencing such loss. Knowing that from God's perspective, not only is there nothing wrong with death, but, rather, there is something very right with death, allows us to surrender to the process. We surrender to "what will be" with a profound faith, a radical trust, that whatever awaits us and our loved ones in the beyond, if anything, is just perfect.

✦ ✦ ✦

Any supposed "faith in God" that does not include trusting that *whatever* happens on the other side of death is just fine is really no faith at all. Fear of a terrifying, hellish after-death scenario *or* hope of a blissful, heavenly after-death scenario are just that: fear or hope—not faith, not trust.

Following is an extract from an "evolutionary parable," written by Connie. Such parables are playful stories for teaching the science of emergent creation in sacred ways. As parables, they teach values. Raw science is nothing without story. As Unitarian Universalist minister Tom Rhodes has said, "Our bodies are made of stardust; our souls are made of stories."

The extract below (full text available on our website) is drawn from the final scene of the parable *Startull: The Story of an Average Yellow Star*. Here the science of stellar nucleosynthesis is coupled with teachings that promote self-acceptance, mentoring, and the understanding that death is naturally creative—even among the stars.

RED GIANT STAR: Carbon is what I am now creating in my core from all the helium I made in my lifetime.

STARTULL: That is a wonderful gift! Will I ever make carbon?

RED GIANT STAR: Yes. But that won't happen for a very long time. During the next five billion years, you will remain an average yellow star—but a star who just so happens to have a planet full of life to take care of.

STARTULL: That is a big responsibility. I will do a good job!

RED GIANT STAR: I believe you will!...Now, my Friend, I must go.

STARTULL: Why?

RED GIANT STAR: It will soon be time for me to recycle back into the galaxy the gift of carbon that I have created. And maybe, just maybe, an average yellow star that is yet to be born will have planets, and perhaps one of those planets will put to use the carbon atoms that I created. Perhaps that planet will evolve Life!

STARTULL: But will you have to die for that to happen?

RED GIANT STAR: Yes. But my way of dying will be much gentler than that of a supernova explosion. Even so, much of what I have created will go back into the galaxy.

STARTULL: [hesitantly] Are you afraid to die?

RED GIANT STAR: I used to be afraid. But now that I am old, I am very satisfied with the star life I have already lived. As a Red Giant, what matters to me most is that this whole amazing process continues. I want new stars to continue to be born. I want life to continue to evolve. I am sure you will feel the same way in another five billion years.

STARTULL: You mean, I am going to die too?

RED GIANT STAR: Such is the way of the Universe. Everything dies eventually. And that is what makes possible this whole grand Circle of Life. Without death, there could be no more birth.

STARTULL: [sadly] Oh...

RED GIANT STAR: [brightly] But you will not die for a very, very long time, my Friend. So, carry on! Carry on as a magnificently average yellow star! And continue with your cosmic task of helping one of your planets evolve life! ... Goodbye!

"In the case of the Sun, we have a new understanding of the cosmological meaning of sacrifice. The Sun is, with each second, transforming four million tons of itself into light—giving itself over to become energy that we, with every meal, partake of. The Sun converts itself into a flow of energy that photosynthesis changes into plants that are consumed by animals. Humans have been feasting on the Sun's energy stored in the form of wheat or maize or reindeer as each day the Sun dies as Sun and is reborn as the vitality of Earth. These solar flares are in fact the very power of the vast human enterprise. Every child of ours needs to learn the simple truth: she is the energy of the Sun. And we adults should organize things so her face shines with the same radiant joy." —BRIAN SWIMME

CHAPTER 6

Words Create Worlds

"As language-using organisms, we participate in the evolution of the Universe most fruitfully through interpretation. We understand the world by drawing pictures, telling stories, conversing. These are our special contributions to existence. It is our immense good fortune and grave responsibility to sing the songs of the Cosmos."
　　　　　　　　　　　　　　　　　　　　　　　—EDWIN DOBB

Few things can enhance our appreciation of divine activity in the world more than an evolutionary view of human language, through which religious ideas necessarily are communicated. Scientists tell us that what separates humanity from the rest of the animal kingdom is not so much our genetic makeup. Nearly 99 percent of our DNA is identical to that of chimpanzees and bonobo chimps, and more than half of our coding genes are indistinguishable from a worm's. Rather, it is symbolic language that sets us apart.

At some point in the not-so-distant past, our ancestors began using words as meaningful symbols. Whether this shift occurred rapidly some 50,000 to 100,000 years ago, as most scientists surmise, or more gradually—perhaps over hundreds of thousands or even millions of years—matters little. What is undisputed by those who accept that we are part of a nested emergent Cosmos is the understanding that at some point in the last two million years we humans added symbolic language to the sounds and gestures born of our primate heritage. Moreover, we then began *thinking* in words.

We are still guided by our senses, experiences, and feelings, as other animals always have been. Even today, gestures and body language dance a conversation in parallel to the movement of our lips and tongue. Try talking with your arms riveted to your sides, and then try to keep your head and eyebrows from moving as well! Over the millennia we have increasingly been guided, individually and collectively, by words. Without the tool of symbolic language, what would remain of our beliefs and stories about how and why the world is as it is, where everything came from, what happens when we die, and what purposes we should devote ourselves to beyond the instinctual drives we inherited from our animal ancestors? Without the tool of symbolic language, how could answers to any of those questions pass from one generation to the next?

Academics use the term *symbolic language* to mean simply this: For our kind of intelligence, *metaphors are not optional.* As Immanuel Kant in the 18th century and countless others since him have pointed out, words as symbols are inescapably metaphorical. They do not directly represent things in the world, or the world itself; rather, they do so indirectly—often by referring to another concept or symbol (in which case the word becomes a metaphor twice removed from the object of reference). A spoken word induces the mind of a listener to associate a particular sound with a specific memory, internal picture, or another word. As Terrence Deacon suggests in *The Symbolic Species*, it is precisely the nonrepresentational nature of words that distinguishes human speech from other forms of animal communication.

One can appreciate the metaphorical quality of language by recalling that whenever we seek to understand something new we do so by analogy. We say, *"This* is like *that,"* and the *that* may itself be a metaphorical expression so long in use that we expect the listener to effortlessly know what we mean. "A clamshell opens like a laptop computer" is something we might say to a child during a walk on the beach. But when laptops were newly invented (I recall we hyphenated the word back then; it was born of two words before it became one), we might have explained our new purchase in the reverse: "A laptop computer opens like a clamshell." Indeed, we Mac users called a previous generation of round-edged laptops "clamshells."

Humans swim not just in a sea of language; we swim in a sea of metaphorical language. "Down" is not just a direction we point to; "down" is how we describe ourselves when we are feeling low. Metaphors thus play a crucial role not only in great literature but also in our everyday speech. As George Lakoff and Mark Johnson note in *Metaphors We Live By,* "Metaphor is as much a part of our functioning as our sense of touch, and as precious." The choice of metaphor, in turn, affects how we perceive the world—and thus what we choose to value and how we choose to act.

Yes, words create worlds. Nonetheless, there is a real world out there, a huge and magnificent world outside our skulls—a world, moreover, from which we arose. Our preconceptions do, of course, affect how we perceive that world; so the practice of science involves a system of checks and balances within a community of perceivers (and their instruments) to ensure that, collectively, we are not deceived. Science is by no means perfect in this regard; assumptions do affect perception, and it may take decades for the errors to come to light, and decades more for the errors to be corrected in the mainstream of science. But science is, if anything, self-correcting. And therein lies another of its great gifts to religion. To again quote Alfred North Whitehead, an early 20th century philosopher, "Religion will not regain its old power until it faces change in the same spirit as does science."

Experiencing God Versus Thinking About God

> *"Thinking about God is no substitute for tasting God, and talking about God is no substitute for giving people ways of experiencing God."* —MATTHEW FOX

Our hominid ancestors experienced Reality as divine. For them, Nature was majestic, mysterious, awesome, benevolent, occasionally severe, all-powerful, nourishing, and more. Virtually every human attribute (the bad, as well as the good) was not only mirrored but also magnified in the mysterious forces of the natural world. Our ancestors experienced Reality this way long before words would label the experience—indeed, before there were verbalized beliefs of any kind. Most beliefs, rational and irrational, spring from the womb of symbolic language.

As our human ancestors began using words to tell stories about why reality is as it is and how it came to be, they naturally used the flora, fauna, climate, topography, and social relationships familiar to them as their source of analogies. The metaphors in use when writing entered the worlds of our cultural antecedents are still with us, for example, in the Judeo-Christian tradition: "The Lord is my shepherd," "the lamb of God," "smaller than a mustard seed," "thy Kingdom come," "gates of heaven," "fires of hell," "the throne of God," "it is easier for a camel to pass through the eye of a needle," "he is my shield and the horn of my salvation," and many more.

Every metaphor or belief about Ultimacy or divinity has its origin in peoples' relationship with the world around them. Each symbolically points to something that was once widely accepted as accurately reflecting what is real and what matters to a particular people in a particular bioregion. Said another way, *all religious stories, metaphors, and spiritual beliefs are true in this sense: they are true to a people's experience of the world.*

If we imagine God as beautiful, gracious, loving, awesome, powerful, majestic, or faithful, it is because we have known or experienced beauty, grace, love, awe, power, majesty, or trustworthiness in the world. As Thomas Berry has said, "If we lived on the moon and that's all we and our ancestors had ever known, all our concepts and experience of the divine would reflect the barrenness of the lunar landscape." Thankfully, we are not confined to a barren moon, but can rejoice as part of a flourishing, intensely creative Earth and a vast and awesomely beautiful Universe that call forth our richest images of God.

For example, we can now understand that a "God's eye view of the world" is not merely the *objective*, transcendent perspective—the view from above or beyond nature. A God's eye view of the world must also include the *subjective* experience of every creature. What dolphins and fishes see, what bats and birds see, what spiders and dragonflies see: all must be included. For me, God is thus not only Love but also Infinite Compassion. God feels the pain and suffering of all creatures—from the inside. Those who think they can love God and trash the environment, or oppress others, must be blind, utterly, to the immanence and omnipresence of divinity. When we truly *get* the nested emergent nature of divine creativity, we know that our love of nature and our love of one another are essential aspects of our love of God.

"What Does God Look Like on the Inside?"

One of the things we love most about our itinerant lifestyle, and about being invited into so many homes, is the occasional factoid, idea, or quotable quote that we acquire from our hosts. Here is a gem I picked up in northern Indiana early in our travels and have been using ever since. Doug Germann, a Lutheran Great Story enthusiast, offered: "What does God look like on the outside? God only knows! What does God look like on the inside? Look around you. Look into another's eyes. Look deeply into your own heart."

✦ *God is the Mystery at the Center* of our amazement that the Universe is here at all, that it is what it is, and that it is always becoming, yet always somehow whole.

✦ *God is the Mystery at the Heart* of consciousness, conscience, compassion, and all the other forms of co-creative, co-incarnational responsiveness of life to life.

✦ *God is the Mysterious Omni-Creative Power* through which the Universe is and ever becomes more intricately and wondrously fulfilled through the interactions of all its parts (each of which contains a spark of the Whole).

Because of the symbolic nature of human language, no one way of thinking or speaking about the Whole of Reality can encompass more than a fraction of all there is to know and express about the Whole. All religious beliefs and stories are metaphorical. They are symbolic statements intended to help us understand the world, our place in it, and how, when, where, and why to cooperate as groups. Such beliefs and stories are all still useful to a degree—and all are limited. Joel R. Primack and Nancy Ellen Abrams offer wise counsel: "Metaphors are powerful and can be perilous, but the danger can't be avoided by locking them in a drawer. Our best defense against their possible misuse is to encompass them in a higher understanding."

"It takes an entire Universe to make an apple pie."   —CARL SAGAN

The Split Between Religion and Science

"We can hardly overestimate the significance of the fact that the scientific and religious propensities were one before they became two different activities. Their fundamental unity precedes their separateness."
        —PHILIP HEFNER

Until the middle of the 14th century in Europe, and throughout the rest of the world, science and religion were inseparable. Natural philosophers (the word "scientist" was not coined until the late 19th century) and theologians were typically the same people, or close colleagues. This cozy relationship was lost when the Bubonic Plague, also known as the Black Death, erupted in Europe. Within a three-year period, from 1347 to 1350, a third of Europe died—some 25 million people—*and no one knew why.* Many communities lost two-thirds or more of their populace. What was going on? Was this divine punishment?

The scourge of the Black Death in Europe fostered conditions that would lead to a nasty divorce, at least in the West. While theologians continued to see the Black Death as punishment, natural philosophers searched for material causes of the disease. Thenceforth, science and religion would regard one another, at best, with polite reserve, and more often with suspicion and annoyance, erupting periodically as culturally wrenching conflict. Not that this was all bad. Without the Black Death, the conduct of science may have remained shackled to assumptions compelled by religious dogma. But that was then. Today, surely, these two solidly differentiated endeavors can amicably mingle once again for the betterment of the human family and the Earth community.

Two trends in Western intellectual history can be traced back to the Black Death. The first is a split in the Western psyche between faith and reason. On the one side were those who devoted themselves to a vision of being saved *out* of this fallen world. On the other side of the divide were those equally committed to understanding everything *within* this world, in order to better the human lot by learning to work with the natural powers.

The second trend born of the Black Death was a shift in Western thinking about our basic orientation to nature. Prior to the plague, nature was often related to as an expression of divine grace. As the widely influential 12th-century theologian Saint Thomas Aquinas wrote in his *Summa Theologica*, "Since God's goodness could not be adequately represented by one creature alone, He produced many and diverse creatures, that what was wanting to one in the representation of the divine goodness might be supplied by another. Thus, the whole Universe together participates in the divine goodness more perfectly, and represents it more fully, than any single creature whatsoever." With the persistence of the plague, however, the face of nature transformed from benign to dangerous—even demonic. It is not surprising, then, that a deep rage against the human condition can be detected in Western thinking from this time forward.

From Clockwork "It" to Creative "Thou"

"The religious conservatives make an important point when they oppose presenting evolution in a manner that suggests it has been proved to be entirely determined by random, mechanistic events, but they are wrong to oppose the teaching of evolution itself. Its occurrence, on Earth and in the Universe, is by now indisputable. Not so its processes, however. In this, there is need for a nuanced approach, with evidence of creative ordering presented as intrinsic both to what we call matter and to the unfolding story, which includes randomness and natural selection." —MARY COELHO

Every name for Ultimate Reality becomes a filter through which life is experienced. If we are to relate to Reality in any meaningful way, we must compare it to relationships that we already know. Humanity's earliest religious expressions were animistic. Nature was experienced as divine, and the "spirits," or energies, of nature were personified as animals. Archeological and anthropological evidence suggest that the first human-like metaphors for Ultimacy were of the Great Mother, or Goddess. This shift in relational experience coincides with the appearance of horticultural villages and a growing reverence for the fertility of the soil and the fecundity of nature.

With the invention of writing and the plow, cultures and civilizations grew more complex and warfare increasingly made sense: "If we conquer those people, we get their land, their women, their harvest, and all their stuff." The emergence of male, warrior images for Ultimate Reality marks a shift in religious expression beginning around 5,000 years ago. Not long after the appearance of humanity's first kings and kingdoms, we find God, or Ultimacy, addressed in the "western" monotheistic cultures as Lord, King, the Almighty. Those metaphors have carried through to this day. Why? Not because we are still governed by kings or earn our living in service (or enslaved) to a feudal lord, but because of the central role played by holy scripture—unchanging scripture—for Peoples of the Book.

The monotheism of Judaism, Christianity, and Islam reflects the dawning within human consciousness that Reality is a unified Whole; that behind the diverse expressions of power, mystery, and grace in this world, there is an underlying Integrity: one Reality, one God. In the West, our one God remained King and Lord while the scientific revolution proceeded, but our sense of what the Almighty actually did in the Universe began to shift. With the invention of the clock, natural philosophers of the era began to view the Universe itself as a kind of clock—a mechanism that would, to a large extent, run on its own. It had to first be created, of course, and set in motion, so the notion of a clockmaker God began to permeate liberal views.

Man, the clockmaker, thus concluded that God must be a clockmaker, too. God fabricated the Creation, just as a clockmaker fashions a watch, as a designer envisions a product, as an engineer manufactures a new tool, as a mechanic fixes broken machines. Seeing God through the lens of this *mechanistic paradigm* was enormously useful for scientists and entrepreneurs of the day. Relating to Creation as a machine made it possible to discover patterns and "laws" within nature, to predict and control our surrounds with ever greater precision and effect, and to treat the riches of the planet as resources to be used. Nothing was off limits. We could poke and prod any aspect of the world in pursuit of knowledge, power, and wealth. Those who adopted this viewpoint concluded that there would be no gods or God to take offense at our intrusion into what had formerly been sacred terrain.

There was, however, a downside. If God was no longer omnipresent, present in everything, if God was effectively banished from the natural world—a Creator divorced from Creation—something important was lost. What was now considered ultimately *real* was a machinelike Universe, composed of parts that were, in turn, machines. What we expected to find, not surprisingly, was exactly what we would encounter. God, now in exile, was nowhere evident in the material realm.

Believers, struggling with this new worldview, thought that perhaps God was still watching over us, but from a distance—away, removed. This *transcendent-only God* might still choose to intervene, to tinker with the clock, especially in response to the prayers of the faithful, or to punish those who were not. But the day-to-day experience of God had markedly diminished. God was no longer revealed everywhere, in all things, at all times. In a world of mechanistic cause and effect, *God's immanence and omnipresence* could no longer be imagined, and thus could no longer be experienced.

Think about it. If your primary metaphor or analogy for "the Universe" is a machine, then not only will you fail to experience the world as an expression of divine presence, or as a revelation of divine goodness or grace; you will naturally and necessarily relate to it as you relate to human-made machines—as lifeless, soulless. A machine is not a He or She. A clock is not worthy of honor or adoration. Like any other machine, the material world is here to be used by... well, us.

Where does creativity reside with respect to a machine? Outside it, of course! Thus, when reality is understood mechanistically, if divinity exists anywhere, it too must reside outside the machine. The birth of this form of transcendent-only theism was thus accompanied by the births of other ways to relate to a desacralized world: deism, atheism, and humanism.

"Two Gods?"

Given the *real* Cosmos in which we live and move and have our being, to imagine God as distant from the here-and-now is, in effect, to posit two gods. If my "God" is not truly immanent, omnipresent, and revealed in the unfolding creativity of matter itself, then how do I relate to a Universe that is so demonstrably, profoundly, thoroughly, and divinely creative? Seeing God in the unfolding of that infinite and wondrous creativity—right where a Supreme Creator belongs—means relating to Nature as a Thou, a Creative Presence worthy of honor and respect. Only then is my perspective congruent with Reality. Only then might I be blessed and guided by that connection.

> How sad that for many people today, God does not exist in any real, tangible way outside their imagination. They cannot experience the Holy One physically. They cannot honor the reality of the divine with their senses. They are cut off from most of God.

Today we are in the midst of a paradigm shift more relationally significant than even the Copernican Revolution and its further developments. This is the shift from perceiving the Universe mechanistically, as a lifeless *it* made by an otherworldly Supreme Being, to seeing the Universe as a creative revelation of divinity—as a *Thou* deserving our reverence. The timing could not be better. In the words of celebrated cultural historian Thomas Berry,

> The world we live in is an honorable world. To refuse this deepest instinct of our being, to deny honor where honor is due, to withdraw reverence from divine manifestation, is to place ourselves on a head-on collision course with the ultimate forces of the Universe. This question of honor must be dealt with before any other question. We miss both the intrinsic nature and the order of magnitude of the issue if we place our response to the present crises of our planet on any other basis. It is not ultimately a political or economic or scientific or psychological issue. It is ultimately a question of honor. Only the sense of the violated honor of Earth and the need to restore this honor can evoke the understanding as well as the energy needed to carry out the renewal of the planet in any effective manner.

The evolutionary perspective restores Nature's honor and restores God to every nook and cranny of this vast Universe. The Creator has been revealed *within* the Creation all along! But for the last few hundred years we couldn't know that. Such was as it had to be. Science in its youth required freedom from religious dogma in order to fully develop its unique gifts. But now, like the prodigal son, we as a species are coming to our senses, and returning home to our true Self. As acclaimed evolutionist David Sloan Wilson writes in his 2007 book, *Evolution for Everyone,*

The prodigal son left home with an inheritance that he foolishly squandered on a profligate life. Destitute and ashamed, he returned to his father's house asking only to be treated as a servant, since he clearly deserved no more. To his surprise and gratitude, he was received with love and forgiveness as one reborn to a new and more sustaining way of life . . . Our conception of ourselves as set apart from the rest of nature is a bit like the prodigal son leaving home with an enormous inheritance. The repeated collapse of past civilizations and uncertain fate of our own is like squandering our inheritance on a profligate life. Before we become truly destitute and ashamed, perhaps it is time to return home to a conception of ourselves as thoroughly a part of nature. Perhaps this can lead to a more sustaining way of life in the future than in the past.

Perhaps we needed the distance to gain perspective. Like an adolescent, we had to strike out on our own and make a few mistakes in order to discover how life works and gain some humility. Speaking mythically, perhaps we needed to leave home in order to appreciate home—and now God can delight in the adventure of wooing us back.

Day and Night Language

> *"I can hear the sizzle of a new-born star and know that anything of meaning, of fierce magic is emerging here. I am witness to flexible eternity, the evolving past, and I know I will live forever—as dust or breath in the face of stars, in the shifting pattern of winds."*   —JOY HARJO

It is vital to remind ourselves, from time to time, of two complementary sides of the one coin of our experience. On one side is the realm of what's so: the facts, the objectively real, that which is publicly and measurably true. Let's call this side of reality our *day experience*. We talk or write about it using *day language*—that is, normal everyday discourse. The other side of our experiential coin I call *night experience*. It is communicated through *night language*, by way of grand metaphors, poetry, and vibrant images. Our attention is focused on What does it mean? This side of our experience is subjectively real, like a nighttime dream, though not objectively real. Night language is personally or culturally meaningful. It nourishes us with spectacular images of emotional truth.

The language we use really does make a difference. Our choice of metaphor will shade our experience of reality, or its portrayal, in a particular way. Moreover, we will always make events mean *something*. Indeed, even if we say something is meaningless, we're making it mean nothing. Humans swim in a sea of meaning no less than fish swim in water. We cannot avoid it. Problems arise when we fail to distinguish the factual, *objectively* real from the meaningful, *subjectively* real—when we mistake our interpretations for what's so. The two are not the same. Facts are delimited; interpretations are manifold.

Building on the distinction introduced in Chapter 4 between public and private revelation, we can now say this: *Private revelation* is grounded in subjective experience and is expressed in traditional, or religious, night language. *Public revelation* is grounded in objective experience; it is measurable and verifiable, and is expressed in day language. (Those familiar with Ken Wilber's four-quadrant model will recognize day language as referring to the right-hand quadrants, and night language as referring to those on the left. Other authors and educational organizations, such as Landmark Education Corporation, also assist people in gaining valuable and empowering skill in distinguishing these realms.) To clarify these distinctions, it may be helpful to imagine a continuum:

Measurable		Non-measurable
Objective	*Subjective*	*Highly subjective*
Day language	Twilight language	**Night language**
A. Event (facts)	*B. Story*	*C. Meaning (metaphors)*
Wide agreement	Some agreement	Strong disagreement
(Public revelation)		(Private revelation)
Reason		*Reverence*

Whenever we think or talk about an event, there is always (A) what happened, (B) the story about what happened, and (C) the meaning we make out of the story of what happened. "What happened" refers to the uninterpreted, measurable, objective facts—the raw data. "The story about what happened" is the narrative context we consciously or unconsciously weave to connect the dots. Central to *story* are the cause-and-effect linkages we effortlessly fashion from a stream of undifferentiated data. As language-using animals, we create stories as instinctually as we seek food when we're hungry. "The meaning we make out of what happened" is even more subjective and nonmeasurable. It's all the things we tell ourselves, and others if they have the patience to listen to us, about how we (usually unconsciously) interpret the story of what happened—that is, what we make the story mean about us, about others, about the world.

A source of anguish at all levels of society (manifesting as conflict between individuals, between religions, among nations) is the consistent and near universal tendency to confuse B and C with A. We assume that what actually happened is not only our story about what happened, but also what we make that story mean. No! What is true is never our story, nor our interpretations, but only the actual, objective, measurable facts. The further we move from day language into night language, the greater the disagreements.

We cannot solve the problems posed by night language disagreements by jettisoning that face of reality. *We need both day and night language in order to have a meaningful experience of life. The important thing is to get the order right.* If we first seek clarity on the measurable facts—which is the very mission of science—the twilight language and night language stories and expressions of meaning that derive from those facts can enrich our lives and support cooperation across ethnic and religious differences. Basing (or reinterpreting) all our twilight and night expressions on a solid foundation of factual, public revelation is our best chance for achieving harmonious relationships at all levels.

"I'm Here, Too"

Driving in rural North Carolina, I noticed a billboard that read,

"I'm in the book." —GOD

Yes, *and* I'd like to have seen a second line:

"I'm here, too. Look around." —GOD

My formal training for becoming a United Church of Christ minister culminated in an ordination paper that I wrote and then presented to a gathering of ministers and lay leaders. Titled "A Great Story Perspective on the UCC Statement of Faith" (available at TheGreatStory.org), my talk stimulated a host of comments and queries. A widely respected minister posed a question I shall never forget. "Michael," he began, "I'm impressed with your presentation and with the evolutionary theology that you've shared with us. However, there's a little boy who lives in me, and that little boy wants to know: Where is Emory?"

Emory Wallace, a well-known and beloved retired minister, had for nearly three years guided me through my ministerial training. He died suddenly, at the age of 85, just a few weeks before my ordination hearing.

"Where is Emory?" My mind went blank. I knew I needed to say something—after all, this was my ordination hearing—so I just opened my mouth and started speaking, trusting the Spirit to give me the words. My response went something like this:

Where is Emory? In order to answer that question I have to use both day language—the language of rational, everyday discourse—and night language—the language of dreams, myth, and poetry. Both languages are vital and necessary, just as both waking and dreaming states of consciousness are vital and necessary. Like all mammals, if we are deprived of a chance to dream, we die. Sleep is not enough; we must be permitted to dream.

We, of course, know that day experience and night experience are different. For example, if you were to ask me what I did for lunch today, and I told you that I turned myself into a crow and flew over to the neighborhood farm and goofed around with the cows for a little bit, then I flew to Dairy Queen and ordered a milkshake—and if I told you all that with a straight face—you might counsel me to visit a psychiatrist. However, if you had asked me to share a recent dream and I told the same story, you might be curious as to the meaning of that dream—but you wouldn't think me delusional.

So in order to respond to your question, "Where is Emory?" I have to answer in two ways. First, in the day language of common discourse, I will say, Emory's physical body is being consumed by bacteria. Eventually, only his skeleton and teeth will remain. His genes, contributions, and memory will live on through his family and through the countless people that he touched in person and through his writings—and that includes all of us.

But, you see, if I stop there—if that's all I say—then I've told only half the story. In order to address the nonmaterial, meaningful dimensions of reality I must continue and say something like: "Emory is at the right hand of God the Father, worshipping and giving glory with all the saints." Or I could say, "Emory is being held and nurtured by God the Mother." Or I could use a Tibetan symbol system and say, "Emory has entered the bardo realm." Any or all of these would also be truthful—true within the accepted logic and understanding of mythic night language.

My response was well received in that meeting of sixteen years ago, and it has shaped my theology ever since. Recently, I blended the core of that distinction into my Great Story talks and workshops. I am sure that my understanding of day and night language—language of reason and language of reverence—will continue to evolve and thus inform my preaching, my teaching, and my personal relationship with the fullness of Reality.

"Read Me a Nighttime Book, Mommy"

After one of my church programs in which I presented the day language versus night language distinction, a woman who was trained as a scientist told me this story: "When my daughter was young, I would read her bedtime stories. I remember trying to read her a book of nonfiction one night. But she protested, saying, 'That's a daytime book. Read me a nighttime book, Mommy.'"

"The most beautiful and most profound emotion we can experience is the sensation of the mystical. It is the sower of all true science. He to whom this emotion is a stranger, who can no longer stand rapt in awe, is as good as dead.... That deeply emotional conviction of the presence of a superior reasoning power, which is revealed in the incomprehensible Universe, forms my idea of God." —ALBERT EINSTEIN

CHAPTER 7

What Do We Mean
by the Word
"God"?

---

*"[Modern] cosmology tells us that the universe encompasses
all size scales, so any serious concept of God must at least do
as much. 'God' must therefore mean something different on
different size-scales yet encompass all of them. 'All-loving', 'all-
knowing', 'all-everything-we-humans-do-only-partially-well'
may suggest God-possibilities on the human size-scale, but
what about all the other scales? What might God mean on the
galactic scale, or the atomic?"*

—JOEL R. PRIMACK AND NANCY ELLEN ABRAMS

D

o you believe in life?

What an absurd question! It doesn't matter whether we "believe in"
life. Life is all around us, and in us. We're part of it. Life *is*, period. What anyone *says* about life, however, is another story, and may invite belief or disbelief.
If I say, "Life is wonderful," or "Life is brutal," or "Life is unimportant—it's
what happens after death that really matters," you may or may not believe me,
depending on your own experience and worldview. What we say *about* life—its
nature, its purpose, its meaning—along with the metaphors we choose to
describe it—is wide open for discussion and debate. But the reality of life is

indisputable. This is exactly the way that *God* is understood by many who hold the perspective of the Great Story—that is, when human, Earth, and cosmic history are woven into a holy narrative. Our common creation story offers a refreshingly intimate, scientifically compelling, and theologically inspiring vision of God that can provide common ground for both skeptics and religious believers. For peoples alive today, any understanding of "God" that does not at least mean "Ultimate Reality" or "the Wholeness of Reality" (measurable and nonmeasurable) is, I suggest, a trivialized, inadequate notion of the divine.

> *The emergence of the Great Story—a sacred narrative that embraces yet transcends all scientific, religious, and cultural stories—will come to be cherished, I believe, first and foremost for enriching the depth and breadth of our experience of God.*

As discussed in Chapter 5, reality as a whole is divinely creative in a nested emergent sense. Subatomic particles reside within atoms, which comprise molecules, cells, organisms, and societies—like nesting dolls of expanding size and complexity. Outward we find planets within star systems, within galaxies, within superclusters of galaxies. Each of these is a holon—it is both a whole in its own right and a part of some larger whole. At every level, each of these holons expresses unique forms of divine creativity, powers that yield emergent novelty. For example, protons and other subatomic particles churning in the cores of stars fuse into most of the atoms in the periodic table of elements. In turn, hydrogen and oxygen atoms merge into molecules of water, with properties that transcend those of the parts. Together, Sun and Earth bring forth fishes and forests, dragonflies and dancers. Finally, out of human cultures come art, music, religious theologies, and scientific theories. Thus, reality understood as "nestedly creative" is not a belief. It is an empirical fact (albeit expressed metaphorically) accepted by religious conservatives and atheists alike.

God, from this perspective, can be understood as a legitimate proper name for the largest nesting doll: the One and Only Creative Reality that is not a subset of some larger, more comprehensive creative reality. God is that which sources and infuses everything, yet is also co-emergent with and indistinguishable from anything. There are, of course, innumerable other ways one can speak about Ultimate Reality and theologize about God. But if "God" is not a rightful proper name for "the One and Only Creative Reality that transcends and includes all other creative realities," then what is?

This way of thinking sheds light on traditional religious understandings of "God's immanence and transcendence." God is the Wholly One, knowable and unknowable. God embraces, includes, and is revealed throughout the entire Cosmos and in all of life. God is the great "I Am" of existence. Yet as the source, energy, and end of everything (which is the very meaning of *ultimacy*), God cannot be limited to the world we humans can sense, measure, and comprehend: Ultimate Reality transcends and includes all that we can possibly know, experience, and even imagine.

This understanding of the divine mocks the question, "Do you believe in God?" Any "God" that can be believed in or not believed in is a trivialized notion of the divine, and certainly not what we're discussing here. Like life, reality simply is—no matter what beliefs one may hold. What we choose to say *about* reality—the stories and beliefs we hold about its nature, purpose, direction, and so forth—*is* open for discussion, and differences among those choices are unresolvable. But who could deny that there *is* such a thing as "Reality as a whole" and that "God" is a legitimate, though not a required, proper name for this Ultimacy? The transparency of this point is surely one reason why, as I share this perspective across North America, it garners the assent of theists, atheists, agnostics, religious nontheists, pantheists, and panentheists alike.

Lately I've even been wondering if this way of thinking about God transcends and includes all understandings of the divine. Whatever any person or tradition might say about the divine, the undeniable fact is: Reality rules! That which is fundamentally and supremely real always has the final word. Everything bows to it, with no exception. Traditional language declaring "God is Lord," and modern expressions like "Time will tell," "Nature bats last," "Your ego does not run the show," and "All creatures evolve by adapting to their environments" point to a similar if not identical understanding and experience. There is something infinitely knowable—and infinitely mysterious—to which everything in the Universe is beholden. This perspective becomes especially relevant when we recall that "the environment" is not just our surroundings. It is our very source.

No Less Than a Holy Name for the Whole

> *"Reality is that which, when you stop believing in it, doesn't go away."*
> —PHILIP K. DICK

Here is a taste of what can be said, using both day and night language, about the Whole of Reality (God):

✦ Source of everything   ✦ Transcends and includes all things

✦ End of everything   ✦ Expresses all forms of power

✦ Knows all things   ✦ Holds everything together

✦ Reveals all things   ✦ Suffers all things

✦ Present everywhere   ✦ Transforms all things

Otherworldly images of the divine notwithstanding, when *God* is understood foundationally as a holy, proper name for "the Wholeness of Reality, measurable and nonmeasurable," everything shifts. New possibilities open for ways of thinking about creativity, intelligence, the Universe, and our role in the evolutionary process. From the perspective of the Great Story, *immanent creativity* may be an ideal way to portray emergent complexity and that would appeal to both evolutionists and proponents of intelligent design. There is no inherent conflict between immanent creativity and a mainstream understanding of biological, cultural, planetary, and cosmic evolution. As well, immanent creativity does not imply (as *intelligent design* does) that the Universe is an artifact, created and manipulated by a remote, and only sporadically interventionist, God.

Although the metaphor of a clocklike Universe helped birth the scientific revolution, scientists working today in virtually all disciplines are moving beyond the constraints of a mechanistic, or nature-as-lifeless-machine, worldview. Creative evolution, self-organization, autopoiesis, cosmogenesis, chaos and complexity sciences, evo-devo—these terms exemplify the shift from a mechanistic to a nestedly emergent worldview. As biologist and philosopher Theodosius Dobzhansky insisted, evolution is neither random nor determined, but creative. Just as water emerges from the union of hydrogen with oxygen, creativity is the child of chance and necessity.

Scientists (using *day language*) regard "the Universe" as evolving in accordance with the dictates of natural law, the happenstance of initial conditions, the unpredictability of chaotic components, and the striking dependability of evolutionary emergence. Theologians (using *night language*) speak of "the Creation" as having been sourced by God's will and maintained by God's grace. Only now can we appreciate that these are different ways of speaking about the same fecund processes. To argue over whether it was God, evolution, or the self-organizing dynamics of emergent complexity that brought everything into existence is like debating whether it was me, my fingers tapping the keyboard, or the electrical synapses of my nervous system that produced this sentence.

Such an understanding of the divine, of course, begs the question: Does this God evoke humility, love, trust, adoration, reverence, and commitment? Is this a God anyone would want to worship, pray to, or devote one's life to serving?

I offer a resounding Yes!

"Finally, a God That Makes Sense!"

Twice in my first six years of itinerant ministry, I faced an audience equally split between the extremes of the science-religion spectrum: atheists and biblical fundamentalists. Give me either group, in its full-blooded richness or interspersed with moderates, and I will do a fair job of making my points in ways that most people can hear and will be led to consider. But the two together? I call that "an audience from hell."

Seriously, in all but the most polarized settings, my reframing of what we might be pointing to when we use the word "God" resonates with ardent atheists and with the scripturally minded. I recall a time when I addressed an InterVarsity Christian Fellowship group of students at a university in eastern Canada. One young man came loaded for bear. Along with his King James Bible, he brought a copy of *Strong's Concordance*. He challenged me from the get-go. "Lighten up, John," someone finally called out to him. "Give the man a chance to present his viewpoint." Well, after my talk and the discussion period, John gave me a bear hug. "Now, I don't go along with everything you say," he began. "But I'm not threatened by your ideas."

Similarly, I have been delighted by responses from most atheists. In Colorado Springs, an older man blurted out in the midst of my presentation, "Finally, a God that makes sense!" In this case, I was presenting exactly the same "Thank God for Evolution" digital slide program as I had used the previous day in my talk to a liberal Christian audience.

I shall never forget the comment made by an elderly woman at a Unitarian Universalist Church in the Midwest. "I'm an atheist," she said. "But I want to tell you how excited I am that you've made it okay to use God language here in our church!"

Prayer in a Nestedly Creative Cosmos

"Praying to an otherworldly God is like kissing through glass."
—PAUL WEST

If we wish to have a meaningful relationship with Reality, we may very well choose to use personal analogies to describe the nature of Ultimacy. Different traditions, necessarily, use different images and metaphors to describe Ultimate Reality and our relationship with Him/Her/It. All such attempts to capture the essence of Supreme Wholeness are legitimate. Most are helpful—and all are limited. Such are the deficiencies of human language and human experience. The map is not the territory; the menu is not the meal.

Spiritual practices that have served many and have stood the test of time, as well as practices born of contemporary psychological research, have this in common: They suggest, at their core, that the path to wholeness and a right relationship to Reality is not complicated. Integrity is the key. The peace that passes all understanding, recovery from addiction, salvation from sin, ongoing transformation, personal empowerment, enlightenment, dwelling in the Kingdom of Heaven, unity with God—each of these can be found in the present moment—and nowhere else.

How? Simply, *get* that you are part of the Whole, and then commit to living in deep integrity—and follow through with it. By being and doing this you will effortlessly express your creativity, take responsibility for your life and your legacy, and listen to your heart for guidance from the source of your existence. You will naturally love Reality (God) with all your heart, soul, strength, and mind. You will love your neighbor as yourself. And, yes, this is "the way, the truth, and the life" that the early Christian gospels portray Jesus the Christ incarnating.

Prayer from this perspective is truly an intimate process, and one that even an atheist might embrace. Prayer is no longer an act of petitioning a far-off supernatural being to miraculously intervene in the world according to one's wishes or desires. With an understanding of "God" as no less than a proper name for Supreme Wholeness, prayer can be understood analogously as a cell in deep communion with the larger body of which it is part.

"The primary prayer is one of awe, and it is probably the most effective prayer because through it we turn to our origins and just behold with a sense of gratitude."   —BRIAN SWIMME

There is a profound difference between *believing* in a personal God and *knowing* God personally, that is, relating to Reality intimately. Believing in a personal God—giving mental assent to the existence of a supernatural entity with a personality—may or may not make a difference in the life of the believer. When belief does not richly transform one's experience, such belief becomes a booby prize.

In contrast, when we relate to Reality personally—knowing that each of us is accepted just as we are, and trusting that it's possible to interpret everything real in one's life as a gift and a blessing in disguise: this will always transform any of us. If we can develop a habit of "conversing" with Wholeness, of quieting our minds, jettisoning all judgments, surrendering to a Higher Power, seeking deep and intuitive guidance, opening to the way of the heart, engaging in contemplative prayer, and many other names and activities that put us in a state of radical openness and receptivity to wider and deeper wisdom, then our *experience* of life will improve enormously—even if the outward conditions of our existence change not a whit.

A Personal, Undeniable God

"When the Bible or classical theology referred to 'God' as an individual person, this was a pictorial way of talking about 'reality in all its fullness'. God is not the greatest or largest of beings. God is the ground of all being. God is that awesome and mysterious Reality in which all things live and move and have their being, and out of which all things emerge and into which all things return."   —GENE MARSHALL

Physicists, philosophers, and others have spoken of a realm of nothingness (no-thing-ness) out of which everything arises. Here are some names that have been offered for this unseen, nonmeasurable, nonmaterial realm: the Implicate Order, the All Nourishing Abyss, the Void, the Vacuum State, the Akashic Field, the Mother Universe, or simply, God. That not all of what is Real is measurable is a fact, not a belief. Any attempt to say something meaningful about this realm of nonmeasurability necessarily evokes night language. Thus any such expressions will be subjectively truthful—or not. They will be believable to some, and not believable to others. The realm of night language would have it no other way.

Again, it may be legitimate to imagine God as far more than a proper name for Supreme Wholeness, but an immanent, omnipresent Creator can be no less than this. Such a way of reflecting on the divine moves God-talk beyond the realm of belief or disbelief. When I say "God," I am not talking about something or someone that can be believed in or not believed in. I'm talking about the Ultimate Wholeness of Reality, seen and unseen—the whole shebang—which is infinitely more than anything we can know, think, or imagine.

This is a simple, nuts-and-bolts distinction. We don't *believe* in things that are undeniably real. We know them. We don't believe in water; we are 60 to 70 percent water. We don't believe in the Universe; we live and move and have our being within this undeniable material and nonmaterial Reality that many today call "the Universe," but which others have for centuries referred to as "Mother," "Father," or "Lord." This understanding does not, in fact, reduce the Creator to Creation; rather, it elevates and REALizes our sense of divine immanence and omnipresence.

I am presenting a God that we cannot deny. This is a God that we cannot help but experience, whether or not we think of God in such a way, and whether or not the word *God* is part of our preferred vocabulary. This God is all around us, among us, within us—overflowing with creativity and abundance at every scale of universal nestedness. And we ourselves are natural expressions, children, of the creativity that suffuses Ultimate Reality. We are no more isolated from God than a tree is isolated from the ground upon which it grows. An evolutionary understanding of Reality can thus bless us with a profoundly personal relationship with God.

One may or may not choose to use the word "God" to refer to this undeniably experiential Reality, the largest nesting doll. It's not necessary to relationalize the Whole of Reality by choosing a divine name as its referent. But there is much to be gained from doing so. The Stoic Greeks, for example, named the Whole "Kosmos." They imagined Kosmos as a vast living being that everything is part of. So while less relational names are both possible and legitimate, it is unreasonable to deny that (a) there is such a thing as the Whole of Reality, and (b) "God" is a legitimate proper name for such Ultimacy.

God or the Universe: What's in a Name?

> "*There is nothing in modern cosmology that requires the existential view, nor anything that requires the meaningful view.... A meaningful Universe encompasses the existential, in the sense that the meaningful can understand the existential, but the existential cannot see the meaningful. The choice of attitude is not a casual one.... Cosmology is not a game; it has the power to overturn the fundamental institutions of society.*"
> —JOEL R. PRIMACK AND NANCY ELLEN ABRAMS

Why call the Supreme Wholeness of Reality "God"? Why is it helpful to give this undeniable Ultimacy a name? And if we do choose to name Ultimate Reality, what metaphors do we find most appealing? More specifically, what do we say "God" is like? And do such choices really matter?

As it turns out, how we name Ultimacy makes a world of difference in our experience of life and one another. We can, of course, refer to Reality in impersonal ways, selecting among the usual secular terms, such as "nature," "the universe," "the cosmos," "the environment." If we choose to use impersonal terms, especially while in the all-too-familiar mechanistic mindset (and thus expecting Reality to show up like a clock), then we may be contributing—albeit unintentionally—to our species' demise. As renowned systems thinker Gregory Bateson has said,

> If you put God outside and set him vis-a-vis his creation, and if you have the idea that you are created in his image, you will logically and naturally see yourself as outside and against the things around you. And as you arrogate all mind to yourself, you will see the world around you as mindless and therefore not entitled to moral or ethical consideration. The environment will seem to be yours to exploit. Your survival unit will be you and your folks or conspecifics against the environment of other social units, other races, and the brutes and vegetables. If this is your estimate of your relation to nature and you have an advanced technology, your likelihood of survival will be that of a snowball in hell. You will die either of the toxic by-products of your own hate, or simply of overpopulation and overgrazing.

If, however, we regard the Ultimate Wholeness of Reality as divine and choose to use the word "God" as a sacred proper name for that Whole, then our terminology itself will incline us (and those who come after us) to honor Nature and to learn all that we can from Her (or Him, or It). As well, we will be moved to do all that we can to prevent further losses of biodiversity and to work for a just and thriving future for all. And we will do this not because we abruptly shed our inborn shortsightedness, but simply as a consequence of seeing the world as it truly is: a material and nonmaterial expression of Ultimate Creativity—the primary revelation of divine love, power, beauty, and grace. We will care deeply for our world *because* of our devotion to God. The two are inextricably related.

"What Does Jasmine Want?"

Over the course of six years of marriage (and life on the road), Connie and I have playfully named a number of holons. In so doing, we have entered into more intimate relationship with each of them. For example, we have chosen a personal name for our home continent of North America: *Nora*.

Jasmine is the name we use to speak of our relationship as a couple—the whole that is more than the sum of the two of us. Naming our marital bond has helped us prevent minor disagreements from morphing into major ones. "I know what you want," I may say to Connie, "and I know what I want, but what does Jasmine want?" Just by posing the question, tensions subside; we might even laugh. Now, why it is that Jasmine usually wants to do what Connie wanted to do in the first place, I haven't yet figured out. ☺

Our van (and the voice of our GPS system) is *Angel*. Earth's seas we call *Oceania*. And what has become one of our own most sacred spots in all of Nora—a cliff on the coast of Maine, where we jumped into the sea nearly every day for two months—we have named *Ziggy*.

Now, of course, we know that Ziggy, Nora, Angel, Jasmine, and Oceania are personifications. Still, our life is far richer because of them. Surely the ancients knew this too.

We are on the verge of the greatest spiritual awakening in history. It took centuries for the Copernican Revolution to transform humanity. Thanks to global satellite telecommunications, the Internet, and all the other technologies that link us, it is quite possible that our own paradigm shift—from seeing nature as an artifact, to seeing Nature as the primary revelation of divinity (and inseparable from that divinity)—will prevail over the course of decades rather than centuries. That the shift will occur eventually is almost certain. How fast it transforms our institutions depends on how rapidly and thoroughly we are transformed as individuals, and where we choose to invest our collective creativity as awareness expands.

Creatheism

"We have all heard some fundamentalist-minded person say something like, 'Don't tell me I'm related to monkeys.' The fact of the matter is that now that we have discovered DNA and its code, we know that we are not only related to monkeys, we are related to zucchini. So let's get over it." —MARLIN LAVANHAR

Occasionally, someone who has heard me speak asks in frustration, "What *are* you, anyway? A theist? Atheist? Pantheist? I can't tell what you are!" My standard response goes something like this: "I'm all of those—and none of them. Actually, my wife and I had to coin our own term. I'm a creatheist (cree-uh-theist), and my wife, well, she's a creatheist (cree-atheist). We spell it the same way. We mean the same thing. We just pronounce it differently." This response almost always evokes smiles or laughter.

Here is why this new word can bridge the theist-atheist divide: *One need not believe in anything in order to be a creatheist.* It's not a belief system. It is based on what we know, not what we believe. I call creatheism a "metareligious scientific worldview" and posit the following three points as core to its understanding:

1. The Whole is creative in a nested, emergent sense.

2. Humanity is now an integral and increasingly conscious part of this process.

3. There are many legitimate ways to interpret and speak about Ultimate Reality.

A creatheistic view of the Universe—whichever way one chooses to pronounce it—celebrates the nested emergent nature of divine creativity. This perspective includes, yet transcends, previous attempts to articulate the relationship of God to the world. The array of "isms" already on the religious menu (including theism, pantheism, deism, atheism, religious nontheism, and panentheism) have all played roles in helping us get to this point, and all offer interesting perspectives on creatheism. Each of these perspectives has a piece of the truth, yet none can deliver the whole truth (nor, of course, can creatheism).

For example, creatheists can believe in God's transcendence—or not. It is a matter of personal preference; it is a belief that can neither be proven nor disproven. Yet both theist and atheist perspectives on creatheism celebrate the awesome, ultimately mysterious cosmic creativity that resides within and everywhere around them.

The most common misconception about creatheism is that it is simply a modern form of pantheism. To be sure, creatheism includes pantheism, and I know some pantheists who find a deep-time, science-based perspective so attractive that they are happily updating and thus enriching their cherished traditions. A creatheistic mindset simply holds that everything is part of the Whole. Pantheism (the term itself was coined in 1705 by John Toland, although the perspective is ancient and indigenously expressed worldwide) is set apart from other theisms by locating the divine entirely within the Universe. Birthed in a time when physicists are discussing the possibility of multiple universes, creatheism simply makes room for the degree to which science reveals the scale of our ignorance. In a nestedly creative Cosmos, we can't possibly know who or what God is in Its/Her/His totality.

Traditional theists tend to focus on divine transcendence. Pantheists tend to focus on divine immanence. Creatheists embrace them both, as does panentheism (or process theology), which is similar. However, panentheism has not sufficiently extended into popular culture, beyond its century-old roots in academia, to engage the full passions of many of us. For conservative Christians, the term has too long excluded their viewpoints, and thus is unlikely to be awarded a fresh start.

Creatheism also embraces yet transcends atheism and atheism's positive expression in the form of 20th century humanism. But as with panentheism, atheism too is evolving—witness the poetry and inclusiveness of today's *religious naturalism* movement. Both creatheism and atheism affirm that all language of Ultimate Reality, without exception, is metaphorical (i.e., there is not an invisible being up there somewhere). But creatheism goes on to say that, if we are going to relate to Reality personally, it is both acceptable and beneficial to use metaphors that engage the heart—metaphors such as Father, Mother, Beloved, Larger Self, Higher Power, and many more. Creatheism is still, of course, a newborn. It is a term that has a long way to go to prove its merit. Nevertheless, for Connie and me it has already done its job: it has bridged *our* differences.

Let us pause for a moment to review various ways of understanding and experiencing the divine:

THEISM: a concept generally thought to mean belief in an interventionist God. Theists tend to imagine nature as a machine-like thing, an artifact made by a Supreme Being residing off the planet and outside the Universe.

DEISM: a concept popularized in the 17th century, meaning belief in a nonin-terventionist God. Deists share with theists a regard for nature as a machine-like thing, an artifact made by a Supreme Engineer residing outside time and space. For both, God is the Creator of the universe but essentially nonexistent within it. The difference is that deists assume that God does not intervene in the workings of the world.

ATHEISM: a concept popularized in the 17th century, meaning disbelief in any sort of god or supreme being, interventionist or not. Atheists, traditionally, also imagine nature as a machine-like thing, which came into being through mechanistic laws, though without a supernatural lawmaker or lawgiver.

PANTHEISM: an ancient concept found throughout the world, repopularized in Europe during the 18th century, meaning belief in a divine Cosmos with man-ifold powers manifesting as distinct forces and most potently in distinct locales. Pantheists typically equate God with the Universe, with nothing transcending Nature.

PANENTHEISM (DIALECTICAL THEISM): a concept originating in the early 20th century with process philosophy, meaning belief that "God is in Creation and Creation is in God." Panentheists see Nature as a revelation, or embodied expression, of divinity.

Finally:

CREATHEISM: a concept introduced in the early 21st century, grounded in an empirical understanding of the nested emergent nature of divine cre-ativity. For creatheists "God" is a holy name for Ultimate Reality—the all-encompassing Wholeness—that which includes yet transcends all other realities. Creatheism regards Nature as a revelation or expression of the divine—particularly in its *emergent creativity*. Creatheism understands humanity as a self-reflective aspect of Creation that allows the Wholeness of Reality, seen and unseen, manifest and unmanifest—i.e., God—to be honored in conscious awareness and to guide our own deliberate manifestations of that divine creativity.

Oh yes. There is one more potent, if also playful, possibility in this list of

godisms. I burst into laughter when I was recently introduced to this term. It is
so aptly named that it requires no definition:

APATHEISM.

> **"Praise God, Brother, So Am I!"**
> Sunday service had ended at a Christian church I was pastoring. While
> greeting parishioners on their way out of the sanctuary, a visitor, a young
> man, extended his hand and declared, "Well, Reverend, I'm an atheist!" I
> shook his hand and responded, "Praise God, brother, so am I!" His face reg-
> istered confusion. I continued light-heartedly, "Tell me about this God you
> don't believe in. I'm quite sure I don't believe in that God either."

The Role of Humanity in an Evolving Universe

*"Evolution means a face set to the future, toward which we
press with faith and high purpose. It means believing in some
better thing, and forever some better thing, for religion, for
humankind, for the world; believing in it so earnestly that we
shall gladly make ourselves coworkers with God to bring the
consummation."*   —JABEZ SUNDERLAND

The meaning and purpose of a person's life is how he or she contributes to the
well-being of any of the larger holons of existence. That includes the holon of
family, of community, and of secular and religious institutions and creative
pursuits that make civilizations possible and persistent. Similarly, the meaning
and purpose of humanity is how we as a species contribute to the larger body
of Life both now and seeded into the future. Traditionally, as in the Westminster Catechism, the issue is addressed this way: "Q: What is the chief end of
man? A: To glorify God and enjoy Him forever." Here is a restatement from the
perspective of the Great Story: "Q: What is our evolutionary destiny? A: To
honor and joyfully participate in the ever-evolving Whole of Reality, with conscious reflection."

The world is almost mind-numbingly dynamic. Out of the Big Bang came the stars. Out of stardust came the Earth. Out of Earth came single-celled creatures. Out of the evolutionary life and death of these creatures came human beings with consciousness and freedom that concentrates the self-transcendence of matter itself. Human beings are the Universe become conscious of Itself. We are the cantors of the Universe.

—ELIZABETH JOHNSON

When considering the role of the human, it is essential to remember that from an evolutionary perspective, we are not separate creatures on Earth, living in a Universe. Rather, we are a mode of being of Earth, an expression of the Universe. We didn't come into the world; we grew out from it, like a peach grows out of a peach tree. As physicist Brian Swimme says, "Earth, once molten rock, now sings opera." And again, "Here's the whole story in one line: You take a great cloud of hydrogen gas, leave it alone, and it becomes rosebushes, giraffes, and human beings."

When the Bible (Genesis 2:7) tells of God forming us from the dust of the ground and breathing into us the breath of life, we can now appreciate that this is a beautiful night language description of the same process—with God presented personally, moving us into felt relationship with the Creative Reality that made it all happen.

As a result of a thousand million years of evolution, the Universe is becoming conscious of itself, able to understand something of its past history and its possible future. This cosmic self-awareness is being realized in one tiny fragment of the Universe—in a few of us human beings. Perhaps it has been realized elsewhere too, through the evolution of conscious living creatures on the planets of other stars. But on this planet, it has never happened before.

—JULIAN HUXLEY

Concerning our evolutionary role in the Big Picture, as well as in the small and immediate, it is crucial to comprehend that human destiny and the destiny of Earth are inextricably linked. If we can know in our bones that everything we are has emerged through billions of years of evolution and that no species can live in isolation from others, then we will finally grasp that the future of our species depends upon the future of this planet—no less than a child in the womb depends upon the mother. This is one of the great lessons of the evolutionary worldview.

> The Universe is an evolving product of an evolutionary process. It is not an accident; it is an enterprise.   —THEODOSIUS DOBZHANSKY

The entire enterprise is integral: soil, air, water, life, stars and stardust. Humanity's Great Work is to further divine creative emergence—God's will—in ways lifegiving for the whole. As Thomas Berry has said, "The human community and the natural world will go into the future as a single, sacred community, or we will both perish in the desert." The time is now at hand to become positive and conscious agents of the next stages of evolution, thereby fostering a future in which the vast diversity of life shall flourish, too. All other issues rest within that overarching context, within that comprehensive sense of collective purpose.

> Our present urgency is to recover a sense of the primacy of the Universe as our fundamental context, and the primacy of the Earth as the matrix from which life has emerged and on which life depends. Recovering this sense is essential to establishing the framework for mutually enhancing human–Earth relations for the flourishing of life on the planet.   —THOMAS BERRY

Because the entire Universe is evolving and we're part of the process, then "knowing, loving, and serving God" really *is* our way into the future! It is, indeed, the only way.

Being "Faithful to God"

> *"How do we choose life for our planet? . . . Life is the entire four-billion-year process of evolution on Earth. . . . To choose life is to nurture and protect this great cosmic process."*
> —JOEL R. PRIMACK AND NANCY ELLEN ABRAMS

From an evolutionary perspective, "idolatry" would entail any instance in which ultimate commitment is grounded in anything less than the Whole. Idolatry is putting one's primary allegiance or loyalty in anything less than Supreme Wholeness—that which transcends and includes all other realities. As the following examples illustrate, the consequences of idolatry are almost always tragic.

+ Countless people and animals have suffered, lives have been lost, and ecosystems ravaged, all because of misplaced allegiance to the materialism of our consumerist culture.

+ Millions of Jews were exterminated because many Nazis were loyal to Hitler but not faithful to anything larger than the Third Reich.

+ Thousands of women were tortured and killed during the Inquisition because of those who were loyal to the institutional Church but not faithful to the feminine experience of life.

+ Whenever two groups go to war, each loyal to its own leader, patriotic ideals, or beliefs about God's will, idolatry culminates not only in mass destruction and death, but also in vengeful attitudes that may transmit across the generations.

Closer to home, when committed partners are devoted to one another and their own children without also being faithful to and responsible for their community and their world, the results are tragic in the aggregate. Few things stress the environment and our health more than millions of us simultaneously trying to survive in the modern world without the emotional and physical support of extended family and community, and without the spiritual support of the body of Life, which gave us birth and which nourishes and sustains us.

Sanity, health, and ecological sustainability (salvation!) all lie in the direction of faithfulness to God understood not as a Supreme Being outside the Universe, but as a holy name for the Whole of Reality. Using traditional Christian language, one might say, *Salvation is to be found in Christ-like evolutionary integrity: honoring the past, being faithful in the present, and taking responsibility for the future.*

> *"We have a new story of the Universe. Our own presence to the Universe depends on our human identity with the entire cosmic process. In its human expression the Universe and the entire range of earthly and heavenly phenomena celebrate themselves and the ultimate mystery of their existence in a special exaltation.... Science has given us a new revelatory experience. It is now giving us a new intimacy with the Earth."*
>
> —THOMAS BERRY

PART III

The Gospel
According to Evolution

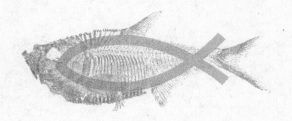

"*Scientific knowledge does limit the imagination, but only in the same healthy way that sanity limits what we take as real.*"
—JOEL R. PRIMACK AND NANCY ELLEN ABRAMS

CHAPTER 8

Growing an
Evolutionary Faith

"How to interpret the Epic of Evolution is neither obvious nor
simple. It requires romantic vision and philosophical rigor. It
requires appropriate metaphysical concepts and inspiring
artistic forms. The Epic of Evolution requires an interpretive
community that seeks to integrate knowledge and wisdom
from across the disciplinary boundaries of our compart-
mentalized modern university and our fragmented postmodern
society. The solution is evolution. Adapt!...Many of the
frameworks best able to interpret the Epic of Evolution are
already present in the world's spiritual traditions. Successful
adaptation is built upon creative replication. We need ancient
wisdom upon which to build this new world."

—WILLIAM GRASSIE

How might we take the core concepts of any religious tradition—
especially those born of a long-ago era—and grow them into an
evolutionary celebration of that faith? This, I maintain, will be the
focus of all religions in the decades ahead. At the outset, resistance to evolving
each tradition will be widespread and adamant: witness the sour relationship
between conservative religious believers and evolutionists in America today.
Yet there will come a tipping point. There will come a time when those so
engaged will find the task no longer cumbersome but beckoning. More, there

will come a time when enough young people are fed on the new and lifegiving interpretations, and at such an early age, that they will gladly take the lead where their elders may falter.

My faith tradition is Christianity. Thus in this book I will periodically explore how flat-earth Christianity might grow into a vibrant, compelling Evolutionary Christianity. My dream is that leaders in other faith traditions will be moved to do the same in their own domains. My approach here is simple: How might the core concepts of our faith be REALized in light of our modern scientific understanding?

What I mean by REALize is to make real—undeniably real—that is, to suggest *day language* referents for traditional *night language* insights. Whether the doctrine be Original Sin or the Kingdom of Heaven, my goal is to make these concepts so real that they touch, move, and inspire everyone, religious and nonreligious alike. I have no interest in merely reconciling the old with the new, or of finding ways to expunge the most egregious incompatibilities between biblical passages and the new cosmology. My aim is nothing less than a deeply fulfilling marriage of faith and reason that leads to empowered lives and healthy relationships—and to vibrant civilizations.

My aim is to show how *what we know REALizes what we believe*. Of course, what I offer here is only a beginning. In a truly evolutionary form of any religion, there will be no single best interpretation for all people and for all time.

Before we begin, let's take a moment to consider the criteria for REALizing traditional night language concepts that I shall use throughout this book. To my mind, a REALized interpretation of a traditional understanding must

1. Validate the heart of earlier interpretations

2. Make sense naturally and scientifically

3. Be universally, experientially true

4. Inspire and empower

As noted in the Introduction, some chapters and sections within this book begin with prophetic inquiries that I then go on to answer. I invite you to join me in considering the answers I offer here, and then to join me and others in exploring more and better answers in a co-creative exploration and celebration on our website.

As communities of Evolutionary Christians, Evolutionary Hindus, Evolutionary Muslims, and the like co-create answers to these questions by way of evolutionary prophetic conversations, each will generate a vibrant collective wisdom about the Great Story teachings for their own tradition—and which will also be of universal value. As for my own first attempt, if the evolutionary interpretations of Christian doctrine offered in this book do not speak to what we all know and experience—that is, if atheists and non-Christians do not also find the interpretations meaningful—then I have failed to satisfy my own four-fold standard for success.

Genesis in Context

> "Our unique attributes evolved over a period of roughly 6 million years. They represent modifications of great ape attributes that are roughly 10 million years old, primate attributes that are roughly 55 million years old, mammalian attributes that are roughly 245 million years old, vertebrate attributes that are roughly 600 million years old, and attributes of nucleated cells that are perhaps 1,500 million years old. If you think it is unnecessary to go that far back in the tree of life to understand our own attributes, consider the humbling fact that we share with nematodes (tiny wormlike creatures) the same gene that controls appetite. At most, our unique attributes are like an addition onto a vast multiroom mansion. It is sheer hubris to think that we can ignore all but the newest room."
>
> —DAVID SLOAN WILSON

The Great Story—evolution experienced in a holy way—can catalyze spiritual and psychological transformation more consistently for modern peoples than can any of the creation stories born of prior ages. That said, as Joseph Campbell so beautifully conveyed decades ago, there will always be power in the ancient stories. What might we discover of continuing value when we regard the biblical epic through an appreciative evolutionary lens? What, specifically, draws our attention within the chapters of Genesis? What of the story of six days of Creation, and then of Adam and Eve in the Garden with the apple, the serpent, and the Fall?

Unlike many others working at the nexus of Christianity and modern science, I have no interest in a passage-by-passage *reconciliation* of the ancient story with today's cosmology. I prefer to start with the best and most up-to-date understandings that have been revealed to humanity, thanks to the disciplines of science and the technological tools that vastly extend human perception and reasoning. This is a story—the unalloyed story given by way of *public revelation*—whose core components could not have been delivered to the ancients in any way that they would have accepted and understood. The ancients could not have envisioned a Universe filled with a hundred billion galaxies, when all they could discern in the night sky was the arc of one grand celestial trace: the Milky Way. The ancients could not have known that "the fixed stars" were suns very far away, and that Earth was in the same class as "the wandering stars" (that is, "planets"). The ancients could not have comprehended the patterning presence of dark matter, the expansive force of dark energy, the chemical cauldrons of supernovas, and the erosive power of raindrops tapping on stone over millions of years. They might have scoffed at a story of crashing continents, of mountains still uplifting, of ocean floors made of rifting and plummeting plates. They could, of course, envision a Great Flood, but they would not have known that enormous rocks falling from the sky had vanquished monstrous beasts long before Noah was assigned the task of preserving the world's biodiversity.

> Tell me a creation story more wondrous than that of a living cell forged from the residue of exploding stars. Tell me a story of transformation more magical than that of a fish hauling out onto land and becoming amphibian, or a reptile taking to the air and becoming bird, or a mammal slipping back into the sea and becoming whale. Surely this science-based culture of all cultures can find meaning and cause for celebration in its very own cosmic creation story.
>
> —CONNIE BARLOW

The ancients most assuredly would have known that green plants require sunlight in order to grow, but they could not have grasped the chemical miracle that we call photosynthesis. They could not have detected the ozone shield nor understood how light-eating bacteria brought it into existence and why that shield is vital to our well-being. The ancients could not have known that God did not aim for the kind of perfection possible through design. Rather, God favored the slow and rambling paths of emergent evolution. As well, the ancients could not have guessed at the stunning improvisations made possible by natural selection, the flamboyant excesses born of sexual selection (as in the peacock's tail), the marvels that would emerge through the push-and-pull interactions of symbiosis, and the role that hundreds of thousands of years of human creativity had played in the evolution of culture—diverse cultures, each with its own creation story.

The ancients could not have known that our own ancestors once lived in the sea and that our umpteenth great-grandparents traversed the trees as gracefully as squirrels and monkeys still do. The ancients were living in the time of the Great Transformation from orality to literacy, and thus from evolving sacred stories to stories set in stone. But they could not have known of the previous transition from the constraints of primate forms of communication to the expansive possibilities of symbolic speech. In all, the ancients could not have known the wonders of deep time.

> In the whole history of human thought, no transformation in attitudes toward nature has been more profound than the change in perspective brought about by the discovery of the past.
>
> —STEPHEN TOULMIN AND JANE GOODFIELD

The ancients cannot, of course, be faulted for their ignorance of deep time. Prior to telescopes, microscopes, computers, and the scientific method of discerning public revelation, no one could have known of the wondrous course of cosmic, planetary, geological, biological, and cultural evolution. Now that we know—now that, for the very first time, we can see who we are, where we came from, and how God made us—we are stunned to discover that our own time in history is no less pregnant with divine possibility than was first-century Palestine.

Don't Throw Out the Apple

"The appropriate stance for scientists to adopt is that traditional societies are likely to embody a great deal of wisdom that remains to be discovered scientifically."

—DAVID SLOAN WILSON

I will now show a way for liberal Christians to safely reclaim the concept of Original Sin, for conservative Christians to hold this doctrine in a REALized way, and for non-Christians to own this insight for themselves. Reframing this particular religious concept is possible only because our culture did such a fine job of casting it aside during the final decades of the 20th century. The "blank slate" (nurture trumps nature) worldview that prevailed in the humanities was part of the impetus. Emancipation also came from within the church. Matthew Fox (former Catholic priest, now Episcopalian) has been a leader in the movement for liberal Christians to pay more attention to what Fox so rightly calls *Original Blessing*. Yes! There is much to be grateful for in our evolved, inherited human nature—our very *animal* nature—that draws from every structure within our brain, as well as in the form and functions of our body.

Our "reptilian" brain—that is, the part that goes all the way back to the time that our own direct ancestors were, in fact, reptiles—handles our heartbeat and our breathing. Our equally ancient cerebellum instantly orchestrates movement to prevent us from falling when we stumble. The part of our brain that we inherited from our early mammalian ancestors uses powerful emotions to ensure that we protect our kin and comrades, and that we pull back from harmful situations and relationships. As well, "gut feelings" are vital complements to rational decision-making. Beyond the brain, we can celebrate that our ability to process food, flush wastes, heal wounds, strengthen muscles, and even grow another human being within—with no conscious effort on our part—is nothing short of a miracle. Nevertheless, we may now have tilted a bit too far in our enthusiasm to proclaim the blessings of our inheritance. I suggest that it is time to wholeheartedly reclaim the concept of Original Sin, but in a contemporary, lifegiving way instructed by the fruits of public revelation—of science.

Original Sin and the New Cosmology

Several years ago, my wife and I were talking with a Catholic nun who had been inspired by Matthew Fox as well as by the Great Story. I shall never forget something she said. We had been talking about one or another of society's pressing problems, when she interjected, "But, of course, I don't believe in Original Sin anymore."

I remember feeling a sudden hollowing inside, like a missing in my belly. Something was deeply wrong. I knew that this religious sister was not alone in her belief, even among Catholics. And from that moment, I began to think seriously about incorporating evolutionary psychology and evolutionary brain science into my talks about our shared creation story.

My fellow advocates of the new cosmology, especially within Christian contexts, have happily drawn from the most up-to-date discoveries in astrophysics, Earth sciences, life sciences, and anthropology. Our movement has been slow to make room for what God has been revealing through evolutionary psychology and brain science. Yet it is here that we shall encounter a fresh take on salvation.

CHAPTER 9

REALIZING "The Fall" and "Original Sin"

> "There is indeed a force devoted to enticing us into various
> pleasures that are (or once were) in our genetic interests but do
> not bring long-term happiness to us and may bring great
> suffering to others....If it will help to actually use the word
> evil, there's no reason not to." —ROBERT WRIGHT

T he ancients had a powerful and still relevant intuition about the mismatch between what I like to call our "inherited proclivities" and the
trials faced by linguistic creatures born into complex societies. Thus
you will see that I attribute great value to the biblical story of The Fall. That
portion of my culture's foundational creation story provides a convenient and
widely understood metaphor that I can call upon while recounting a key aspect
of our modern creation story—the aspect that spotlights the challenges of the
human condition.

This is my claim: *The evolutionary sciences, especially evolutionary brain
science and evolutionary psychology, provide a more realistic and universally
relevant picture of the human condition than was possible when the Hebrew
people acquired their creation story.* There is no better way to constructively
deal with our quirks and pathologies than to begin with the best understanding possible of what exactly we are up against.

Especially when we learn about "human universals"—innate tendencies we all share, thanks to our evolutionary heritage—we begin to understand why we are the way we are and why we struggle with the things we do. With blinders removed, we can move beyond denial. We lighten up. We can forgive ourselves and others. From a platform of self-acceptance we can begin the constructive task of improving our ways of being, interacting, and living. An evolutionary understanding thus provides the perspective needed to develop practices for actually achieving peace and lasting victory over that which may have caused us to stumble in the past. Halleluiah!

Lessons from Evolutionary Brain Science

"We are rag dolls made out of many ages and skins, changelings who have slept in wood nests, or hissed in the uncouth guise of waddling amphibians. We have played such roles for infinitely longer ages than we have been human." —LOREN EISELEY

Our modern understanding of the human brain offers scant evidence to suggest that a designer God engineered a perfect mental toolkit for the human endeavor. Indeed, the scientific evidence is overwhelming that the human brain is an emergent phenomenon in which physical structures and neurological connections developed in an additive and exploratory way over millions of years. This is how God, the Creator, made "Adam and Eve" and the rest of us. It is now beyond dispute in the scientific community that our deepest and most basic brain structures were shaped within the skulls of our reptilian ancestors who ate, survived, and reproduced in an era that long preceded the dinosaurs.

A half century ago, Paul MacLean introduced the idea that the human brain consists of three main parts, and that the parts of our "Triune Brain" correlate to the time sequence of their evolutionary emergence: reptilian, paleo-mammalian, and neo-mammalian. Since then, scientists have recognized that there is a fourth and far more recently evolved mammalian structure, which is profoundly manifest in the human brain: the prefrontal cortex, or frontal lobes. Let us begin with a brief overview of each of the components of our *Quadrune Brain.*

REPTILIAN BRAIN: Lizard Legacy. The cerebellum and brainstem together handle our involuntary breathing, basic bodily movements, and acquired "muscle memory"—most notably, our learned ability to throw a spear, ride a bike, drive a car, kick a ball, type on a keyboard, or play a musical instrument without having to think about it. Our ancient reptilian brain is also the seat of instinctual drives that are least subject to conscious control. I call these primordial drives "the 3 S's" of our *inherited proclivities.* They are Safety, Sustenance, and Sex. Or, we might call them "the 5 F's". In times of acute danger, our reptilian brain responds with hormones that instantly generate fear or hostility, culminating in action or rigid inaction that entail the first 3 of the 5 F's: *Fight, Flight,* or *Freeze.* As with its expression in reptiles, the fight/flight/freeze response in humans is involuntary—beyond the range of conscious choice, especially in the face of sudden and unanticipated danger. Indeed, so powerful is reptilian control of our behavior when we are traumatized that the other parts of our brain have a difficult time overriding these automatic and instantaneous reactions. Our Lizard Legacy is also the seat of deep territorial defensiveness and aggression when our boundaries are threatened, which is why arguments and wars tend to escalate.

The remaining two F's that define our Lizard Legacy are *Food* and…*Copulate.* If individuals are challenged by food, sex, drug, and other physical addictions, it is the reptilian brain that is the deepest (though not only) source.

Importantly, while this part of our brain is called reptilian, its structures and fundamental drives hail from even earlier times, when our ancestors still lived in the sea and sported fins instead of limbs. Indeed, researchers at the University of Michigan reported in 2006 that even nematode roundworms can become addicted to nicotine and will suffer effects of withdrawal.

OLD MAMMALIAN BRAIN: Furry Li'l Mammal. This is the limbic system, consisting of the amygdala, hippocampus, thalamus, hypothalamus, and insula. Reptiles do not have a limbic system, but all mammals do. It is the seat of deep emotions, and its health and well-being seems to require periodic entry into the dream state. Mammals, in fact, must dream or they will die. This is why sleep apnea and menopausal night sweats can be so destructive of emotional health. In contrast, no reptile or fish seems to dream. Now here is an interesting piece of biological trivia: Monotremes (the duck-billed platypus and spiny

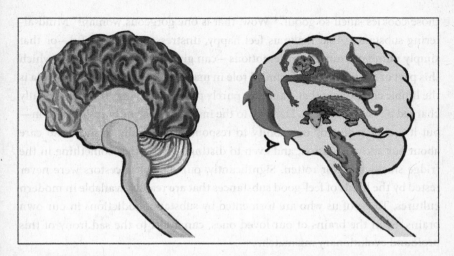

Our Quadrune Brain. Our deepest, oldest brain components (and behavioral drives) reflect our ancient reptilian heritage—what might be called our Lizard Legacy. Next, and wrapping around the reptilian core, is our paleomammalian brain, the limbic system, which is the seat of emotions—our Furry Li'l Mammal. Superimposed on those two structures is our newer, neomammalian brain: our neocortex, which is our incessantly talkative Monkey Mind. Last to evolve is the section of neocortex at our forehead. With a left side and a right side, these are our frontal lobes—the seat of our higher purpose, our Higher Porpoise.

echidna) are considered mammals because they produce milk (a kind of nutritious sweat in their case, as they do not have nipples). Yet the monotremes lay eggs just like reptiles do. They are also the only milk-producing creatures that seem to have no need for dreams. Truly, they are a transitional form in the evolutionary story of life.

It seems that the limbic system evolved as a way to provide more nuanced behavior and experiential learning than reptiles are capable of. As well, because only a few reptiles (e.g., crocodiles) nurture their young posthatching, but all mammals do via the provision of milk, emotions for familial bonding have always been crucial for mammals. For those mammals that evolved the further adaptation of social groups, familial bondedness is supplemented by nonkin bondedness, which prompted new emergent drives: status-seeking and reciprocal cooperation. These are all parts of what I like to call our Furry Li'l Mammal heritage.

Our Furry Li'l Mammal is also what ramps up the reptilian drives into emotionally powerful, and thus consciously experienced, imperatives. "Oh those cookies smell so good!" "Wow, that is one gorgeous woman!" Mind-altering substances that make us feel happy, unstressed, or powerful—or that simply numb out unwelcome emotions—can give rise to addictions for which this part of the brain plays a central role in mammals. Specifically, the insula is the limbic component that translates purely physical cravings into emotionally charged physical cravings. Damage to the insula can abruptly end addiction—but it can also destroy our ability to respond emotionally to music, to care about our social situation, and even to distinguish whether something in the fridge smells fresh or rotten. Significantly, our distant ancestors were never tested by the kinds of feel-good substances that are readily available in modern cultures. Those of us who are tormented by substance addictions in our own brains, or in the brains of our loved ones, can attest to the sad irony of this profound evolutionary mismatch.

NEW MAMMALIAN BRAIN: Monkey Mind. This part of our brain, the neocortex, could be called our chatterbox, calculator, or computer brain because it is incessantly talking to itself (fretting about the past and worrying about the future), performing rudimentary cost-benefit analyses, and computing the balance of favors and debts in each of Furry Li'l Mammal's social relationships. *Monkey Mind* is the nickname I shall choose here, because that name is already in use by Buddhists, who use it to refer to one key aspect of the mental phenomena that I listed above: it is the condition of mentally being anywhere *except* in the present moment. The story goes that our Monkey Mind is akin to a monkey who leaps from tree to tree, taking a bite of just one fruit, before moving on to the next tree, and the next. Similarly, our new mammalian brain has a tendency to endlessly toss up bits of thought—leaping haphazardly, wastefully, from one thought to the next. Our Monkey Mind will continue to ramble until (a) it is disciplined internally by conscious focus (including meditation); (b) it is called into service by an external situation that commands our immediate attention; or (c) it is engaged in a creative or physically demanding task that generates a mental state of flow.

On the plus side, the neocortex is a fabulous tool for using symbolic language, and for listening to and generating a continuous stream of information with seemingly no effort—that is, engaging in conversation. It is also absolutely terrific for thinking about things rationally, engaging in logical analyses.

This new mammalian brain (highly developed in primates and dolphins; meagerly expressed in rabbits and tree sloths) also offers extra room for storing memories and associations and for accessing genetically inherited skill sets. It also provides two powerful functional advantages.

First, there is the scenario-building function. "If I do X, then Y might happen." Scenario building and imaginative testing make it possible for actions to be "selected" within the brain. Thus actions can be tested safely within the mind before one actually makes a choice that is tested (and selected) by the world at large. As philosopher Karl Popper noted, "Ideas die in our stead." Scenario building has obvious advantages, but the downside is that the process can be emotionally draining—that is, when we remain in a state of indecisiveness. And even after we make a decision and take action, so long as we keep wondering whether we may have made the wrong choice, scenario building is no comfort. Thus, a by-product of evolutionary advance is that the new mammalian brain generates an internal source of stress capable of magnifying the external stresses that the world sends our way.

Second, one has to choose between competing drives. As Paul R. Lawrence and Nitin Nohria hypothesize in their 2002 book, *Driven,* the imperative of having to choose between multiple, independent drives is what gives birth to free will. An understanding of evolutionary brain science thus demonstrates, far more compellingly than can any philosophical treatise of the past, that free will is real—very real. Big-brained mammals have the capacity to consider alternatives and thus choose among the often-competing reptilian and paleo-mammalian drives. For example, imagine that you are an elk or an antelope. Do you choose to go down to that succulent patch of grass near the thicket, or do you stay in the open where you can easily spot approaching predators but where the food is less appealing? (Competing reptilian drives: sustenance versus safety) Do you try to sneak a copulation with a female in the herd, even though that would put you at risk of injury by the big-antlered male who has claimed all the females for himself? (Competing reptilian drives: sex versus safety)

Social mammals have additional drives that complicate their choices. Imagine you are a monkey and that you are a member of a large monkey troop. Exploring on your own, you come upon a luscious patch of ripe fruit. Do you call out to your comrades to join in the feast, or do you decide to eat your fill in

silence—and risk being caught as a defector? (Competing reptilian and mammalian drives: sustenance versus status)

Humans living in a world co-created by symbolic language face additional
dilemmas unique to our species: Do I keep my sexual infidelity a secret and
thus risk being found out (and live in fear of being found out), or do I confess
to my spouse and risk being shamed, shunned, or divorced? Indeed, all
the competing drives and the complexity of our lives in a civilized world
can propel us into a state of incessant worry and despair: Monkey Mind
and Furry Li'l Mammal team up in an endless loop of negativity that can
escalate to disastrous ends: depression and even suicide. To quell our hyper-
active Monkey Mind, practices have emerged—some healthy, others not—to
give us respite from this compulsion to fret about the future and revisit the
past. Practices include meditation, chant, speaking or thinking in tongues,
ecstatic dance, drumming, playing a musical instrument, attending a sym-
phony, sports and watching sports, immersion in hot water, massage, tantra,
computer games, television, mind-numbing drugs, shopping, and much, much
more.

PREFRONTAL CORTEX: Higher Porpoise. Evolutionary brain science has
revealed that the prefrontal cortex, also known as frontal lobes, (the part of our
neocortex right above and behind the eyes and which is the most recently
evolved), are crucial for humans to live successfully in any culture today. As
Elkhonon Goldberg states in his book, *The Executive Brain: Frontal Lobes and
the Civilized Mind,*

> The frontal lobes perform the most advanced and complex functions in all
> of the brain, the so-called executive functions. They are linked to intention-
> ality, purposefulness, and complex decision making. They reach significant
> development only in humans; arguably, they make us human.

It is here in our prefrontal cortex where a new drive can emerge that is strong
enough to help us choose among competing mammalian and reptilian drives—
and to do so with less stress and far more conviction. It is here that we can
create and nurture a higher purpose, which I delight in calling, "Higher Porpoise." A higher purpose will, of course, act in concert with some of the deeper

drives. A teenage boy who devotes himself to grueling hours of athletic training has a Higher Porpoise that directs him to rise before dawn in order to practice for a swim meet, refuse dessert in order to lose weight for a wrestling match, or say no to drugs that might cause his expulsion from the football team. At the same time, excelling at the higher purpose endeavor will gain him immense status—and more opportunities to date.

A higher purpose need not always be a good thing individually or collectively. A teenage girl might choose a higher purpose of becoming a fashion model, risking depleted bone mass and anorexia. Any young person might choose a career for the primary purpose of becoming wealthy, at the expense of developing more deeply satisfying goals. In the extreme, one's higher purpose might be to martyr oneself (metaphorically or actually) for a particular cause. All in all, however, it seems that those who are least troubled during their teenage years have dedicated themselves to some sort of higher purpose.

Higher purpose will surely change during one's life. Some cultures even have norms and rites of passage that encourage such transitions. A variety of developmental models (e.g., Integral Consciousness, Spiral Dynamics) comport well with an evolutionary understanding of how individuals (and cultures, too) develop, and how our most powerful drives may shift with maturation and with life's opportunities and challenges. Rita Carter in her book *Mapping the Mind* explains,

> The frontal lobes are where ideas are created; plans constructed; thoughts joined with their associations to form new memories; and fleeting perceptions held in mind until they are dispatched to long-term memory or to oblivion. This brain region is the home of consciousness—the high-lit land where the products of the brain's subterranean assembly lines emerge for scrutiny. Self-awareness arises here, and emotions are transformed in this place from physical survival systems to subjective feelings. If we were to draw a 'you are here' pin on our map of the mind, it is to the frontal lobes that the arrow would point. In this our new view of the brain echoes an ancient knowledge—for it is here, too, that mystics have traditionally placed the Third Eye—the gateway to the highest point of awareness.

Our higher purpose is intimately connected with our sense of the future. Like all other animals, human beings are motivated by instincts. But we are also motivated by desire, yearning, longing, hope, spiritual aspiration—what renowned evolutionary theologian John Haught calls "the lure of the future." This dimension, our vision of what is possible, or what seems likely, is every bit as important as our past and in some cases may even be more influential as a motivating drive in human life.

Lessons from Evolutionary Psychology

"Our true ancestry is the emergent creativity of the Universe. Our forebears were those who learned how to coalesce hydrogen and helium into stars, to form planets, to sustain life first from mineral nutrients in the sea and later to capture delicious photons, to exploit oxygen for energy rather than be exterminated by it, to diversify via sexual reproduction, to form social groups for greater security and protection of offspring. We are the beneficiaries (and, admittedly, also the victims) of this narrative of emergence. Our 'companions' are all of these progenitors. Indeed, they are more than companions; they are family. From them we have inherited our corporeal shapes and movements, our body chemistry, and even some of our behavioral agendas." —JOHN BREWER

It is a truism today, both in science and in popular culture, that who we are as individuals is profoundly influenced by how each of us has been shaped by the environments we were born into and that influenced us as we developed and matured. Evolutionary psychology is a body of research that suggests that the same holds for us as a species. Who we are collectively has been profoundly shaped by the environments of our ancestors—the drives that still play out in our reptilian, paleo-mammalian, neo-mammalian, and distinctly hominid brains. These are the drives that are now a genetic, inherited part of our human nature.

While far from explaining everything, evolutionary psychology helps us gain insight into the human condition: why people behave similarly all over the world, no matter what their doctrines of belief; why we all struggle with one or more items on the same list of unwanted tendencies or addictions, and thus why we think, feel, and act in ways that can harm ourselves and others. Evolutionary psychology also goes a long way toward explaining many of the differences we commonly observe between men and women. Here are some concepts culled from a deep-time understanding of our instincts:

✦ MISMATCH THEORY. Cultural evolution has occurred at such a fast pace and has so impacted natural environments that the brain structures and behavioral proclivities we humans have inherited are adapted to conditions that are as out of sync with those of today as riding horseback on a freeway or throwing a spear at a freight train. The mental equipment we are born with is attuned for surviving, adapting, and reproducing in a bygone era. In the parlance of evolutionary psychology, this is known as *mismatch theory*. Consider the ridiculous excesses of road rage and the media's penchant to focus on news of fallen heroes and heinous crimes, while ignoring collectively threatening trends that lack a compelling human face. In the days when our instinctual priorities were shaped, there was no such thing as global warming and no possibility for forward-looking remediation. But there were many instances in which our survival depended on curiosity about the human drama, on our drive to understand the causes and consequences of human fallibility and maliciousness.

✦ NATURE VERSUS NURTURE. Which is more important for shaping the personalities and behavioral choices of individual humans? Is it our genes or our environment? Is it nature or nurture? The answer is a resounding, "Both!" Evolutionary psychology and evolutionary brain science are helping social scientists, psychologists, parents, and teachers move away from the 20th century assumption that we each start out as a blank slate and that culture is by far the major determinant of behavior. The new, more nuanced view explores how and when nurturing practices can be most effective, while acknowledging that there are genetically based behaviors and inclinations that no amount of nurturance is going to eliminate. Psychopharmacology adds a whole new layer to this work, given that our moods and behavior affect our internal chemistry— and that our internal chemistry affects our moods and behaviors.

✦ MALE/FEMALE DIFFERENCES. Evolutionary psychology begins with the biological fact that men and women differ profoundly in how they go about the business of procreation. Men have the physical ability to father hundreds, even thousands, of children in a lifetime, and sometimes with less than an hour of personal investment. But they have no absolute assurance that any particular child is genetically their own. In contrast, a woman has only a limited supply of eggs. Moreover, for every successful fertilization, a nine-month pregnancy, hazardous birth, and years of physically demanding nurturance will follow just to raise one child. On the plus side, a woman does definitively know that every child she births is her genetic offspring.

This basic male/female reproductive dichotomy has fostered correlative differences in physical forms, hormonal outputs, sexual urges, emotional tendencies, and cultural practices between the genders. Characteristics do, of course, play out over a range (a bell curve) within both genders and thus do not mean that every individual male will exhibit more "male" physical and emotional traits and sexual urges than every individual female, and vice versa. But on average, there are real biological and behavioral differences between the genders that we express as a species because of our evolutionary heritage and that trump cultural and familial variability in the specifics of how we are raised. These gender differences are thus *human universals*. These very real differences have been captured in folk wisdom and expressed through cultural traditions and expectations that, for example, put a premium on female (but not necessarily male) coyness and choosiness about sex partners—and thus on the importance of virgin brides and faithful wives.

Given the devastating legacy of "social darwinism" and the equally devastating legacies of constraints imposed upon the female gender by cultural and religious traditions, this aspect of evolutionary psychology could emerge in a beneficial way only in our postmodern world—a world in which cultural rules have already been substantially reshaped by feminists and technological innovations of birth control. Our culture has been exploring the implications of decoupling sexual activity from procreation, because this is the first time in human history that women and men have access to reliable contraceptives. The explorations venture far beyond sexuality and into ways in which procreative choice opens possibilities for women to enter traditionally male domains of work and leadership.

✦ SELF-DECEPTION. Because trustworthiness is so highly valued among social animals, advanced primates (like us) have evolved superb "gut instincts" for detecting deception. An ill-timed blink of an eye, tension in the lips, or perhaps a slight stiffening of the neck will be noticed by our peers, though perhaps not consciously. Suspicion will arise, and they will become more wary and watchful. Add to that the vocal nuances of tone and pacing that we verbal primates are on the alert for, and it is easy to see why evolution has balanced the equation by equipping us with marvelous powers of self-deception. After all, the best way to deceive another is to first deceive oneself. Then an untruth or partial truth can be communicated with heartfelt conviction. Evolutionary psychology thus teaches us that *we cannot necessarily trust even our own thoughts and memories.* We cannot expect to always and accurately know what we are feeling, what is driving us, and how our actions are likely to play out. Self-deception exacerbates the troubles that stem from the mismatch of our ancient drives with today's cultural expectations.

✦ APPRECIATING THE SHADOW. Virtually every aspect of our specieswide psychological inheritance that seems troublesome today is part of a package that evolved to serve individual and collective well-being in ancestral environments. Forcefully trying to eliminate the shadow side of our reptilian, mammalian, and hominid instincts—what could be called our "unchosen nature"—is neither realistic nor desirable. Rather, evolutionary psychologists invite us to channel those troublesome energies in safe and productive ways, while consciously strengthening drives that promote our individual and collective well-being. More to the point, we can call upon our mammalian drive for bonding (our Furry Li'l Mammal) to keep our reptilian drive for sexual adventure from unraveling our marriages. We can call upon our mammalian obsession with status to trump the Lizard Legacy's quest for sex and sustenance. Equally, because our reptilian brain holds the portfolio for movement, our Higher Porpoise can direct our Monkey Mind to figure out a way, and then pull the strings of voluntary movement, so as to trick our Lizard Legacy into sensing that the danger has passed. For example, when a fear response has been triggered, but our neocortex assesses that we'd be better off if we calmed down, Monkey Mind can direct the muscular apparatus in our chest to slow down our otherwise involuntary breathing. Monkey Mind can also instruct the body to go for a stroll, put on a smile, take a hot bath, stretch out in the sun—all actions that confound our body's detection of stressful circumstances. Finally, we can cultivate a higher purpose powerful enough to step in and insist, "No!" The Higher Porpoise will whisper that it is time to do some yoga, put on a motivational CD, read some scripture from the Good Book or the Big Book, pick up the phone and call our 12-step sponsor, integrity partner, or friend, or otherwise engage in some activity that will dissuade or distract us from acting out in emotionally, physically, or socially harmful ways.

Resurrecting "The Fall"

"Habituation to any goal—sex or power, say—is literally an addictive process, a growing dependence on the biological chemicals that make these things gratifying. The more power you have, the more you need. And any slippage will make you feel bad, even if it leaves you at a level that once brought ecstasy."
—ROBERT WRIGHT

PROPHETIC INQUIRY: Where did sin come from, anyway? How does the evolutionary epic help me deal with the fact that I didn't play a role in "The Fall," yet I have to live with its consequences?

The theological doctrine of the Fall is a biblical concept embraced by Jews, Christians, and Muslims alike. It was not until the 4th century C.E, however, that the concept of Original Sin was introduced into western Christianity, by St. Augustine. A holy view of evolution REALizes both.

While I was writing the first draft of this chapter, Internet blogs were abuzz with a sad and stunning example of what can happen when we fail to acknowledge that our most troublesome inherited proclivities are natural—indeed, instinctual. Appreciating that unwanted inclinations are part of our heritage doesn't mean we must do their bidding. But it does help us accept that the yearnings themselves need not be judged as shameful—and thus we don't have to be in denial about their existence. A man choosing to live in a committed, monogamous relationship with a woman, for example, can accept that sexually promiscuous thoughts (heterosexual and/or homosexual) are natural and to be expected from time to time. This is true even for those who are completely happy with their partner. Only from the stance of acceptance can one effectively notice and then seek peer support and accountability to remain in integrity when unwanted urges do arise.

In November 2006, Ted Haggard, then-president of America's National Association of Evangelicals and head of a megachurch in Colorado Springs, was outed by a homosexual man whom he now admits having paid for sexual services. And before Pastor Ted, there were the falls of other prominent evangelical leaders, Jimmy Swaggart and Jim Bakker, for scandals that also were kindled by reptilian drives mismatched with the demands of their religious culture. These scandals played out in the larger arena in which public lives subject the famous to extraordinary scrutiny. In the case of Pastor Ted, his sin was not just one of deceit but hypocrisy, because he actively preached that homosexual acts under any circumstances are immoral.

For me, personally, there is no more poignant example from which to ponder the possible benefits that an evolutionary theology might afford its members, including its leaders. First, through an understanding of the Quadrune Brain, we can acknowledge that *we all share inherited proclivities that, left unmanaged, can harm not only ourselves but also the larger and smaller spheres, or holons, of our existence.* Whatever "demons" we may experience are indeed not so shameful, nor so exclusively ours, that they must be guarded as deep, dark secrets. Precious few of us are entirely free of unwelcome sexual thoughts and impulses. Yet each of us can move beyond denial and the futility of trying to secretly make them go away, once and for all. At the Sunday service in Colorado Springs immediately following the allegation, the congregation was read a letter of confession written to them by their pastor. Ted Haggard wrote, in part,

> The fact is, I am guilty of sexual immorality, and I take responsibility for the entire problem. I am a deceiver and a liar. There is a part of my life that is so repulsive and dark that I've been warring against it all of my adult life. For extended periods of time, I would enjoy victory and rejoice in freedom. Then, from time to time, the dirt that I thought was gone would resurface, and I would find myself thinking thoughts and experiencing desires that were contrary to everything I believe and teach. Through the years, I've sought assistance in a variety of ways, with none of them proving to be effective in me. Then, because of pride, I began deceiving those I love the

most because I didn't want to hurt or disappoint them. The public person I was wasn't a lie; it was just incomplete. When I stopped communicating about my problems, the darkness increased and finally dominated me. As a result, I did things that were contrary to everything I believe.

"There is a part of my life that is so repulsive and dark that I've been warring against it all of my adult life." Had Ted Haggard been able to learn from the evolutionary sciences about the naturalness of homosexual/bisexual inclinations and acts—not just in humans but in many species, including our nearest primate relatives—perhaps he could have more effectively lived with his Lizard Legacy. Perhaps he could have lived in integrity as a monogamously married heterosexual man, or perhaps not. But, surely, the gift of perspective afforded by today's evolutionary brain science and evolutionary psychology would have offered him better odds than would unquestioned adherence to simplistic cultural admonitions of long ago now frozen in scripture.

Gordon MacDonald, a fellow evangelical leader who also experienced a fall when a sexual impropriety became public, wrote this commentary for *Christianity Today* a few days into the Ted Haggard saga:

I am no stranger to failure and public humiliation. From those terrible moments of twenty years ago in my own life I have come to believe that there is a deeper person in many of us who is not unlike an assassin. This deeper person (like a contentious board member) can be the source of attitudes and behaviors we normally stand against in our conscious being. But it seeks to destroy us and masses energies that—unrestrained—tempt us to do the very things we "believe against." If you have been burned as deeply as I (and my loved ones) have, you never live a day without remembering that there is something within that, left unguarded, will go on the rampage.

Tellingly, MacDonald speaks of a "deeper person" within each of us, a kind of "assassin" that "left unguarded will go on a rampage." Evolutionary brain science confirms how right he is! Any of us whose lives have been damaged by slipping in our commitments and thus following our deep impulses knows what Gordon MacDonald is talking about. Evolutionary brain science helps us comprehend why: the deepest and most difficult to control urges are those whose territory resides within the fortress of our ancient reptilian brain. When those drives take over, "we" are no longer in control. Something else is. And it can feel like an assassin; it is destroying our lives against our will. This sense that something not-us nevertheless tempts and even controls us can be seen throughout history, though it is given different names. It has been called Satan or the Devil. Freud called it the Id (German: "It").

Our reptilian brain truly has its own agenda, a set of three ultimate goals: sustenance, survival, and sex. Evolution found ways to make sexual fulfillment extraordinarily pleasurable in order to ensure procreation. But the penis, in particular, doesn't remember the "in order to" part of the deal. Moreover, we are not alone. There is ample evidence in the natural world that nonprocreative sexual acts (homosexual or otherwise) are common among social animals. Respected theories abound as to the selective value of nonprocreative sex in such circumstances, most notably, among the peaceful bonobo chimps, our closest genetic cousins. But it is only for the human, the "moral animal" immersed in a sea of civilization with ethical codes of conduct and spoken commitments—it is only for us that the choices around sexual expression can become problematic.

Sexual drives that would lead to marital infidelity may, of course, be quiescent in some individuals, but there is a well-established link between high levels of testosterone and how insistent and relentless the sexual drives become. Moreover, there is a well-established link in mammals between gains in status and elevated levels of testosterone. Either can cause an upswing in the other. So even if we begin our social climbing with our internal "assassin" adequately restrained, once our status exceeds a threshold, without support and accountability our elevated hormones may be our undoing. As our secret indiscretions become public, our paleomammalian drive to maintain high status kicks in big time, such that we are tempted to violate other moral principles as well—by lying, blaming others, covering up, perhaps even blackmailing possible informants and threatening them with physical harm.

The Ted Haggard story is only one recent scandal involving a public figure of high status and significant influence whose Lizard Legacy managed to outfox Higher Porpoise. Before him, of course, there was President Clinton. And let us not forget the several thousand Roman Catholic priests in America who have recently been charged with sexual abuse, alleged to have occurred over the past several decades. Confession or judgment of guilt has been established for more than a few of them—and this, in turn, has driven many parishioners away from the Catholic Church, and even away from any faith at all. Have we learned anything from these personal and public tragedies? I would venture this:

> *So long as religious and political leaders continue to ignore our evolutionary heritage, and thus do not put in place structures of internal and external support that can withstand the high dosages of testosterone that high status and power necessarily confer, then there will be no hope for a less calamitous future.*

Understanding the unwanted drives within us as having served our ancestors for millions of years is far more empowering than imagining that we are the way we are because of inner demons, or because the world's first woman and man ate a forbidden apple a few thousand years ago. The path to freedom lies in appreciating one's instincts, while taking steps to channel these powerful energies in ways that will serve our higher purpose. Even so, "demonic possession" is a traditional night language way of speaking about someone who is compelled to act in harmful ways. "Demonic temptation," in this sense, is anything that would have us disregard the well-being of the larger holons of which we are part (our families, communities, world), or the smaller holons for which we are responsible (our bodies, minds, principles). It is my hope that—however evolutionary theologies manifest in the future—there will be room for traditional language (demonic possession), scientific language (reptilian brain), and metaphorical night language born in our own time (Lizard Legacy).

Your Brain's Creation Story

"Most of the open-ended processes that we associate with human uniqueness, from flexible brain development to symbolic thought and cultural diversity, reflect fast-paced evolutionary processes that take place within an architecture created by genetic evolution. We have not escaped evolution. We experience evolution at warp speed. The starship of evolution is not like the starship Enterprise, *however. Unless we understand how it works, it will take us to places we don't want to go."*
                                        —DAVID SLOAN WILSON

I have found that the perspective afforded by evolutionary brain science and evolutionary psychology can be salvific, offering hope and real possibility for transforming not only lives, but *my life and your life.* This is especially so for those of us struggling to understand why our best intentions have not always led to best behavior.

By way of evolutionary psychology, God, the Supreme Wholeness of Reality, has revealed to us through public revelation (science) how and why our Quadrune Brain evolved, why men and women can seem utterly alien to one another, why public figures so regularly "fall" to sex scandals, and why all of us sometimes fail to make the right choices and to honor our commitments. These lessons are especially lifegiving for teens and young people struggling to find their way to self-determination and fulfilled lives under the influence of powerful hormones—with precious little life experience of their own to give them guidance. Who among us can learn of our evolved human condition without feeling as if the scales have dropped from our eyes?

"There must be something wrong with me…" Relax. We're all that way. Now let's understand why. Let's appreciate how those same urges once served our ancestors. Let's take a square look at where they cause us trouble today. And then let's move ahead and see how we can support one another in growing in deep integrity and fulfilling our evolutionary purposes—God's will for our lives.

The Parable of the Pickle Jar

During her sophomore year in college, Connie's niece Halsey Barlow, who grew up with no understanding or appreciation of evolution, read Robert Wright's classic book on evolutionary psychology, *The Moral Animal*. Here are extracts from the email that Halsey sent us immediately afterward:

I now put *everything* into an evolutionary context. It's almost become a reaction that when I encounter a situation I think, "OK, why am I feeling this way? How and why would my 780th great grandmother react to this?" It doesn't even have to be anything significant. For instance, if I see one of my roommates snacking on my jar of pickles, I get this feeling in my stomach like, "Hey, back off. Those are mine, mine, mine!" I want to be laid-back and not care, but my instincts are so darn uptight. But knowing that I am a selfish, status-seeking hornball only motivates me to do things that say, "F you!" to my evolutionary roots—even though I do love them.

Halsey concludes her email with wisdom astonishing for her age (then, only 18 years old). Is it possible that knowledge of evolutionary psychology can nurture a spiritual capacity that normally requires years of disciplined meditation or contemplative prayer? That is, is it possible that this scientific perspective can help one *witness* one's feelings rather than simply act on them? She writes,

I absolutely love how imperfect and clumsy we are as humans. I mean, almost any situation can be laughed off if we only step back and take a look at *why* we feel the way we do. I love it! I wouldn't want it any other way.

The new evolutionary worldview offers blessed membership in the club of the less-than-perfect. Finally, there is good reason to hope that, yes, here is something undeniably real, something that can help me and my loved ones with our difficulties. This sense of new possibility, of profound relief, and of genuine freedom is not unlike the Christian experience of being saved. Seeing my most troublesome urges through the lens of evolutionary psychology has indeed felt that way for me. More, I shall boldly claim that *the salvation of the liberal Christian churches, and the grounding in reality for conservative churches, depends on their adopting for everyday use and translating into appropriate religious language these and other findings of evolutionary brain science and evolutionary psychology.*

When I speak about the Quadrune Brain, especially to young people, I like using the names Lizard Legacy, Furry Li'l Mammal, Monkey Mind, and Higher Porpoise. The names are personal and comforting, and they evoke light-heartedness even while discussing the most befuddling and serious issues of our lives. These four brain components and their associated drives are what we all have to work with. They are our unchosen nature. They are human universals. None of us can expect our mental equipment to function flawlessly, given their origins in life conditions that are now ancient history, and especially because natural selection aims for the happiness of individuals only to the extent that happiness affords the genes a better chance of propagating into the next generation—and the more widely the better. Our inner urges are confounded whenever life presents us with situations in which the basic drives necessarily conflict with one another. And yet, in an emergent Universe, God would not have done it any other way.

Not all of us may be inclined to give our wholehearted assent to this evolved human condition. We might not honestly feel we can thank God for *all* of the heirloom instincts and urges that the ancestors have passed on to us and to our loved ones. But surely we can be grateful for the public revelations of evolutionary brain science and evolutionary psychology that so powerfully REALize the story of the biblical Fall and provide us with the self-acceptance, insights, and tools that are essential for leading more serene and fulfilled lives.

The public revelations of science teach another comforting truth: emotions evolved for many good reasons. Emotions are the means by which the parts of our brain that are unconscious communicate with the parts of our brain that "we" can readily access. Emotions are the way that our paleo-mammalian brain, with its powerful drives for bonding and status, communicates its wishes and its fears to our conscious awareness. Emotions are also the way that aspects of the reptilian drives (for safety, sustenance, and sex) are translated within the old mammalian limbic system into emotionally charged signals that our conscious brains then take as directives for acting in the world. *Emotions are also*

And we wonder why we occasionally struggle
with issues related to food, safety, and sex!

the means by which a consciously chosen higher purpose is invested with the
energy it requires to become a beacon in our lives. Indeed, without the deep
motivations of these two ancient, unconscious realms—our Lizard Legacy and
our Furry Li'l Mammal—we would have no drive to do anything at all.

Thank God for emotions! And thank God for our increasing ability to comprehend emotions and to have them serve our lives and the lives of those who
are touched by our actions.

We are, of course, responsible for how we act upon our emotions and for
how we choose between competing drives. We are responsible for what we choose
to say (or blurt out) as well as what we do. We are also responsible for how well
we clean up the messes we make (and have made) just by following our instincts
and doing what comes naturally. We are responsible for whether we put in
place structures of support and accountability that will strengthen the positive
drives that we pray will prevail. Nevertheless, progress begins with full acceptance of where we really are. So whenever we are challenged by our inherited
proclivities, or when we are disappointed with the choices we make and in the
mistakes we seem to keep repeating, we can take some comfort in this: Each
and every aspect of our behavioral repertoire in some way served the survival
and reproductive interests of our hominid, mammalian, or reptilian ancestors.

Reclaiming "Original Sin"

"Natural selection never promised us a rose garden. It doesn't
'want' us to be happy. It 'wants' us to be genetically prolific....
Understanding what is and isn't pathological from natural
selection's point of view can help us confront things that are
pathological from our point of view."   —ROBERT WRIGHT

Okay, so now what do we do? How is the story of the evolutionary emergence
of our human brains liberating? How does it call forth better strategies for living one's life? How, too, might core theological concepts inspired by the Bible
support and enrich the evolutionary understanding by giving it a mythic
frame—a mythic frame that is already embedded in our culture? And might
an evolutionary gloss on aspects of the Genesis story universalize that story,
making it meaningful and instructive not just for Peoples of the Book, but also
for peoples of other traditions and of no religious tradition?

Let us approach this inquiry by introducing a core insight: *The Fall and Original Sin are both trivialized when we interpret the events in the Genesis story as historically factual.* I envision that traditional, flat-earth religious believers will eagerly shed the last vestiges of literal interpretations of ancient scripture once they understand how our evolutionary story corroborates the central teaching about the Fall; once they *get* that Original Sin (our unwanted, unchosen, "sinful" nature) is, itself, literally true and thus is fully supported by scientific facts.

Here is another potent example of how the ancients sensed and recorded in mythic language a phenomenon now known to be physically true. The Bible says that the fruit that God forbade humans to eat was fruit from the Tree of Knowledge of Good and Evil. This is a superb night language description of the day language process through which our ancestors evolved the frontal lobes—the seat of our human ability to sort among and override competing impulses and drives. The frontal lobes indeed made possible *conscious* awareness of Good and Evil and conscious awareness that we are, in fact, making choices all the time.

Thus, from an evolutionary spiritual perspective, liberals as well as conservatives can find enormous value in the concept of Original Sin and in the story of The Fall, for this is the reality: *Even the most innocent babe carries the seeds of "original sin" in its brainstem and limbic system.* Unchecked, any of us can and will fall. Thankfully, because it can be a powerful corrective, almost all of us will feel guilty and ashamed when we fall, because we know that such actions are wrong; we are aware that there is good and there is evil.

Who among us chooses to crave sugar, fat, caffeine, and other foods and substances that are unhealthy if taken in excess? Who among us wants to become addicted to alcohol or nicotine or painkillers—and to suffer the judgment of others and the agony of withdrawal whenever we try to quit? Who among us asks to explode in rage, or be possessed by anger that festers and wounds? I don't know anyone who really wants to lie when confronted by an angry or suspicious and thoroughly self-righteous spouse, yet we almost all do (at least in our lies of omission). Then too, who among us wishes to suffer resentment and an overpowering urge to retaliate? Who asked to be burdened by the sins of envy or lust or jealousy that consume our goodness and better judgment? Who among us decided that what we really want is to feel the allure of power, status, greed, and selfishness that not only hurts others but also diminishes ourselves?

From a sacred evolutionary perspective, the mythic story of The Fall of Adam and Eve carries a profound meaning. Something in our evolutionary history caused a mismatch. Our inherited proclivities are not ideally suited for the demands of living an honorable life in a world of words, complex social relations, and technologies (such as the Internet) that provide temptation in ways no animal, human or otherwise, ever had to deal with before. In this way The Fall can be seen as universally, eternally true. That is, to be human is to be burdened by our *unchosen nature*. This is as true for Hindus and Buddhists and Taoists as it is for Jews and Christians and Muslims.

Our western religious story of humanity's fall into sin might also remind us that as cultural conditions change, the instinctual urges that once served us and our communities may begin to play out in ways that are no longer benign. When humans learned to create artifacts and began to accumulate possessions, the sins of envy and greed were born. When the complexities of culture necessitated that children be given many years of nurturance and education, which gave rise to the legal institution of marriage, the sins of adultery and lust were born. When agricultural and economic systems made it possible for wealth to be accumulated and inherited, the sins of laziness and pride were born. These sins are all cataloged in our sacred scriptures. As human awareness and technology continue to evolve, new sins and new salvations begin to appear.

There is profound relief in knowing that the inclinations we most dislike in ourselves and others are often not of our or their own doing. In a way, our flaws are not, at base, our fault. We didn't choose them; nor did others. We all, to some extent, inherited them. Our inherited proclivities were shaped by the particulars of our human, mammalian, and vertebrate evolutionary journeys, nuanced by the developmental journey each of us navigates from womb to tomb. This gift of understanding is the foundation for any lasting transformation. It encourages us to move beyond denial or condemnation and simply accept that there are powerful drives within all of us that we did not choose. Once relaxed and accepting, we can begin to forgive self and others for past transgressions. This forgiveness, in turn, clears the board and gives us the courage to look full-square at our current situation and from that vantage to embark on realistic paths for bettering our lives, enriching our relationships, and blessing our world.

CHAPTER 10

REALIZing "Personal Salvation"

"There is a wolf in me . . . fangs pointed for tearing gashes . . . a red tongue for raw meat . . . and the hot lapping of blood—I keep this wolf because the wilderness gave it to me and the wilderness will not let it go."
—CARL SANDBURG

PROPHETIC INQUIRY: What are the blessings of evolutionary spirituality that can save us from the dark forces within that threaten our physical well-being, our mental health, the quality of our relationships, and our spiritual integrity?

From a science-based, evolutionary perspective, there is no place for belief in a literal Satan—an otherworldly being with demonic intent—just as we no longer find helpful the notion that God is divorced from, less than, and residing somewhere outside the Universe. Nevertheless, personalizing or relationalizing the forces of evil—especially those within us—can be helpful, whether or not we choose to use the words Satan or the Devil.

When I need to muster extra resolve against my inherited proclivities, especially regarding the lure to lie for the sake of status, sexual attraction, or the temptation to indulge in feel-good substances, it occasionally helps for me to see those tendencies as something other—as not me. That sense of otherness makes it easier for me to "witness" my unchosen nature—my instincts—and thereby gain the calm objectivity that distance affords, rather than being ruled impulsively by it. These inherited proclivities are *not me*, and yet they are *within me*. I shall never be entirely free of them.

"I Don't Know That Guy!"

Folksinger Greg Brown wrote a song titled "I Don't Know That Guy." It is a funny and poignant reminder of how challenging our Lizard Legacy can be. Here are the first two verses:

> Me, I'm happy-go-lucky—
> always ready to grin.
> I ain't afraid of loving you—
> ain't fascinated with sin.
> So who's this fellow in my shoes—
> making you cry?
> I don't know that guy.
>
> Who took my suitcase?
> Who stole my guitar?
> And where's my sense of humor?
> What am I doin' in this bar?
> This man who's been drinking,
> and giving you the eye—
> I don't know that guy.

Evangelical opinion leader Gordon MacDonald, as already mentioned, referred to these tendencies as a kind of assassin within. In saying that we feel "tempted by Satan," we mean exactly that. For many Christians today, the words "tempted by Satan" may still be helpful in dealing with the most troubling aspects of our unchosen nature. For me, "Satan" is still a useful term, but with this proviso: From the standpoint of evolutionary faith, "Satan" points to nothing that can be believed or disbelieved. Rather, *"Satan" as the great Tempter is something that every human experiences by virtue of having an evolved brain.* Why? Because the human brain was not *designed* by an all-knowing, otherworldly engineer God. It was *evolved* by the living, immanent, omnipresent God, and the world of today is a far cry from the world of our prehuman ancestors.

For me to publicly use the word "Satan," however, would shut down the listening of those toward the liberal pole of Christianity—not to mention anyone outside the Christian or Islamic perspective. But what if we begin talking about our "reptilian brain" or, better yet, "Lizard Legacy"? What a light-hearted, playful way to get real about the most serious challenges that we, as individuals, face in right living; that is, abiding in integrity!

The Challenges of Our Lizard Legacy

"There is a hog in me . . . a snout and a belly . . . a machinery for eating and grunting . . . a machinery for sleeping satisfied in the sun—I got this too from the wilderness and the wilderness will not let it go."   —CARL SANDBURG

Inclinations toward excess with regard to food, sex, and feel-good substances are deeply rooted in our reptilian brain. The brainstem and cerebellum work great in their environment of origin. When our reptilian ancestors were hungry, they looked for food. They didn't waste energy looking for food when they weren't hungry. Instead, they basked in the sun (it helps digestion). As to sex, the reptilian brain's simple drives made a lot of sense several hundred million years ago. When our ancestors felt the urge, they searched for a mate until they found someone who could be induced or forced to copulate. There were no emotional complications. And, for our umpteenth great-grandfathers in reptilian times, there were no consequences to that compelling act; no eggs to guard, no babies to tend.

Even for our early mammalian ancestors, the natural world imposed constraints on the drives inherited from reptiles. Because it took effort to find food, especially food with a high content of sugar, salt, or fat, and because we exposed ourselves to becoming food every time we left the safety of our burrow, we did not overeat on a habitual basis. Our mammalian ancestors who lived in temperate climates, in which winter hibernation made sense, did experience a shift in instincts, drives, and metabolism. Seasonally we were willing to go the extra distance in our foraging habits in order to fatten up, and our metabolisms shifted so that we could actually do it. Periodically, too, our environment forced us to fast, during which time we shed the extra weight. As to drugs: other than occasionally stumbling onto naturally fermented fruit, that temptation was foreign to our mammalian great-great-grandparents.

Our prehuman ancestors were, by and large, not troubled by debilitating and injurious acts of excess in consuming substances or engaging in sex. But the absolutely essential elements of the brains they used and that they passed down to us *are* capable of leading us into temptation. Our Lizard Legacy and Furry Li'l Mammal worked just fine in their contexts of origin—but there is a mismatch in how they function in a culture of fantasy foods and mind-altering drugs easily acquired and in fabulous supply.

And then there is sex. Sex, no matter how pleasurable in the moment, did not evolve to serve the interests of any individual—reptile or mammal. Sex evolved to serve the interests of the genes, of genetic continuity and propagation. Evolutionary biologist Richard Dawkins wrote a popular book in 1976, titled *The Selfish Gene*, which popped our bubble of self-importance in that regard. Genes are long-lived; they are potentially immortal. One legitimate way of thinking about biological evolution is to imagine ourselves as "survival machines," compelled by instincts to do what it takes to successfully transmit the precious DNA we carry into the next generation. And then we die.

This evolutionary shift in perspective makes sense of some of the procreative cruelties of the natural world: Picture salmon hurtling through rapids and leaping over falls, eating nothing, all with the aim of living just long enough to ejaculate sperm and eggs far upstream. Well before the goal is reached, decomposing flesh is falling off fin and flank. Consider, too, our arthropod cousins: Tiny male spiders do the deed and then acquiesce to becoming a protein-rich snack for their larger mate, who in turn will offer her own body to cannibalistic hatchlings. Male praying mantises keep thrusting while their mate swivels just enough to calmly consume any section of her partner's head and thorax within reach. As for us mammals, males among herding animals (such as elk and bison and sea elephants) will spar mightily for access to harems of females. The victors will be battered, exhausted, and possibly too depleted of fat reserves to survive into the next mating season.

This time of sexual craziness among mammals is disruptive, often dangerous—but at least it doesn't last forever. During other months of the year, males are not distracted by a passing female. Think of an unneutered dog, who behaves very well around other dogs of both sexes until a female in heat is detected. That female looks no different than before, but her body is releasing chemical pheromones into the air, which, when they then enter the nasal passages of the male, elicit a highly predictable response. Chimpanzee females supplement pheromones with bright pink buttocks to ensure that, when ovulating, they attract the attention of males. But humans evolved what is called *concealed ovulation*. In theory, women as well as men are available for sex 24/7. On the practical side, this means that unlike elk and bison and dogs, an average human male (and those females with unusually high levels of testosterone) can be distracted into pursuing sex, or thinking about pursuing sex, or wishing they could pursue sex, virtually all the time. None of us chose to be this way; it is our unchosen nature.

Especially with respect to our inherited sexual proclivities, evolutionary psychology reveals the wisdom (though not always fairness) of widespread cultural tendencies for ensuring that adolescents spend at least some of their waking minutes learning what is needed to become functional adults. Cultural fixes today include modest dress codes, school uniforms, or even gender-segregated schools.

"Do You Want Them to Gaze at Your Belly?"

Several years ago I was stunned on more than one occasion by the attire some teens wore to church. The fashion at the time was deliberately provocative. Adolescent girls in America wore outfits that bared their bellies and hips. I'm not necessarily proposing dress codes. What I do suggest is that in sex and love education classes we teach our youth frankly about our evolved sexual condition, about the range of ways that various cultures have traditionally dealt with gender-differentiated sexuality. Their attire will still be their choice, but at least we can do our best to ensure that it is an evolutionarily informed (and perhaps even wise) choice. "Do you want people to hear what you say when you stand up, or do you want them to gaze at your belly? It is your choice."

Guidance for nurturing teens (and our staying sane in the process) is one of the biggest blessings offered by evolutionary psychology. My wife read Robert Wright's *The Moral Animal* when it was published in 1994, and she still remembers thinking, "I wish every 13-year-old girl would read this book, because then they would know *why* boys are different!" With my last child right now in the throes of teenage angst, I have learned not to expect my logical exhortations to be decisive. From an "unintelligent design" perspective, one could argue that God blundered by arranging for the human body to develop its secondary sexual characteristics a full decade before the frontal lobes (Higher Porpoise) of the human brain become fully operative.

Even beyond the teen years, we never outgrow the challenges of the sea of symbols we swim in. Prior to language, we could not have vowed to love one another exclusively and "till death do us part." Our inherited proclivities, now as ever, are loyal to the interests of our genes, not our pledged commitments. If sexual infidelity happens on the sly in a nonhuman mated pair, the lapse is likely to remain undetected. But in the human, these questions are possible: "Where were you last night?" "Why is there lipstick on your collar?" "Whose number is that on your cell phone?" "But you told me a month ago that you were at Ralph's that night!"

"Satan" can, indeed, bring temptation by way of our Lizard Legacy. Nevertheless, our brainstem and cerebellum are vital for life. Without our Lizard Legacy, we would starve and leave no offspring. Without our Lizard Legacy, every stumble would result in an injurious or fatal fall, and we would not have learned to walk in the first place. Without our Lizard Legacy, we would have to remember to breathe. Finally, there would have been no physical impetus for our Furry Li'l Mammal to have evolved the bliss of romance. Yes, there *is* Original Blessing, in abundance. But, oh, the challenges!

Furry Li'l Mammal to the Rescue

> *"There is a fox in me . . . a silver-gray fox . . . I sniff and guess . . .*
> *I pick things out of the wind and air . . . I nose in the dark*
> *night and take sleepers and eat them and hide the feathers . . .*
> *I circle and loop and double-cross."* —CARL SANDBURG

The "inherited proclivities" of our paleo-mammalian brain include our deep drive to be in bonded, nurturing relationships and to acquire and retain high status in our social groups. These drives can be called upon to restrain the most disruptive of our Lizard Legacy's inclinations. Fear of losing our partner if we are caught in sexual infidelity can be real and powerful. Our drive to be in nurturing relationships can also prevent us from overeating if we think that obesity will cause those bonds to deteriorate or reduce our chances of entering into new partnerships.

Evolutionary psychology helps us understand why it is that *accountability, or the lack of it, is the single best predictor of long-term integrity in individuals and groups.* The supreme value that we (and most other highly social mammals) place on status, thanks to our paleomammalian brain, can help us hold our reptilian drives in check. Obesity in a culture of profligacy (though not in a culture of scarcity) may well cause us to lose status. If colleagues discover that we have a problem with substance abuse, our status in the working world will decline. And if we are in a position of moral authority (pastor, teacher, therapist, politician) and are exposed as having indulged in any form of sexual impropriety, then our status will plummet. Lower status brings fewer rewards—and reduced possibilities for fulfilling our reptilian drives for safety, sustenance, and sex.

Fortunately, we can call upon our Furry Li'l Mammal when our Lizard Legacy would otherwise lead us down the path of temptation. And our neomammalian heritage, our Monkey Mind, can be called upon to conjure up dreadful scenarios of loss of love and status should we follow through on the urges that our rational minds tell us will not serve the larger and smaller holons of our existence. We thus can tap into powerful energies born of fear of consequences. And we can escalate that process by seeking counsel and support from the paleomammalian and neomammalian brains of those who are already bonded to us and want to see us thrive—for their sake as well as ours. Family members and friends will step forward, if we give them half a chance, and support us in staying the path, perhaps volunteering to spend time with us in healthy pursuits.

Our Furry Li'l Mammal can be a vital force in helping us channel our Lizard Legacy energies in integrous ways. But even our Furry Li'l Mammal has a "demonic" side. It adds its own challenges into the mix, for there are downsides to the uniquely paleomammalian drives. Many of us who have lived with an alcoholic or addict of any kind know the torment of codependence. We can't live with them the way they are, and yet we cannot, or cannot imagine, living without them. So we all too easily become "addicted" to trying to control them. Our lives are warped by the urge, the seeming necessity, to control—to somehow make the intolerable, tolerable; the unbearable, bearable. And we isolate; we cover up for our partner's or parent's or child's addiction. To do otherwise would be to risk demotion in the eyes of others; we would lose status. Our family would no longer be worthy of high esteem. And that might lead to a loss not only of love and nurturance but also of material essentials and the safety that material essentials afford.

Thus, the paleomammalian drives that can be so helpful in fending off addiction can be debilitating if one has to live with an active addict. Our Furry Li'l Mammal willingly drags us into the hell of codependence. And there, too, resides "the Devil."

Thank God for Our Higher Porpoise!

> "There is a fish in me ... I know I came from saltblue water-
> gates ... I scurried with shoals of herring ... I blew waterspouts
> with porpoises ... before land was ... before the water went
> down ... before Noah ... before the first chapter of Genesis."
> —CARL SANDBURG

Our Furry Li'l Mammal's drive to maintain a primary bonded relationship, especially when supplemented by compassion born of an evolutionary understanding of our instincts, can help hold a marriage together in difficult circumstances. To this mix we can add another form of assistance: our higher purpose, or Higher Porpoise, which is the fourth and most recent endowment of our Quadrune Brain. Higher purpose is, in fact, what kept Connie and me together early in our relationship—especially when I betrayed her trust. Something bigger than our embattled self-interests demanded reconciliation.

Connie and I each have a robust Higher Porpoise—a clear and unmistakable sense of calling. Important for our marriage, it is precisely the same purpose: we have dedicated our lives to bringing a holy evolutionary awareness into our culture in the most interesting and inspiring ways we can. What is more, our talents and energies are so complementary, we each know that the very best way to pursue that purpose is by staying in intimate, joyful, mutually beneficial partnership. We are mission partners as much as we are wedded partners. Indeed, the former is the reason why the latter was even possible, given our personality differences.

For many couples, the shared higher purpose is raising children. Shared interests, such as gardening, skiing, and dancing, are also important in holding a marriage together. But interests are not higher purpose; higher purpose is commitment to something that transcends the happiness of either spouse. Raising children is a supreme higher purpose. Concern for the children, and a deep parental desire to remain in the same home as the children, keeps many a marriage together, for good or ill.

Married or not, a higher purpose can be a lifeline. In a time of crisis, it can help us stay the course—or muster the courage to make big changes. Ultimately, it is an important part of a life well lived, and thus of a death that can be approached with equanimity, knowing that one's life was, in fact, worth something: we were of service to our world in some way that will carry on after we are gone. We made a positive difference.

"Using the Sexual Impulse to Evolve"

Barbara Marx Hubbard, a leader in the conscious evolution movement, has a playful way of portraying three distinct ways in which the sexual drive of our Lizard Legacy can manifest:

> In procreative sex there is a higher purpose, which is to create new life. Recreational sex, on the other hand, is for intimacy and pleasure. It's enhancing in many ways, but it doesn't have the higher sacred purpose of procreational sex. In evolutionary sexuality, or what I call "co-creational sex," rather than reproducing the couple, or engaging in intimacy and sexual pleasure for recreation, the sacredness of the intimacy is compelled by a vision of the couple evolving through their union. In that sense, evolutionary sexuality is comparable in its sacredness to procreational sex. While nature's purpose is to reproduce the species through procreation, in co-creational sex, we are using the sexual impulse to evolve the species for the highest purpose.

Your higher purpose may well shift and evolve as the decades pass. If you're not clear what yours is at this stage in life, I invite you to do the following "Discerning Your Calling" exercise. It's an excellent way to begin finding out. Do it in increments, perhaps over the course of several days. Better yet, recruit a friend to do it with you.

Discerning Your Calling Exercise

1. *Take a piece of paper and make three columns. Then, while you're in a non-judgmental, accepting frame of mind...*

2. *Breathe deeply; pay attention to your body and focus on your heart.* Why the heart? Because the heart, scientists have discovered, is as much neuronal tissue as it is muscle tissue—and it has relatively direct neuronal connections with the frontal lobes of the brain. It is not surprising that many cultures have intuited the importance of involving the heart in one's decision-making. Indeed, when indicating oneself, we instinctively point to our heart—not our head. Now, after a few minutes of noticing, rather than thinking...

3. *Bring to mind those activities, creative projects, passions, or interests that bring you lasting joy and deep satisfaction.* Think, too, about times when you have offered a helping hand or contributed to others or to your community in ways that made you feel great. Basically, what do you love doing? What lights you up, gives you energy, or ignites your imagination?

4. *Title the left column "My Joys" and begin to list the words and phrases that articulate what you've just brought to mind.* Be sure to include whatever you're good at and what other people would say you are good at. Periodically stop thinking and writing just to notice—notice your breathing, the sensations of your body, the beating of your heart, extraneous sounds. Then, as more possibilities come to mind, add to your list. When you feel complete...

5. *Close your eyes again, breathe deeply, and ask yourself: "Where do I hurt over what is happening to others—what is happening to my community or my world?* What troubles me or causes my heart to ache? Where do I get angry or frustrated or depressed about what's going on around me? What causes my heart to open with compassion?"

6. *Title the column on the right "World's Needs" and begin to create your new list.*

7. *Keep your lists handy, and add to them over the next few days.* Don't worry about "getting it right" or putting everything down initially. Add to both lists as ideas spontaneously arise. Periodically revisiting and adding to your lists is a spiritual practice that can span a lifetime.

8. *Return to a contemplative state and let your imagination roam while you begin to creatively mix and match, guided by your heart.* Ask yourself: "What are some possible avenues (not just the practical, but also the outlandish) where my joy and the world's needs intersect? How might I contribute my time and energy in ways that would make a difference to at least one other person or creature, and that would also give me great joy?"

9. *In the middle column, begin to list these intersections.* Perhaps draw diagonals to the items in the surrounding columns that would thus be connected. Don't censor or judge the possibilities yet; write freely, periodically stopping to notice yourself breathing. Then, study the connections you have drawn and prayerfully imagine...

10. *Your calling:* Where your joy and the world's needs intersect will indicate the directions of your calling, your mission, your vocation—God's will for you at this time and place. This is where you can join with the impulse of evolution, with the flow of Life, and thus participate consciously in what God is doing in the world.

A powerful way to conclude the Discerning Your Calling exercise is to *articulate your life purpose in a single sentence.* It may take days or weeks of revising your draft sentence before you come to a time when you know that you've got it: you will feel in your gut that this sentence is it.

Halleluiah! Feel the energy, feel the joy of possibility, feel your intimate connection with the Way of Life, with Supreme Wholeness, with God. And then pause for just a moment to remind yourself that, while this is so today, your life purpose will likely evolve as the years pass and that such evolution is a good thing (albeit potentially disruptive and painful). Know, too, that you can revisit this practice when it is time to discern what your next stage might be.

"My Life Purpose"

Here is my life purpose in a sentence: "I serve the future by living in deep integrity, evangelizing evolution, and heralding as God's will the aligning of individual and group self-interest with the well-being of the whole planet."

Last, but by no means least, begin to take the steps necessary for your life purpose to reach into the world. This may require time for study or apprenticeship with someone already walking that path. Or you may be ready to launch your purpose into the world right now, assisted by the Higher Porpoises of those you enlist as colleagues, advisors, or accountability partners.

"My portion of the great work, like that of any other person, creates its own synergy. My job is to be vigilant to the fact that the world creates neither coincidence nor accident—only opportunity. I need only have the courage to ask for what I need and the valor to accept it once it appears."   —ED COLLINS

A person's great work, or higher purpose, is more often co-created than it is discovered. A common mistake made in the effort to discern God's will is assuming that one should wait until the path or direction is revealed from on high. Sometimes it is revealed; sometimes it even grabs us by the collar and drags us into our great work before we've given our assent. But often that is not how a calling manifests. I like to think of it as God waiting to see if we are ready and serious about this—if we are willing to make sacrifices, take risks, jump into the void—trusting, ever trusting. Are we *really* ready to accept a higher purpose that may require us to shed unhealthy habits and to restrain the excesses of our Lizard Legacy? Do we have the courage to proceed, even when our Furry Li'l Mammal wants to scamper back into its burrow? Will we be able to enroll our Monkey Mind in collaborating with our higher purpose, so that it will stop fussing about all that might go wrong and instead get to work imagining positive, empowering scenarios, while figuring out ways to actually manifest our new purposeful path? And who are "we" anyway—if not all of them, if not the entire *menagerie of the mind*. Can we all work toward consensus, give and take, rather than control, domination, and intimidation?

Co-creating your life purpose means being guided by your heart, noticing, listening to what life's circumstances are whispering, and then making it up—all the while trusting the Universe, having faith in God. In the days and weeks and years ahead, you will know when adjustments need to be made, when a temporary tangent becomes the most alluring path. The important thing is to begin. Trust that by saying yes to the invitation to participate, doors will open and signs will guide your way.

"The moment one definitely commits oneself, then Providence moves too. A whole stream of events issues from the decision, raising in one's favor all manner of unforeseen incidents, meetings and material assistance, which no man could have dreamt would have come his way. I learned a deep respect for one of Goethe's couplets: 'Whatever you can do or dream you can, begin it. Boldness has genius, power, and magic in it.'"

—W. H. MURRAY

Being on purpose in the flow of God's will is allowing the wisdom of Wholeness to guide us toward action that will bless the present and the future. Each of us is a part of the Whole, allured by the Whole, to serve the Whole. By serving the body of Life of which we are part, we are really serving our true Self—our Great Self.

Salvation Through Evolutionary Integrity

"There is an eagle in me and a mockingbird . . . and the eagle flies among the Rocky Mountains of my dreams and fights among the Sierra crags of what I want . . . and the mockingbird warbles in the early forenoon before the dew is gone, warbles in the underbrush of my Chattanoogas of hope, gushes over the blue Ozark foothills of my wishes—And I got the eagle and the mockingbird from the wilderness." —CARL SANDBURG

PROPHETIC INQUIRY: How does "salvation" relate to our everyday struggles? What kind of salvation can we find through a holy understanding of evolution? What does it require of us?

"Satan" can and does use the most seductive of disguises—from sex, friend-ship, and righteousness, to power, profit, and patriotism—in order to tempt us away from concern for the common good. Most dangerously, Satan can kidnap our higher purpose, as when a zealously religious young man straps on a chestful of explosives and boards a bus, or when leaders of a nation-state react to a terrorist act at a scale that escalates the problem, all the while fanning the fears and invoking the patriotic assent of its citizens. Where is salvation to be found under these circumstances?

My experience of Evolutionary Christianity suggests that as our under-standing of the Wholeness of Reality (God) expands and evolves, so too, natu-rally and inevitably, will our understanding of the meaning and significance of salvation. From a holy evolutionary perspective, *salvation is not something that can be believed in or not believed in. It simply is.* What we call "salvation," like "sin," is an undeniable part of the human experience.

To know the joy of reconciling when I've been estranged; to experience the relief of confession when I've been burdened by guilt; to taste the freedom of forgiveness when I've been enslaved by my resentments; to feel passion and energy when I've been forlorn; to once again see clearly when I have been self-deceived; to find comfort when I've been grieving; to dance again when I've been paralyzed by fear; to sing when I've been short on hope; to let go when I have been attached; to embrace truth when I've been in denial; to find guid-ance when I've been floundering—each of these is a precious face of salvation. No matter what our respective beliefs, we all have experienced salvation in these and other ways.

Given an evolutionary understanding of the dark side of our inherited pro-clivities, and the universality of salvific experience, does it make sense for *evo-lutionary* Christians to insist that the path to salvation, the road to freedom, individually and collectively, runs through Jesus?

Absolutely!

The only way we can move from alienation to wholeness, from denial to recovery, from bondage to freedom, from guilt to empowerment, is by follow-ing the path that the early Christian scriptures report as the deeply integrous one Jesus incarnated: humble faith, courageous authenticity, compassionate responsibility, and loving service to the Whole. *The doctrine of salvation is powerful when it is held as eternally true—true for everyone, everywhere, and available at all times.* This central Christian doctrine has something to teach everyone, Christian and non-Christian alike. To do so, however, our theology of salvation must be freed from otherworldly and unnatural interpretations. Only in this way can its transformative powers bless our lives and the lives of those around us here and now.

"What the Hell Are We Preaching?!"

In mid February 2007, as my wife and I were in the final editing stages of this book, Connie approached me, her face distraught. She had taken time out from the tedious task of line editing to do something different—working on our 2006 tax forms. As it turns out, her angst had nothing to do with money or mathematics. She told me that she had been multitasking, listening online to the most recent sermon by one of her favorite ministers. The sermon was on love; it was a practical sermon about how couples—and church congregations—might work to sustain love over the long haul. Connie was exasperated. "It was a good sermon, as far as sermons go," she said. "But without a grounding in evolutionary psychology, what the hell are we preaching?!"

The doctrine of salvation may be assisted in modern times by incorporating what I like to call *evolutionary integrity,* or *deep integrity*—integrity as Jesus taught by word and deed, but now set within the perspective of deep time—future as well as past.

Deep time in the past tense means we value the contributions and sacrifices of countless ancestors and others, human and nonhuman, who have made this moment possible. It also includes an appreciation for even the most unwelcome of our instincts, fed by our understanding of what ends those urges and drives served in by-gone times. *Deep time* in the future tense means we accept full responsibility for ensuring a just, healthy, beautiful, and sustainably life-giving future for Planet Earth and its diverse species.

Christian leaders and laity alike have long recognized that it is not *beliefs about* Jesus that will save Christians. It is, rather, *faith in* him (i.e., trust in the values he incarnated, the integrity he enfleshed). *The key to salvation is committing to Christ-like integrity.* Being "in Christ" and being "in evolutionary integrity" (or, deep integrity) are different ways of saying essentially the same thing. One uses night language; the other, day language. To speak traditionally, deep integrity is the way, the truth, and the life that Jesus embodied. "Christian," after all, originally meant "little Christ." When I trust like Jesus, love like Jesus, live my truth like Jesus, take responsibility like Jesus, and serve the Whole like Jesus, I know heaven—even in the midst of the chaos and crucifixions of life.

As I walk an evolutionary path of personal salvation, humility is a requisite, for *I cannot save myself by myself. Original Sin runs deep. My Lizard Legacy is too powerful, and my Furry Li'l Mammal lives in a world of small and often selfish concerns. I need others and, in fact, I need the Whole of Reality.* Here we see another face of deep-time grace in and through evolutionary psychology and brain science. How much more workable to accept ourselves, in all our flaws, rather than to resist those inborn aspects as if they shouldn't be! Instincts simply *are*, and we can see how they served our ancestors. Now, how do we go about the task of channeling those energies in integrous ways, with equanimity and insight rather than white-knuckle horror?

Salvation is found in saying *yes!* to the possibility that my self-interest is best served by serving others and the world around me. God's will for individuals today is essentially what Jesus said it was two thousand years ago: "Love the Lord your God with all your heart, soul, mind, and strength, and love your neighbor (including your enemy) as yourself. All the Law and the Prophets hang on these two commandments." Translated into modern-day speech: "You want to be saved? Love the Whole of Reality (immanent and transcendent) passionately, with everything you've got—in every person, part, and detail—and know that by pursuing the well-being of the larger and smaller holons of your existence, you are not only furthering your own well-being but also fulfilling the promises of the past and manifesting the possibility of a glorious future!"

Recalling the circumstances in which we've personally tasted the fruit of salvation—all the times we've known healing, reconciliation, forgiveness, wholeness, empowerment, freedom, and so on—we can be sure that God's grace (the grace of Ultimate Wholeness) was our ground. The Christian notion of being saved from eternal hellfire by grace through faith does make sense (in a this-worldly way) when seen in an evolutionary context. Freedom from the hell we create for ourselves through pride, arrogance, deception, resentments, hatred, and so forth is available only when we accept that some larger Wholeness is at work, and that this undeniable Reality is trustworthy and ultimately on our side. By turning from our narrow self-centered ways and fears and giving ourselves over to the care and guidance of *this* Reality (however we may choose to name It/Him/Her), we really can experience salvation, the peace beyond understanding.

From an evolutionary perspective, a salvific experience that is deep and lasting shows up in daily life as an abiding, almost effortless state of integrity and trust in time, and in life—including trusting the chaos and challenges that inevitably come our way. And once we attain this state, we can manifest the creativity of divine wholeness flowing through us. We are released from the paralyzing grip of resentments, what-ifs, rehearsals, and futile anxiety that otherwise can so easily consume our time and energy and distract us from our calling.

Being saved *from* behaviors destructive of ourselves and others is only half the story. There is this, too: What are you saved *for*? Now that you are free, how will you serve God? The poet Mary Oliver challenges us to consider what we will do with our "one wild and precious life."

To be "saved" surely includes growing in evolutionary integrity and pursuing one's higher purpose. It is to have sufficiently dealt with our shortcomings and put into place inner and outer structures of support such that we can stride into the world of action undistracted by imperfections in ourselves and in those with whom we are in relationship.

The DNA of Deep Integrity

"Many of our impulses are, by design, very strong, so any force that is to stifle them will have to be pretty harsh. It is grossly misleading to talk as if self-restraint is as easy as punching a channel on the remote control." —ROBERT WRIGHT

Millions of people around the world can personally attest to having found freedom from their addictive nature (or freedom from the desperate habit of trying to control others) by way of 12-step and other recovery programs. Twelve steps are difficult for anyone to remember, but the aim of this approach is easy to describe. It is to move individuals to a habitual place of what I'm calling deep, or evolutionary, integrity.

The four "letters" (chemical bases) in what I call *the genetic code of evolutionary integrity* (or DNA of deep integrity) not only express the essence of recovery programs; they also comprise the heart of religious morality. This, of course, should not be surprising. Much that we are now learning scientifically was intuited long ago and has been rediscovered time and again, as the same process of testing and selection at work in nature also shapes culture. Here is the DNA of deep integrity:

1. TRUST/HUMILITY/FAITH: Surrendering to the wisdom of divine Wholeness— that undeniable physical and nonphysical Reality beyond thought, belief, or denial, which is at work in the world and to which each of us is ultimately accountable.

2. AUTHENTICITY/HONESTY/SINCERITY: Getting real with oneself and others, owning the painful truths about one's life, and grasping the comforting truth that God loves us anyway. Then making commitments that will cultivate healthy habits and supports for living in integrity.

3. RESPONSIBILITY/ACCOUNTABILITY/COMPASSION: Stepping into the shoes of those we have harmed, and then making amends—while cultivating compassion for ourselves and others. Enlisting the support of others, too, as integrity is a team sport.

4. SERVICE/GENEROSITY/PURPOSE: Supporting others in maintaining integrity and providing lifegiving service in additional ways. In so doing, we not only bless the world; we support our own growth and fulfillment, while boosting our chances for long-term integrous living.

Genuine self-interest is thus served by cultivating evolutionary integrity in a four-fold way: growing in trust, authenticity, responsibility, and grateful service. To the extent we are guided in ways that fulfill this standard, God guides us.

STAR Clusters

Trust, Authenticity, Responsibility, Service . . . Trust, Authenticity, Responsibility, Service . . . each letter in the genetic code of evolutionary integrity that we make the effort to practice feeds into the next, round upon round, spiraling our character in positive ways. For some, the process begins with *trust,* with the recovery movement's first step of surrendering to the wisdom of a beneficent Higher Power or to one's community of support. For others, the spiral may begin with stepping up to a life of *service,* for which trust, authenticity, and responsibility are vital supports. But of course, who of us would trust or serve if we did not believe that it was in our self-interest to do so? So let's start there.

Self-interest is a powerful biological instinct, and it can dance with service if we develop and expand our sense of who we are and why we're here. What looks like service to some may actually feel like self-interest to those who have an expanded sense of self (or, Self). The mother who defends her child is identifying (consciously or not) with her family and her instinctual drive to care for her legacy. The activist who supports environmental action may be identifying with the larger body of Life of which he or she is part. The person handing out sandwiches in a soup kitchen may be identifying with all of humanity. Whether prompted by self-interest (perhaps by a desire to reduce personal suffering or by a desire to grow) or whether sensing into a greater Self-interest and thus acquiring an urgency to serve, or to serve in an enlarged capacity, the way of evolutionary integrity offers a viable and flexible path to the fulfillment of our evolutionary mission.

I've come to think of the acronym STAR as summarizing this work to increase and deepen our evolutionary integrity...

> **S** elf-interest/Service (and their dance)
> **T** rust
> **A** uthenticity
> **R** esponsibility

I imagine people coming together in groups, locally or online, to support one another's growth in evolutionary integrity. We might call these groups "STAR clusters." And as we develop connections among such groups—email lists, conference calls, websites—for community-building, information-sharing, networking, support, action, we can imagine bridging the distances between the STARs, reaching through "interStellar space."

When I mentioned this idea to my friend Tom Atlee, he joked, "Right now, we humans are like the Big Bang early in its career: a lot of hot air. In the case of the Big Bang, the hot air was hydrogen gas; in our case it's mostly egoic gab. Out of the hydrogen gas eventually coalesced stars and everything that stars, in turn, make possible. Now, in our case, by coalescing into STAR clusters we can assist our evolution, and to the benefit of the whole planet."

What I am calling *the DNA of deep integrity* consists of four groups of character traits, each of which pertains to a primary stance toward some aspect of life. The following chart distinguishes evolution-furthering traits from those that are deevolutionary.

The DNA of Deep Integrity		
Primary Stance Toward...	**Evolutionary ("Christ-like") Character Trait**	**Deevolutionary Character Trait**
1. Known and unknown	trusting/full of faith	fearful/anxious
	humble/unassuming	proud/arrogant
	open/interested	closed/conceited
	curious/respectful	know-it-all/judgmental
	childlike/blameless	jaded/guilty
2. What's so in & around me	authentic/earnest	inauthentic/fake
	sincere/genuine	insincere/false
	honest/truthful	deceptive/misleading
	true/forthright	unfaithful/devious
	real/straight	false/crooked (slimy)
3. All my relations	responsible/accountable	irresponsible/unreliable
	compassionate/benevolent	unfeeling/mean
	loving/caring	self-absorbed/cold
	thoughtful/considerate	thoughtless/insensitive
	sympathetic/understanding	heartless/unconcerned
4. Present & future	purposeful/mission-driven	aimless/directionless
	serving/helpful	self-centered/useless
	grateful/appreciative	ungrateful/unthankful
	generous/munificent	stingy/miserly
	inclusive/expansive	dogmatic/contracted

Integrity is not a solo sport; it is a community undertaking. For this reason, I dream of the day when *baptism* in a Christian church comes to mean this: We know that Original Sin cannot be washed away by a daub of water. Rather, the baptismal act is a commitment by the community to lovingly guide the baptized individual through all of life's stages and through every challenge, using the awareness and tools that God has revealed and will continue to reveal through the time-tested wisdom of our cultural and religious inheritance and the public revelations of science. The religious community would provide structures of education and support that would acquaint us with our evolved Quadrune Brain: our Lizard Legacy, our Furry Li'l Mammal, our Monkey Mind, and our Higher Porpoise.

Values teaching would begin in early childhood, not only through Bible stories, but also through fun and playful *evolutionary parables* derived from our shared creation story. "The Lucky Little Seaweed," "Ozzie and the Snortle-fish," "Startull: The Story of an Average Yellow Star," "Earth Had a Challenging Childhood" are some of the evolutionary parables that I, and many others, are already using to educate and delight adults and children. You can download them from our Great Story website.

Then, in adolescence and continuing throughout adulthood, our baptismal community would be counted on for peer counseling, recovery work, and encouragement of our Higher Porpoise through participation in evo-integrity groups, or STAR clusters.

Growing in Deep Integrity

1. Lizard Legacy

   a. What do your reptilian instincts want that helped your ancient ancestors survive and reproduce but that now have negative consequences if you act on them indiscriminately, habitually, or in ways that are out of integrity? (List everything related to food, substances, safety, and sex that occasionally cause you problems or challenge one or more of your relationships.)

   b. What do you appreciate about your reptilian instincts? How do they serve your life and your relations?

2. Furry Li'l Mammal

   a. What does your Furry Li'l Mammal want that would have served your ancient ancestors but that now have negative consequences if you act on them indiscriminately, habitually, or in ways that are out of integrity? (List all your issues and challenges related to love, parenting, sibling relations, status, security, and wanting to look good or be right.)

   b. What do you appreciate about your mammalian instincts? How do they serve your life and your relations?

   c. How have your mammalian instincts helped keep your reptilian drives in check?

3. Monkey Mind

   a. What challenges does your chatterbox mind cause you? (List anything that typically takes you out of the present moment—things you habitually worry about, anxiety for the future, guilt or regrets about the past, whatever it is that clutters your mind with preoccupations.)

   b. What do you appreciate about your rational mind? How does it serve your life and your relations?

   c. How has your Monkey Mind's propensity to conjure worst-case scenarios helped you hold the excesses of your reptilian and mammalian instincts in check?

4. Higher Porpoise

   a. What is your life history of higher purpose? (List the pursuits that have given your life focus, beginning with your earliest memories of childhood hobbies and fascinations.)

   b. Reflect on the times in which you have lost focus, when you have floundered, or when life felt meaningless. How did you regain a sense of purpose?

   c. How has your higher purpose helped you lead a life you can be proud of, and helped you hold in check excesses of your reptilian and mammalian instincts?

   d. Where do your joy and the world's needs intersect?
      (See pp. 178-79)

Christ-like Evolutionary Integrity

"Here's a new commandment for you: Love one another as I have loved you. By doing this everyone will know that you are my disciples."

—JESUS

The picture of Jesus we find in the early Christian scriptures can be a model for all of us, no matter what our particular faith tradition or philosophy. There we read about the life and teachings of one who saw and honored God in all things, boldly lived his truth, loved more broadly and deeply than his contemporaries, nonviolently challenged the stagnant institutions of his day, and took responsibility for his world. And here, too, we see that one person can change the course of history.

I will explore two questions that often arise when I speak on the subject of evolutionary integrity. While the answers may not be obvious, they will make sense upon reflection. The first question concerns the character trait clusters that I suggest are "the DNA of deep integrity." The second question centers on my claim that these same traits are central to what it means to be "Christ-like" and "Christ-centered."

Question 1: *What do humility, authenticity, responsibility, and service have to do with evolution? Why are these particular traits necessary for evolutionary integrity?*

Humanity is poised at the cusp of a great transformation. As we awaken to the nature of the larger Creative Reality that is our source and sustenance, we also awaken to our own fabulous—and frightening—powers to set the future course, not only for our own descendents but also for the entire community of life. Such powers can be exercised prudently only with humility.

Planet Earth comes first. The health and well-being of the body of Life must take precedence over the health and well-being of any single species, including our own. Indeed, as Thomas Berry reminds us, "Human health is a subsystem of the Earth's health. You cannot have well humans on a sick planet."

Our species emerged out of the naturally divine processes of this planet in the same way that apples grow out from an apple tree. That is how God made us. And this fact must now guide us as we move into the future. Ignoring it will lead inexorably to our estrangement from Earth and quite possibly to our extinction. In the words of Thomas Berry,

> It is time to step back and find the human place in the natural world and not think that we can make the human world primary and the natural world secondary. We have got to say to ourselves, 'Let's begin to try to understand the natural world and find a way of prospering the natural world first.' Then find our survival within that context. Because if we think we can put ourselves first and then fit the natural world into our program, it's not going to work. We have got to fit the human project into the Earth project.

Ever since Darwin, we have been learning how planetary, biological, and cultural evolution have produced the stunning diversity of life and the complexity of human societies. It is only by coming to know and align with these patterns and principles that our kind will not merely persist into the future but do so with gusto and grace. Now taking these twin realities—first, that Earth is primary, and second, that we must align ourselves with the ways of the Universe— let us return to the first question: What do humility, authenticity, responsibility, and service have to do with evolution?

To see why these four trait clusters are essential to evolutionary integrity, imagine the impact of their opposites on the evolution of an individual, a group, or humanity as a whole. When we think and act out of fear or arrogance, rather than trust and humility; out of inauthenticity or deception, rather than honesty and sincerity; out of blame or self-righteousness, rather than compassion and responsibility; or out of apathy or self-centeredness, rather than generosity and service, our life sucks (to put it bluntly). Problems abound. The same is true for groups, corporations, nations, and our species. When things don't work, when breakdowns occur, when relationships sour, personally or institutionally, it's a sure bet that fear, arrogance, inauthenticity, irresponsibility, blame, and narrow self-interest are the cause. It's also a sure bet that humility, honesty, responsibility, and consciously attending to the needs of the larger whole will transform the situation and put it back on the path of healthy evolutionary emergence. The four traits are central because only by growing in these areas are we growing in our relationship with Reality, with God, with the way things really are in a nested emergent Cosmos.

Question 2: *What does "Christ-likeness" have to do with evolution? Why are the traits clustered around humility, authenticity, responsibility, and service central to Christ-likeness and Christ-centeredness?*

From an *evolutionary* Christian perspective, the focus is the Christ enshrined in our sacred texts and creeds. What stories and teachings were spiritually important enough to survive two thousand years? It is here that we find the mythic, Cosmic Christ in all his glory. And it is *this* picture, what people felt compelled to write or say about him long after he lived, and what hundreds of millions still feel compelled to say about him, that offers insight into how trust, honesty, responsibility, and service entail the essence of "Christ-likeness" and "Christ-centeredness."

If we take an appreciative look at the night language of the ancient scriptures, we detect in Christ Jesus the superhuman capacity to incarnate the ideals of humility, authenticity, responsibility, and service. The good news of the Jesus story is that when these possibilities are brought fully into the world, even in the life of just one individual, those around him or her will be transformed, and the stories they in turn tell will transform others, continuing to work miracles in the lives of any who aspire to be "Christ-centered" for generations to come. Ponder the following:

+ What speaks more powerfully of *humility* and *trust* than God Almighty, the Creator of the Universe, humbling himself to be born in a barn, apprenticing as a carpenter, walking as one of us, speaking in defense of the poor and the maligned, then suffering and dying at a young age, with no assurance that his teachings would make any lasting difference?

+ What evidences *honesty* and *authenticity* more perfectly than Jesus courageously living and speaking his truth, nonviolently challenging the unredeemed powers that be, and regularly exposing the self-righteousness and hypocrisy of the religious leadership of his time and their all-too-narrow circles of care and commitment?

+ What better picture of divine *compassion* and *responsibility* could there be than God taking full responsibility, despite excruciating personal cost, for healing a relationship that was broken by the arrogant self-centeredness of our kind?

+ Finally, one would be hard pressed to find anywhere in the world's mythic-religious literature a better example of generous *service*, or a superior instance of someone devoted to a *higher purpose*, than that of a simple carpenter doing his best to catalyze human transformation by way of self-sacrificial love.

Jesus' contemporaries could not have known empirically the truth of evolution that is now so available today. Nevertheless, the dynamics of evolutionary emergence surely were at work all the while. As a creATHEIST, I choose to regard as no coincidence that the mythic stories of Jesus the Christ so well match what we now know both experientially and experimentally through the public revelations of science.

"Getting right with God," "coming home to Reality," "abiding in Christ," and "growing in evolutionary integrity" are different ways of saying the same thing. Thousands of years before humanity could have rationally known about the salvific power of deep integrity, wise humans in each and every culture came to the same understandings by experience and heartful reflection. Mythic (meaningful) stories were told and embellished to inculcate these values, to offer peoples across the globe opportunities for salvation, here and now. In my own tradition we speak of it this way: the Lord Jesus Christ is our redeemer, savior, and divine-human reconciler. He is the incarnate expression of God's love and forgiveness, and the very embodiment of the ultimate transformative powers of the Universe.

"I've Never Told Anyone"

More and more I feature evolutionary brain science and evolutionary psychology in my public presentations. Why? Because the telling of our brain's creation story evokes deeply spiritual and lifegiving responses. For example, at a church in Louisiana a woman rose and told the gathering: "I am studying neuroscience, and the more I learn about how the human brain evolved, the more compassion I have for myself and others. Isn't it amazing, and rather ironic, that science leads to this religious virtue."

Elsewhere in the South, an older man took me aside after a church service. I had illustrated the brain science segment of my sermon that morning by describing how a rise in status will boost baseline testosterone levels in humans and other primates. I explained that this kind of hormonal response had, in fact, served our ancestors—but we live in a very different world. In our world there is language, there are social strictures, and there are spoken commitments, including "till death do us part."

"If we are ignorant of the science," I proposed, "if we deny our evolutionary heritage, then promotion at work or election to public office will catch us by surprise—and possibly wreak havoc in our marriages and other relationships."

As it turns out, such had indeed been the case for the gentleman who pulled me aside after church. "I've never told anyone this before," he began. "Many years ago, when I was working for a large corporation, I got a big promotion. Immediately, my life began to fall apart. I couldn't understand it. Within a year, I had multiple affairs and my marriage was in ruins. It cost me my job, too." He grasped my hand and said, "Thank you for helping me understand how this could have happened."

That's not the end of the story. The man was able to forgive himself. With shame lifted, he set about making amends to those he had harmed long ago. A month later I learned that his long-standing depression was not the only thing that had lifted, and that he and his beloved were reaping the benefits.

No matter your religion or philosophy, making a commitment to grow in deep integrity will offer you much the same experience as those who have repented of their sins and asked Jesus to be their personal Lord and Savior. The two are day and night language reflections of personal transformation. That's the beginning. Then there's the great adventure of living a victorious and deeply fulfilling existence (i.e., dwelling in the Kingdom of Heaven) and doing so for the rest of your life. That's where the DNA of deep integrity provides a more useful blueprint, and more detailed and universally accessible map, than could have possibly been revealed in a prescientific world of private revelation.

You can live a life of "Christ-like" integrity with or without an evolutionary worldview. It can be done either way. It's just much easier with an evolutionary worldview.

REALIZING "Saving Faith"

"Faith is the strength by which a shattered world shall emerge into the light." —HELEN KELLER

PROPHETIC INQUIRY: How can we possibly have faith in God in a world that seems so imperfect and beset with problems? What does evolution tell us about how to live in such a world?

It is no surprise that faith, or trust, is at the heart of every religious tradition. Given the nested emergent nature of divine creativity, trust just makes sense. We can't know for certain what the larger and smaller holons—the inner and outer spheres—of our existence are up to. Worrying or fretting about the future doesn't get us anywhere: neither will change a thing—except, of course, us, by diminishing us. From this evolutionary perspective we can begin to appreciate how rational faith really is! In fact, other than doing all we can to ensure that inner and outer conditions will thrive, trust is the wisest choice we can make. Without it, there is no peace.

Faith in God is trusting the wisdom of the Whole. It is having faith in time. It includes trusting the intelligence and creativity of the larger holons of which we are part and the smaller holons of which we are composed and for which we are responsible. It also means trusting that bad news and breakdowns are not cosmic mistakes. They can help us grow. Faith in God is also trusting that our intuitions and feelings can be guides for action that will serve our larger interests, our higher purpose, and glorify God—that is, be a blessing to the Whole.

Faith and trust are synonyms; faith and beliefs are opposites. (I suggest you reread the previous sentence; it's important.) Trust is what most other animals do pretty much instinctually. When some difficulty occurs in nature—a drought, flood, or hurricane—other creatures don't make it mean anything. They don't say to themselves, "Damn, why did this have to happen to me?" They just accept the situation as it is (trust) and then make the best of it by being in action. Faith helps us reclaim our birthright as animals. Through faith we experience communion with God, the same intimacy that our ancestors enjoyed "in the garden" for hundreds of thousands or perhaps even millions of years, as hominids prior to language.

Now that we live in a world shaped so extensively by language, faith is not automatic. It needs to be nurtured. Faith enables us to move beyond the beliefs born of language, and thus to receive God's guidance directly as we pursue our

Great Work: doing our part to better the world, in ways big or small, to leave a positive legacy that lives on after we are gone.

Earlier generations cannot be faulted for not knowing what we know. God (the Whole) reveals truth to each and every generation and culture in ways that make sense at that time and place, given the psychological, technological, political, economic, and social realities of the day, as well as the landscape, climate, and other natural factors that shape the awareness of, and metaphors available to, the inhabitants. In a diverse, developmental Cosmos there is ultimately no privileged position theologically—no place and time from which we can be sure to have God's Absolute Truth. The insights revealed to the Buddha, Confucius, Moses, Paul, Mohammed, the Hebrew prophets and other biblical writers, early church leaders, reformers, counter-reformers, and others through the ages and in other cultures throughout the world were only those that could have been received then and there. The same is true for us. Future generations will surely have larger, more comprehensive understandings of the meaning and magnitude of each of our faith traditions than any of us can imagine today. Regularly reminding ourselves that *facts are God's native tongue* is a spiritual practice that can encourage this humble, hopeful stance.

If someone were to call a 2,000-year-old medical treatise or legal document "new," we'd all have a good laugh. Now consider the term *New Testament.* Okay, no laughs there. Perhaps some shock, or sadness. What keeps many of us from noticing how outdated it is to continue calling the early Christian scriptures the *New* Testament is the erroneous belief that God stopped revealing truth vital to human well-being and destiny when people thought the world was flat, only a few thousand years old, and at the center of the Universe.

This is not to say that the Bible contains no wisdom relevant for today. It does. What I am saying is that we have new ways of understanding truth. Just as we now test ancient herbal remedies with double-blind experiments reported in peer-reviewed journals, we now can explore the Bible for the deep truths of Christianity through the God-given eyes of science, and with the assistance of more recent facts publicly revealed in the 19th, 20th, and 21st centuries.

> *If you are wholeheartedly committed to growing in deep integrity and have*
> *no resentments, no guilt, no shame, no regrets, and no unfinished business,*
> *you are saved no matter what your religion or philosophy.*

From a creatheistic perspective, to think that "getting right" or "being right" with God requires one to hold a particular set of beliefs implies that God is beset with human limitations. If a person is expected to give mental assent to word-based propositions in order to be "saved," then God's love is hardly unconditional, nor is God's wisdom infinite. To interpret "faith in God" as meaning that one must subscribe to a particular way of seeing the world in order to go to an unnatural or otherworldly place called "heaven" is to miss out on the this-world saving grace of the Gospel. Those who cling to flat-earth theological positions on biblical grounds diss the Holy One. To imply that the best guidance available today for interpreting salvation or any other doctrine is to be found in 2,000-year-old texts is to declare God as cruel, uncaring, and impotent. I don't think people mean to say this about God, but consider what would be thought of a human father who denied his children ongoing instruction and guidance as they matured.

"Why Do We Think Differently About God?"

Imagine hearing about a father with a troubled and confused teenage daughter. If you were then told that the father had not communicated with his daughter since she was a toddler, you'd blame the father for how screwed up the daughter was as a teenager, right? Of course! Certainly, "kind, loving, or generous" are not words you'd use when speaking about a father who stopped guiding his daughter when she was a toddler. Why do we think differently about God?

Chaos, breakdowns, and difficulties are often God's greatest gifts. Bad news is, in fact, the primary driver of creativity and transformation—in our own lives as well as throughout the Cosmos. Only faith offers the "peace that passes all understanding." Believing certain things about the Bible or Jesus, or anything else, is all well and good. But if your heart doesn't trust that chaos is grace in disguise, beliefs alone will not sustain you when chaos comes your way.

Evolutionary faith extends much farther in time and space than flat-earth faith could ever hope to. Real life is full of pain and disappointment. To have faith regarding the past means refusing to play the blame game, while nurturing respect for everything (good and bad) that has made this moment possible. To have faith in the present is to be mindful that every breath and every touch is a gift of grace. To have faith in the future means trusting that whatever happens can facilitate your growth and learning. As the saying goes, if it doesn't kill you it'll make you stronger—or softer, more compassionate. Trusting time means experiencing the flow of real life with an open heart and a deep-seated attitude of curiosity, generosity, and responsibility.

Faith has a spatial as well as a time dimension. I know how important it is for me to trust my inner nature: my dreams, intuitions, failings, and my life energy. I trust my outer nature, too: the natural and social contexts within which I am embedded.

✦ To trust what is *inside* is to listen to the subtle voice of the Spirit within. It is to follow only those external authorities that align with my heart's wisdom, in service to my calling. It is also to accept that even my faults and shortcomings serve a purpose.

✦ To trust what is *outside* is to appreciate the natural world for what it has always been—teacher, healer, provider; revealer of divine mystery, majesty, and power—and to accept my condition as an earthling, as a gloriously mortal expression of this creative planet. It also means having faith that the faults and shortcomings of modern society serve a purpose; they reflect a necessary, though immature, stage in the evolution of consciousness and culture.

In religious terms, in night language, trusting time and space means having faith in God. It is choosing to stay open to the possibility that we are being allured by the same mysterious Reality that guided the process of evolution for billions of years and birthed the prophetic promise that God is Love. It is also knowing that nothing in our lives or in the world is an irredeemable mistake.

Trusting in these ways does not mean passivity. I trust that our Western consumer society is not a cosmic mistake. Even so, I am doing all I can to help our culture mature beyond its present short-sighted and Earth-destructive practices.

Looking within, I trust that my shadow serves a purpose, but I'm also doing all I can to remain in deep, evolutionary integrity. Likewise, we can trust that those who oppress others are less evil than they are ignorant or unenlightened, and at the same time we can do everything within our power to ensure that freedom and justice prevail. We can trust that everything we do in creative response to the seemingly dark side of the world is God's way of bringing light into the world.

"Trusting the Universe" or "having faith in God" means trusting that *everything is right on schedule*. But it also means trusting that the anguish and anger we sometimes feel for what is happening around us, and our yearnings for a just and sustainable society, are part of the Universe too—and right on schedule as well.

> When you come to the edge of all the light you know, and are about to step off into the darkness of the unknown, faith is knowing one of two things will happen: There will be something solid to stand on or you will be taught how to fly.
> —PATRICK OVERTON

REALizing "the Gospel"

> *"If what we mean by the 'Gospel' today is the same as what Christians two millennia ago meant when they used the term, we do our tradition a terrible disservice."*
> —PAUL WEST

PROPHETIC INQUIRY: Given what we now know about the deep-time creativity of the Whole, what does the Gospel have to teach us about the nature of God's grace and the great good that is possible for us—individually, relationally, and collectively? What is the saving good news revealed in our common creation story?

The meaning of *the Gospel* is infinitely rich. No generation can possibly exhaust its depths. Every generation has the privilege and responsibility of reinterpreting the core insights of its faith tradition for its own time, as the Spirit leads them.

Thomas Aquinas, one of the greatest theologians in Church history, wrote nearly a millennium ago, "A mistake about Creation results in a mistake about God." What Aquinas knew then is even more consequential today. As our understanding of the Cosmos expands, so must our view of God and, for Christians, our appreciation of the meaning and significance of the Gospel. Seen as a sacred story of nested creativity in which life becomes more complex, more aware, and more intimate with itself over time—this epic of evolution can revitalize the meaning and magnitude of the Christian gospel.

The disciples and early Church leaders, reflecting on Jesus' ministry within the context of their own first, second, and third century CE political, judicial, religious, and cosmological understandings, formulated creeds and doctrines about the significance of his life and mission. Since then, our view of reality has grown enormously. If the Christian tradition is correct in its assertion that Jesus truly did incarnate God's great news for humanity, then

The meaning, grandeur, and this-world relevance of the Gospel today must reach far beyond what any previous generation, including the biblical writers themselves, could have known.

A REALized understanding of salvation urges each generation to do all that it can to support the human venture in *this* world and for ages to come. A REALized understanding of salvation propels 21st-century Christians to grow beyond any residual fascination with literalistic understandings of the so-called End Times—a fascination that may have made sense in some communities of Christians in previous centuries but which is now not only dated but dangerous. Let us allow that Christians of the next century—indeed, Christians of the next decades—will be wise to scrutinize interpretations of core doctrines born of our own time in history.

Crucially, ongoing efforts to reinterpret and revitalize the Gospel in no way need diminish or demote core doctrines of the faith. By my own fourfold standard (introduced at the beginning of Chapter 8), reinterpretive efforts would be applauded only if they (1) validate the heart of earlier interpretations, (2) make sense naturally and scientifically, (3) speak experiential truth both to those within and to those outside the faith, and (4) inspire and empower.

This imperative to refresh religious doctrines as times and conditions change is more than just a matter of staying relevant. The religious leaders who have most inspired me are those who proclaim that *the greatest and boldest creedal assertions are in the future, not in the past.*

When we become accustomed to seeing God's will, God's love, and God's transforming power operating on the scale of billions of years and embracing all of Creation, our understanding of the Gospel opens and magnifies. Its greater realization, however, will take time and will never be exhausted. The Protestant Reformation, made possible by the printing press, did not happen overnight. Similarly, the "evolution revolution," made possible by advanced telescopes, computers, and the Internet—as well as by ever-increasing knowledge of how living systems function—will take decades before its implications are fleshed out theologically, politically, and economically. Nevertheless, we can say this:

> Given what we now know about deep-time creativity and grace, we can no longer in good conscience continue interpreting the story of Jesus' birth, life, teachings, passion, death, and resurrection as primarily having to do with saving a select group of human beings from the fires of a literal hell when they die.

Such cannot possibly be the truth of the Gospel for our time. That interpretation may still appeal to millions of Christians (sadly, who quibble or fight over just who is in the select group), but it is in no way Good News for most of humanity. Indeed, for all other forms of life on Earth, such an anachronistic interpretation is far more a curse than a blessing. How can we continue to think that this is what God wants?

What we call "the Gospel" will be experienced as good news only if it is a saving response to the bad news that people are actually dealing with. Said another way,

> If what we mean by "the Gospel" does not address in a hopeful, inspiring way what people themselves regard as their greatest personal and collective challenges, then for them the Christian message will not be salvific. It will be irrelevant.

This mismatch, between what people in fact experience as bad news and what our church liturgies present as the Good News, is a big reason why those under thirty are largely unchurched—and why the epic of evolution told in God-glorifying ways is gaining wide appeal. To be frank, most young people are not preoccupied with concerns about whether heaven and hell really exist. The difficulties in their lives today, as well as their concerns for the well-being of the world at large, trump any such otherworldly preoccupations.

> Salvation must be available here and now. And of course this is precisely what an evolutionary gospel proclaims.

Salvation must be available, as well, to those who cannot give their assent to literal interpretations of the Christian doctrines of virgin birth, resurrection, ascension, and so forth—even in the broadened ways offered here (also see Appendix B). Indeed,

> *The good news must also be made available to those who do not find the biblical story of creation credible, nor helpful even when interpreted metaphorically. The Gospel, thus, must no longer be restricted to those for whom the Bible is the centerpiece of God's revelation.*

"The Joy of Watching Young and Old Alike Light Up"

For Connie and me, the occasional hardships of living on the road are more than offset by the joy of watching young and old alike light up when we have done nothing more than give them a creation story. Through this Great Story, people see that their own lives are part of something grand and majestic—a creative process that encompasses their inner thoughts as well as a hundred billion galaxies. What a relief to know that God is no less active in the Universe today than thousands, millions, or even billions of years ago! Yes, time can be trusted, and Grace is at work *right now*—calling, inspiring, freeing, and enabling all of us to participate in the Great Work. This is a REALized God. This is a REALized Gospel.

My experience on the road has taught me that, especially for teens, the Gospel must highlight the lessons of evolutionary brain science and evolutionary psychology. Time and again, I have watched young people experience salvation by learning about their evolutionary heritage—that they are the way they are because those drives served their ancient ancestors. Halleluiah! Our instincts are to be worked with, appeased in moderation, and rechanneled integrously—not condemned and repressed.

From a meaningful evolutionary perspective the Gospel includes the Great Story of God's love and saving grace as revealed in the Bible, on the cross, and throughout the entire 14-billion-year epic of evolution. The Gospel, as such, is transformative on three levels: individually, relationally, and globally. To ignore or discount any one of these is to miss the meaning and magnitude of them all.

+ *Individually,* the Gospel can free a person from addiction to sin and self-absorption, enabling each of us to savor the fruit of the Spirit in the midst of the never-ending challenges of life, and empowering all to be blessings to the world regardless of our shortcomings. It can also enable one to know peace, even in the midst of difficult circumstances and in the presence of difficult people.

+ *Relationally,* the Christ story shows us how reconciliation is possible with virtually anyone. When I take responsibility, let go of thinking I'm right and the other is wrong, step into their experience, and communicate with love and compassion from *that* place, miracles occur.

+ *Collectively,* the Gospel can free us from species pride, arrogance, and human-centeredness by revealing the trajectory of divine creativity and how we as a human family can fulfill our role in furthering what God has been up to for billions of years.

"I Don't Merely Believe...I Know!"

I don't merely believe in the fall of Adam and Eve, in Original Sin. I know that the reptilian and mammalian parts of my brain have drives of which my conscious mind is clueless—and that these inherited proclivities, my unchosen nature, evolved to serve my ancestors in life conditions far removed from those that govern my life today. The story of Adam and Eve reminds me of this.

I don't merely believe that I am saved by grace, through faith, and that someday I'll go to heaven. I know that every time I have been enslaved then freed, estranged then reconciled, lost then returned home, it was a gift of God that gave me a peace beyond description. The early Christian scriptures remind me of this.

I don't merely believe in the Resurrection. I know that for billions of years, chaos, death, and destruction have catalyzed new life, new opportunities, and new possibilities. I know, both from my own life and from Earth's history, that Good Fridays are consistently followed by Easter Sundays. The story of Christ's death and resurrection reminds me of this.

I don't merely believe that someday Christ will return and I'll fly away with him, I know that wherever trust, authenticity, responsibility, and service reign supreme, Christ has already returned. And as long as I remain in deep integrity and grow in these qualities, I experience, right here and now, rapturous joy. The theological promise of the Rapture reminds me of this.

"The blessings of religion do not require departures from factual reality."
                                        —DAVID SLOAN WILSON

The practice of evolutionarily REALizing core doctrines of religious faith, as demonstrated in this chapter, is no mere conservatism—nor is it liberalism. Evolution seen through "God-glorifying, Christ-edifying, scripture-honoring" lenses transcends the glorious diversity of any faith. This effort sanctifies science, REALizes religion, and shows that our way into the future, "God's will," is obvious and universal.

PART IV

Evolutionary
Spirituality

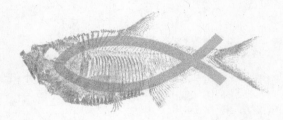

"Our evolution has been an awesome journey of fourteen
billion years. Every entity that ever moved or swam or crawled
or flew, every being that lived to reproduce itself, all the vast
numbers of species now extinct and presently living who have
invented the amazing capability which we have inherited as our
eyes, our ears, our organs, our very atoms, molecules and cells—
all of those preceding us are represented in our emergence now.
We bow down in awe and gratitude for the past. Without all
that came before us, none of us would be awakening now!"
—BARBARA MARX HUBBARD

PART IV

CHAPTER 11

Evolutionary
Integrity Practices

"The most extraordinary fact about public awareness of evolution is not that 50 percent don't believe it but that nearly 100 percent haven't connected it to anything of importance in their lives. The reason we believe so firmly in the physical sciences is not because they are better documented than evolution but because they are so essential to our everyday lives. We can't build bridges, drive cars, or fly airplanes without them. In my opinion, evolutionary theory will prove just as essential to our welfare and we will wonder in retrospect how we lived in ignorance for so long." —DAVID SLOAN WILSON

I concluded the previous chapter with the claim that a sacred, meaningful view of evolution sanctifies science, REALizes religion, and shows that our way into the future, God's will, is obvious and universal. What *is* this way—perhaps our only way into a glorious future? Christ-centeredness! Evolutionary integrity!

Once we accept that God's Word is not confined to ancient texts, and that God is still speaking through the public revelations of science, how can we personally tap into that guidance for living joyful and on-purpose lives? What tools can help us apply the wisdom gleaned from billions of years of cosmic and biological evolution, millions of years of primate and hominid evolution, thousands of years of human cultural evolution, and hundreds of years of scientific, technological, and methodological evolution?

For those of us who celebrate evolution from a Christian perspective, how can we live the Gospel and grow "in Christ"? What practices can help us walk more gracefully the path of deep, evolutionary integrity, so that our actions will be a blessing now and for generations to come? For those who celebrate evolution from a non-Christian or nonreligious perspective, how can one enjoy the fruit of humanity's collective wisdom for leading happy and fulfilling lives?

REALizing "the Centrality of the Cross"

Any doctrine, in any faith tradition, can be understood in at least two ways. Traditionally, theological concepts such as "heaven," "hell," "the rapture," "the second coming of Christ," and so forth, have been thought of in abstract, imaginary, otherworldly ways—as unnatural places, events, or experiences. That's how I, too, interpreted these doctrines for many years. Today, however, I interpret all such theological concepts from a sacred evolutionary perspective—as religious symbols that point to what is concrete, universal, and undeniably real. For example, "the centrality of the cross" is often taken to mean that only Christians who believe that God's Son suffered and died on the cross for their sins will ascend to a place somewhere outside the universe called heaven. Everyone else will be tortured forever in hell. This is an abstract, imaginary, otherworldly interpretation. From a deep-time perspective, however, this same doctrine, the centrality of the cross, can be seen as universal, as experiential, and as undeniably real.

Today, I see the cross as, indeed, a symbol of our only path to salvation (rightness with God), individually and collectively: the path of vertical and horizontal integrity. *Vertical integrity* is getting complete with the past and being responsible for the future. It includes honoring the contributions of my human and nonhuman ancestors and the divine earthly processes that made my life and this moment possible. *Horizontal integrity* is being in right relationship with my nested world. It means carrying no secrets, holding no resentments, having no unfinished business, and doing what I can to leave a positive legacy. The cross also, importantly, reminds me to trust that pain, suffering, difficulty, and even death are never the end of the story.

So now, whenever I see a cross, with or without Jesus on it, I feel deepest gratitude for this holy reminder of my way home, my sure road to heaven, the way of Christ.

There is no path to integrity; integrity is the path!

The following exercises will help you grow in deep integrity. They are universal spiritual practices. If I can reference a particular source, I will; but most are now just part of the commons. Each of these practices has evolved in the selective environment of my personal life choices and challenges. Thus my own quirky signature will be evident. I am sure that any practices you adopt will continue to evolve through you.

Regardless of your religious beliefs, philosophical orientation, age, gender, sexual orientation, political views, or anything else, if you practice these exercises you will experience transformation, growth, and a joy that can only come from getting right with Reality (God) and remaining in such right relationship.

Taming Our Monkey Mind

How can we be guided by the divine within and around us, rather than distracted by our ceaselessly chattering Monkey Mind? How can we develop a habit of periodically silencing the thinking, interpreting, judging part of our brain so as to pay attention to the still small voice of the Holy One to whom we ultimately belong? The following two exercises are elegant means for quieting the rational, verbal part of the brain and thereby facilitating a state of joy and peace. They will help you develop the habit of distinguishing thinking from noticing. It doesn't require years of dedicated practice to begin to achieve stunning results. Even if you have never meditated before, these exercises will help you develop a habit of attending to what's real within and around you at any given moment.

PRACTICE: Noticing Two or More Sensory Stimuli at the Same Time

John Selby, in his acclaimed meditation guide, *Seven Masters, One Path*, speaks of a simple and effective practice used in retreat centers across the world, as well as in hospitals and research centers where the healthful aspects of meditation are studied. The practice requires no more than this: notice two or more sensory stimuli at the same time and maintain that attention. For example, you might choose to sit comfortably and close your eyes, and then begin to notice your breathing, slowly, gently—in…out…in…out. After a few cycles of this, and before your mind begins to wander, also notice and pay attention to any sounds you can hear. Remarkably, if you pay attention to both breath and sound at the same time, there is no room in your conscious awareness to think about anything else. I invite you to close your eyes and sample this technique before reading on. It is very soothing.

You can repeat this exercise with eyes open, even while walking. You just need to soften your gaze so that you see everything ahead of you—broadly, with no focus on anything in particular. Notice how much you can see without shifting your eyes or moving your head. Now also pay attention to anything you can feel or smell or hear. It really doesn't matter what the two sensory stimuli are, so long as you focus your attention simultaneously on both. Effortlessly, the verbal, rational, assessing part of the brain is calmed. Monkey Mind relaxes, and our awareness fills with the eternal present moment.

PRACTICE: Speaking in Tongues

Speaking in tongues has been a significant part of my spiritual practice for half my life. Speaking in tongues has its detractors, but there are sound evolutionary reasons for its effectiveness. The following practice will REALize the act of speaking in tongues, because it doesn't require you to believe anything. It's an experience available to anyone who tries it.

How I speak in tongues is simple. I pretend I can speak a foreign language; vocalizing nonsensical sounds in a gentle, melodic, or rhythmic way. I encourage you to try it, right now. Do it in whatever way comes naturally, for a few minutes or longer, until it becomes effortless. Now speak in tongues again, but this time inaudibly, though perhaps still moving your lips. Then continue this "speech" without moving your lips; have it happen just internally. Whichever form suits you best, you should notice almost immediately that your awareness expands. You are more aware of what you see and hear and feel—without trying.

Just as a person who speaks a foreign language can also think in that language, if you can speak in gobbledygook, you also can think in gobbledygook. Because you cannot think in made-up syllables and in English at the same time, this practice effectively silences the verbal part of your brain. It gives your Monkey Mind a banana to chew on. Speaking in tongues (outwardly or internally) makes it easy to attend to *noticing* what's real in the present moment, rather than falling back into distraction. It's no coincidence that many report feelings of ecstasy when speaking or thinking in tongues.

When speaking in tongues first came to me a few months after my born-again experience, it truly was *baptism in the Holy Spirit,* as my Pentecostal Christian tradition had taught me. "Baptism in the Holy Spirit" is a resonant way to describe this experience using night language. Speaking in tongues is immersion in the holiness of this moment, this time and place. I often do it intentionally, to quiet my mind while driving, for example. Or it may arise on its own, especially when I am overcome with gratitude or overwhelmed by beauty. On such occasions, emotions take control of my body: arms lift skyward and I babble away in gentle ecstasy.

"Who Wants to Be Filled with the Holy Ghost?"

I had my first experience of what Pentecostals refer to as "baptism in the Holy Spirit" while at an Assemblies of God retreat in the German Alps in the late 1970s. The minister concluded the worship service with an invitation for those who wished to be "filled with the Holy Ghost" to come forward. I knelt at the altar, and with closed eyes stretched my arms heavenward. Soon I felt a warm hand on my shoulder and the comfort of hearing someone praying that I would open up and allow the Holy Spirit to speak through me in wordless syllables. It was then that I began "praising God in other tongues," as I had witnessed so many others do in Pentecostal church services.

Almost immediately, I experienced an ecstasy beyond thought or belief. It was the pure experience of divine presence, unmediated by words. I felt I was back with Jesus' disciples, in the upper room depicted in Acts 2:1-4.

> When the day of Pentecost came, they were all together in one place. And suddenly from heaven there came a sound like the rush of a violent wind, and it filled the entire house where they were sitting. They saw what seemed to be tongues of fire that separated and came to rest on each of them. All of them were filled with the Holy Spirit and began to speak in other tongues as the Spirit enabled them.

While there may be documented cases of people "speaking in other tongues" who were actually speaking in a language that they had not yet learned (e.g., Acts 2:8), for most Pentecostals the experience is an incoherent babble—*as if* they were speaking a foreign language. The emotional, psychological, and spiritual benefits are the same either way. When I speak in tongues or quietly think to myself in tongues, even for a few moments, I usually feel a connection to God and to everyone and everything around me—a connection that is difficult, if not impossible, to experience when my Monkey Mind is doing its thing. My conscious mind is released from the bondage of words.

Speaking in tongues helps me give voice to emotions too difficult to express any other way. I thus often pray in tongues. Early on in our relationship, Connie and I occasionally relied on this gift of the Spirit during difficult times. I could express my anger, frustration, or disappointment to her, and she could hear it and respond similarly, and neither one of us had to deal with the aftermath of cleaning up hurtful words or compounding the problem by misstatements or misinterpretations.

Recently, I have begun to rely on the gift of tongues not only for emotional expression in times of great feeling, or while in prayer. I now regularly think in tongues simply to still the otherwise constant conversation in my head, quieting the jabber of opinions and insistent trivialities that otherwise isolate me from the presence of the Holy Spirit. Quietly speaking and thinking in tongues, at will, has thus become my preferred form of meditation. The Great Story helps me understand how this gift of tongues is both a natural outgrowth of the human developmental journey (day language) and a gift of the Holy Spirit (night language). The Great Story thus helps me receive the blessings of an ancient spiritual practice, while living fully in our contemporary world.

Taming Our Lizard Legacy

Given the pull of instincts, I have found it helpful to write out and share with my STAR cluster, my deep integrity group, a three-fold list: *my integrity circles*. These identify which behaviors are to be avoided and which are to be encouraged. Each circle is represented in two ways. First, the colors of a traffic signal: *Stop! Try to stop* (or proceed with caution if I can't). And *Go!* The second label is a spatial referent: all the good and nurturing actions and habits in my outer circle are intended to wrap around and contain, or hold in check, the circles within.

RED / INNER CIRCLE: *Behaviors that qualify as a violation of my integrity.* Doing anything in my red circle means I must reset my "date of evo-integrity" (DEI). My commitment is to say "No!" to everything listed in my red circle of behaviors, and to keep doing so one day at a time.

YELLOW / INNER CIRCLE: *Behaviors that may set me on a dangerous path.* Acting in such ways do not constitute violations of my integrity, but I must proceed with caution because they may lead to something that clearly is. These behaviors I liken to a baseball field warning track. If I am an outfielder running back to catch a long fly ball, when I detect a change in footing, I know that pain and possible injury could result if I keep going and hit the wall. My commitment is to report any yellow circle occurrence to someone in my STAR cluster or accountability team within 48 hours.

GREEN / OUTER CIRCLE: *Behaviors that promote my integrity and bless me and those around me.* These are actions that deepen my walk with God and enrich my relationships and bliss-quotient in life. They are to be encouraged, enjoyed, celebrated, and practiced religiously. Sustained deep integrity depends on my spending as much time as possible in this domain. Within this realm is my salvation, transformation, victory over temptation, and success in life. It is here, too, that I nurture my Higher Porpoise (higher purpose) and cultivate a positive evolutionary legacy.

PRACTICE: Integrity Circles

Create your own tripartite list of Integrity Circles. Share it with someone you trust. Consider it a spiritual practice to regularly reread your list and revise it, if necessary. Whether or not you are challenged by addictions or codependence, everyone can benefit from taking the time to envision, and then commit to pursuing, Green Circle activities.

Growing in Trust: Nurturing Humility and Faith

In the biological world, nothing contributes more effectively to healthy evolution than does feedback. Feedback—learning how one's choices and actions affect others—is the way that organisms survive and thrive and the way that species (albeit, unconsciously) adapt and evolve. In the human world, unfortunately, very few of us have had decent models or instruction in how to compassionately offer and graciously receive feedback. Yet nothing will more effectively help us grow in humility and further our own evolution and the evolution of those with whom we are in relationship than soliciting honest feedback.

I've used a variety of means for soliciting feedback over the years, but none more elegant than the practice described by Jack Canfield in his 2005 book, *The Success Principles*, as I summarize that practice here:

PRACTICE: Soliciting Feedback

To succeed in the world and to foster thriving relationships, one needs to secure an accurate picture of how others perceive the effects of our actions. The boldest way to do this is to outright ask those so affected—but to do so in a way that feels safe for both parties. Jack Canfield, in *The Success Principles*, suggests a two-step method for soliciting feedback—from spouses, children, friends, students, clients, indeed anyone with whom we are in ongoing relationship. Step 1 is to ask,

On a scale of 1 to 10, how would you rate the quality of our relationship?

Now here is the grace of this exercise. Because giving and receiving criticism can be uncomfortable, for any response that is less than a 10, Canfield suggests this follow-up question:

What would it take to make it a 10?

Notice that you are not asking for criticism per se. Rather, you are looking for positive suggestions for improvement. By no means will you challenge or debate your respondent's assessments. Rather, you are looking for feedback and suggestions—and you can think about their validity at a later time. Right now, the only thing you are doing is seeking out the data.

Pose this set of questions on a regular basis to all the important people in your life. Eventually, your respondents will begin to trust that they can be honest with you—that they can say what they really feel without your reacting harshly, incredulously, or in some way requiring them to justify their responses. Thanks to this exercise, you will have the information you need to keep improving, all the while strengthening your humility muscles.

REALIZING "Love Your Enemies"

Few things impact our experience of the world and the quality of our life more directly and profoundly than our habits of meaning-making and self-talk. Why? Because none of us experiences the world as it is. We experience the world through the filter of our interpretations. The meaning we make of an event, the story we tell ourselves about it, is generally far more consequential over the long run than the experience itself. The good news is that habits of self-talk and meaning-making are just that: habits. They are not hardwired. They can be changed.

Nothing is more corrosive to our own well-being and relationships than blaming others. This is true for everyone. Arguably, Jesus' most important admonition was to love your enemies as yourself. The following two self-talk exercises use the insights of evolution to help us fulfill this directive. They will assist you in letting go of blame, while growing in faith, trust, and humility.

PRACTICE: Reframing the Past

Pick a mildly challenging memory (you can work up to a major trauma later). It can be something that happened long ago or that occurred recently. Now imagine, and journal if you care to, three very different ways of interpreting this event mythically (be playful and dramatic, exaggerate, make it big and obviously silly):

First interpret this experience from the stance of a victim, as if you really had nothing to do with it—it was someone else's fault entirely. Make sure that you describe the experience as much worse than you have ever described it before (to yourself or others). Get into the gory details; don't by-pass anything. When you think you have finished, notice the judgmental and victimlike thoughts and feelings that naturally arise. Notice how your chest and belly and neck and extremities feel, and your breathing too, when you interpret an event as though anyone but you is to blame—someone else is obviously bad, wrong, stupid, or malicious. Maybe even move around the room feeling this way. Encourage any shoulder-slumping victim to transform into the persona of vengeful victim. You may need to begin by willing angry movements, and then perhaps they will begin to come on their own, creating their own little dance of vengeance. Yes, indulge in it! Encourage your Furry Li'l Mammal to feel as angry and self-righteous as it can, all the while encouraging your Lizard Legacy (your innate bodily control) to be influenced by the rush of hormones. "I hate my enemy! *Grrrrrrrrr!*"

Before proceeding to the next interpretive stance, you will need to shake off the emotion and bodily sensations wrapped up in the victim stance. So flop your arms aimlessly and roll your head. Whoosh out some breath again, and again. Perhaps do some aerobic dance or exercise. Sing with abandon! Do something, anything, to free your mind and body from the grip of self-righteous blame.

Next, take that same mildly challenging memory, but now interpret it as a victor, as if this event were exactly what you needed to help usher you into greatness. Imagine that it was no one's fault; it just happened. Allow yourself to feel compassion for yourself and whoever else was involved. What might have been going on inside each of you before and after this experience? Allow yourself to feel tenderhearted and even grateful for this event and what you've learned from it. How interesting that it turned out, in a strange but real way, as a blessing in disguise. How amazing that you could not have seen this before! Stay with this mythic possibility in your imagination and notice what feelings and thoughts naturally arise. Notice how your chest and stomach and neck and extremities feel, and your breathing too, when you interpret an event as though it was an experience of God (Reality) conspiring on your behalf—though you can hardly be blamed for not having recognized this great good fortune in the moment. Again, if you are moved to get up and move: *move!* Move like a victor, like a champion! Move with gratitude and appreciation and confidence. Move with the easy flow of one who is entirely right with the world, who can accomplish anything! Sing, shout. Proclaim the great news! "I am on my way!"

Okay, use your breath and move your body again in ways that will release the effects of this emotional state. Return to neutral. Take your time.

Finally, take the same memory and interpret it comically, as though this event had been orchestrated by God, or the gods, as a practical joke—and you and everybody else involved were mere pawns. Imagine that this episode exposes something ironic or funny or just plain ridiculous about human nature or the nature of the Universe. Thinking back on the experience from this perspective, you now can't help but smile, even laugh. Allow yourself to feel lighthearted about the episode, and appreciative of what it taught you about yourself, about others, about life. Notice what thoughts and feelings naturally bubble up when you imagine this event, without judgment, as if a cosmic trickster or comedian were just having a good time at everyone's expense. Notice how your chest, belly, neck, and extremities feel, and what your breathing is like, when you interpret an event as though it was obviously an experience of life's absurdity or goofiness. Do move around on this one: allow the divine comedy to penetrate your being. Begin by intentionally moving in comical, slapstick ways, and soon they may generate on their own. Intentionally erupt in a hearty laugh—and see if it carries on its own. If you can, keep laughing until you are laughing uncontrollably—until your sides hurt or the tears are flowing.

Okay...shake it out. Rest. Perhaps on another day, revisit this threefold exercise, but take on a more challenging memory. Patiently build up your self-talk/interpretation muscles before you attempt to work with the really biggest hurts of your past. Perhaps enlist a friend to undertake this exercise with you. Coach each other as if you were movie directors coaching actors. If you really take this on, I promise, it may be one of the most spiritually transformative exercises you'll ever do.

The goal here is not just to know, but to actually experience that you can use your imaginative Monkey Mind to fashion an interpretation of your choice of any memory. More, you can do this well enough so that any emotion you choose will spontaneously emerge from the depths of your Furry Li'l Mammal and then continue to run on its own. Experience how you can direct the Lizard Legacy control over body movement well enough that the movement itself will feed back on the feelings of your Furry Li'l Mammal, amplifying them such that the body moves even more. Know, too, that it is your higher purpose of self-healing that makes any of this possible. It's your higher purpose that directs your Monkey Mind to weave an interpretation couched as victim, victor, or pawn such that your Furry Li'l Mammal will be lured into believing it is true, thus generating the intended emotion. It is your Higher Porpoise that calls your Lizard Legacy into play and then steps back when the fur and scales start flying, as emotional mammal and embodied lizard drive one another into a frenzy through positive feedback.

Here is the self-healing: You will experience the lovely rush of pent-up emotions, judgment, and guilt finally getting *cleared*. You will know that you do in fact have choice over how the actual events in your life are interpreted in your life story—what meaning you make of them. You will understand that your life story is not something that exists independently of your interpretations, and you will know that you have choice regarding how you interpret your past and present. No memory need be repressed in this way; just fully and artfully reinterpreted and reinvested with healing emotions, although several rounds over a period of weeks or months may be necessary to recolor your most painful memories. You may also notice other subtle or not so subtle changes. For example,

+ You may have less resentment about your past, and more forgiveness and trust than ever before. If so, you will have discovered that forgiveness is a selfish act.

+ You may feel more compassion for those who are still caught up in blame and who have not yet had the opportunity to see for themselves that there is another way.

+ You may begin to habitually interpret events in the present moment in a less victimlike and more empowered way—or not. There is a lot to be said for living spontaneously, fully feeling whatever emotions naturally occur. Then, when it is time to reflect (that is, when your Furry Li'l Mammal is no longer scampering on high alert), do this tripartite exercise again—victim, victor, pawn—and feel freshly empowered to clean up whatever mess you might have contributed to.

PRACTICE: Looking for Opportunity

Now that you have a way of moving beyond our instinctual tendency to blame others when life's difficulties inevitably arise, develop the habit of asking,

What is the opportunity here? What is possible now that wasn't before? How might this (event/experience/emotion) be a gift and blessing in disguise?

If your emotional state would make that task about as difficult as climbing Mount Everest, then try on this junior varsity set of questions:

When did I feel this way before? Was it possible for me to envision a good outcome at that time? If not, did things somehow work out anyway? Can I simply trust that the same might be unfolding right now?

Just asking ourselves questions like these can help us relax into life and trust that (no matter what we're dealing with and though we may not be able to envision how in the moment), there *will* be a solution. There *will* be a time when life resumes a smooth course, and perhaps something amazing will be gained along the way.

I expect miracles to occur when variations on the two preceding practices are used in community conversations during crises or to resolve institutional or societal conflicts. *"What's possible now that wasn't before?"* and *"What are the most inspiring and empowering ways we can think of to interpret this issue or event?"* I can almost hear a new world cracking through its shell as I imagine hundreds, thousands, even millions of people asking these questions together, and then acting on what God reveals.

Growing in Authenticity:
Realizing "Remove the Plank"

Removing the plank in one's own eye before attempting to remove the splinter in another's isn't easy when you can't see the plank. Virtually all of us have an innate propensity to shade the truth, to engage in deception big or small, especially when we're afraid that otherwise we would be judged harshly. And it gets worse. Evolution has quite effectively selected for skill in *self-deception*, too. We stand the best chance of deceiving others to our benefit if we first deceive ourselves.

Given our natural instincts, then, it takes courage and support from others to consistently grow in honesty. Twelve-step programs have developed approaches for taking a hard look at oneself. Following are three exercises that will help any of us "remove the plank" and thus grow in authenticity. (The first two are distilled from the 12-step approach, as well as other empowerment methods.)

PRACTICE: Inventory Your Character (Step #4 in 12-step programs)

Write out your strengths and growing edges. Be as thorough as possible. List what you, and others, would consider your assets *and* your liabilities, your positive characteristics and your challenges. Also inventory your impact on all the smaller holons for which you are responsible and the larger holons of which you are a part. Extra credit: Ask your kids and your partner (or ex-partners!) what you've left out.

PRACTICE: Come Clean (Step #5 in 12-step programs)

Read your inventory aloud to someone you trust. Most important: tell this person every embarrassing, shameful, arrogant, hateful, self-centered, harmful thing you've ever done. Confess everything. Don't hold anything back. When something comes to mind after you've completed your recitation (it surely will), then tell that too. Also report all your self-righteous judgments and resentments. Be thorough and fearless in this.

Why is this practice so life changing? When you're keeping just one secret, or nursing just one resentment, it's very easy to keep more. When you're holding on to no secrets or resentments whatsoever, it's relatively easy to remain in integrity.

We all carry around rich, smelly compost from the past that impairs our attitudes and actions today. Most of us also suffer, unknowingly, from stingy judgments and smoldering embers of resentment. Yet nothing so consistently robs us of joy as unexpressed resentments. Living free of guilt, shame, and judgment is truly heavenly, and hugely empowering.

PRACTICE: Light-Hearted Integrity Checks

When you catch yourself, after the fact, as having lied, or when you notice that you just made yourself right by making someone else wrong, tell on yourself in a playful way. Practice light-hearted integrity checks. I do this with Connie all the time. It's humbling, for sure, but it also builds courage, honesty, and intimacy like nothing else. Those on the receiving end will likely give you a doggie bone for your courage and honesty—and perhaps a kiss!

Growing in Compassion/Responsibility: REALizing "Judge Not"

Another message from Jesus that is universally applicable is this: "Don't judge, and you won't be judged; don't condemn, and you won't be condemned; forgive, and you will be forgiven; give, and it will be given to you." I can think of few better ways of articulating the evolutionary imperative to grow our circles of compassion and responsibility.

We naturally hold in high esteem those who put themselves in the shoes and experience of others and who then speak and act from that place. Similarly, we admire those who refuse to play the blame game, those who don't look elsewhere when things go wrong but always assume there's something they could have said or done differently that would have made a difference. We feel safe around compassionate people; they nurture trust and evoke in the rest of us a desire to be more like them.

Growing in compassion and responsibility, we find it easier to connect with others in loving ways. Because we have compassion for ourselves, too, we no longer need to play the victim. We become serene and whole, spontaneously free, and we are held in Grace. And here is where evolutionary psychology can be of great help. Of course we find it natural to blame others! Of course we shirk from taking responsibility! Of course we want justice: "I will admit to my share of the problem if only *they* will admit to theirs!" There is nothing wrong with us in having those tendencies; they are part of our evolutionary heritage and surely served our ancestors. So let's lighten up about it and get on today with what works.

PRACTICE: Taking Responsibility for Your Wake
(Steps #8 and #9 in 12-step programs)

First, accept that you've left a wake, for good or ill. You have said or done things and not said or done other things, and have interpreted events, in ways that have left their mark. As well, others have interpreted what you've said or done, or not said or done, in ways that now cause them to harbor good or bad feelings toward you, accurately or inaccurately. Accepting this fact as normal and natural is the foundation for taking full responsibility for your life and your effect on others.

Second, make a list of those you think you may have harmed and those you know you have harmed, and also a list of those who may not think well of you for whatever reason.

Third, select one person at a time on your list and imagine generously what you could have said or done differently for a better outcome. Write it out. (If you have difficulty with this, simply imagine what someone you admire might have said or done in your shoes). Think and feel from inside the offended person's experience. We're not talking about the truth here; we're talking about their perception—remembering that it is oh-so-natural to view ourselves as victims and others as transgressors. Imagine what they would say if they were asked about you.

Fourth, now think about what you could say or write to that person that would make a real, transformative difference for them. This is not about justifying your actions. It is only about letting the aggrieved party know that you understand their response, and that if you could go back to the situation you would do things differently. Do get support from someone you trust, someone who's been through this process, to ensure that your intended communication is as generous as possible, free of stinginess and blame. Don't go it alone! Role-play your intended communication and have your friend give you feedback. Or write a draft letter of apology, and have a friend review it. We will naturally be stingy rather than generous, so this will require effort on your part. We tend to apologize and then follow up with a ruinous "But I was only trying to..." clause. Not helpful!

Fifth, now go ahead and do it. Make your apology (in writing or in real time) in a responsible, humble way. You may find it helpful to include something like, "If I could go back and do it all over, I would have..." and describe yourself saying or doing something that would have left a positive wake. This process, of course, requires a great deal of intention. Someone whom you have harmed may find it painful to be confronted with your apology, as it forces them to revisit a memory they may have long ago tucked away. This is not the time to ask for forgiveness. Asking for forgiveness is not only uncharitable; it could be counterproductive. If pressed to give you a response, emotions may compel them to respond hastily, in a way they later feel bad about. Then they are stuck with not only a painful memory but a new memory that disturbs them. Meanwhile, hey, you've come clean; one more name crossed off your list, and you feel great! Again, the key here is to take full responsibility, standing in their shoes, imagining what it must have been like for them, and recounting your offense so thoroughly from their perspective that they feel you really do understand the magnitude of their hurt and why they hurt—and that you completely understand if they cannot forgive you. This may sound like a lot of work, and it is. Why bother? Because it's a stairway to heaven!

"Magic in Any Relationship"

In 1997 I enrolled in a transformational education program called The Landmark Forum. The Forum Leader suggested something toward the end of the training that made a lasting impression on me. In the context of discussing relationships where love or warmth used to be present but now is missing, he said, "If you want to see magic occur in any estranged relationship, do the following. Give the other person full credit for everything that did or does work in the relationship. You take full responsibility for everything that didn't or doesn't work in the relationship. Then just shut up." I took his coaching, and he was right. I've never seen this approach fail to produce magic. Typically, it is life changing on both sides. The simple act of my coming forward and taking responsibility often prompts the other person, no matter how wronged they may feel, to also acknowledge some degree of culpability—thus not only freeing them of resentment but also giving them a chance to feel good about their own generosity. One good deed calls forth another.

Growing in Gratitude:
REALizing "Love God and Your Neighbor"

Volumes have been written on the importance, even the necessity, of nurturing an attitude of gratitude if one wishes to grow spiritually. It's impossible to love God without gratitude. Gratitude strengthens trust and expands compassion. As Meister Eckhart put it, "If the only prayer you say in your whole life is 'thank you,' that would suffice."

Gratitude manifests in the midst of everyday life when we pause to take account of how much we ourselves have been given. We are present to the wonder of the simplest gifts: a glass of water, a bite of food, a breath of fresh air, the scent of a flower, the touch or kiss of a loved one. At such times, our hearts are full. Thomas Berry has movingly written that, for humans, it is ultimately our role, our calling, to become "celebrants" of the Great Story. Affirmations of gratitude we speak as individuals in our own reflective moments are one form of celebration. So too are our comings together in community to celebrate a holiday (holy day), a life passage, or the memory of a moment of transformation in the immense journey of life. Celebrating life—as it is, not as we wish it would be—is an essential part of deep integrity. In fact, singing and dancing may be one of the more important things you can do to help usher in the Reign of Reality (Kingdom of God), if for no other reason than that such practices will transform *you* and those around you.

Many years ago I watched a videotape of evangelical teacher Winkie Pratney speaking on the importance of expressing gratitude to the people in our lives who deserve it. As I soon discovered, virtually everyone deserves it and benefits enormously from being on the receiving end of authentically expressed thanksgiving. Pratney told a powerful story that I have been reciting to audiences ever since. I remember it this way:

A woman had worked at a newspaper for more than a dozen years. One Sunday morning, her pastor spoke on the subject of thanksgiving. After church, the woman felt led to express gratitude to her boss—a gruff, curmudgeonly fellow. The next day she walked into his office and said, "You know, it dawned on me yesterday that I've never told you how much I appreciate you for being my boss. You're hardworking, fair, you pay me well, and I really enjoy my job." She then went on to thank him for particular things he had said or done over the years. When she finished, he replied incredulously, "Is that it? You came in here just to thank me? You're not buttering me up for something?" "No," she said with a smile, "I just felt you deserved to hear how grateful I am for my job and for the fact that you're my boss. That's all." She turned and walked out.

Ten minutes later he came to her cubicle and asked if he could see her back in his office. He shut the door and motioned for her to sit. Voice wavering, he began, "As you know, I've been the editor here for forty years. In all that time, no one has ever thanked me like you just did." He fell silent, clearly fighting back his emotions. After a few seconds, he said, "You just validated what I've been doing here for four decades." His eyes moistened and he could say no more.

When I heard this story, I thought to myself, "Who do I owe a debt of gratitude to?" My father immediately came to mind. I sat down and wrote my dad a sixteen-page letter. I started with the earliest memory I could recall and went through my entire life, thanking him for all sorts of things—some general, some specific. "Remember when..." and I would mention a particular memory. "I never thanked you for..." and I would write whatever my heart led me to say. What I was unprepared for was the impact that writing this letter would have on *me*. It was bathed in my tears. I soon learned it had the same effect on my dad.

Three years later, I had a conversation about gratitude with the business administrator of the seminary I was then attending. I retold the story of the woman and her boss, as background for telling about the letter I wrote to my father—and how that letter transformed our relationship. Three days later, as I was entering the building, the administrator called out to me and asked that I come into her office. As she shut the door and invited me to take a seat, I was feeling nervous. "What did I do?" I wondered. She began, "Remember the stories you told me the other day? Well, yesterday afternoon I wrote my husband a letter of gratitude and sent it to his office. It turned out to be twelve pages long, typewritten single-spaced." "Wow!" I replied. She continued, "We've been married for 47 years. Once I started writing I discovered there was lots I was thankful for." She paused, and then, "I just got off the phone with him a few minutes ago. It's been over 25 years since I've heard my husband cry. Thank you for telling me those stories!"

PRACTICE: Letters of Gratitude

Who is or was important in your life? Especially, who is (or was) important in your life who may not yet know how grateful you are for the difference they made? These people need not have been entirely helpful or nurturing, but the point here is to call up only the positive memories. As a spiritual practice, take one person at a time and write a letter of gratitude. In what ways did their influence make you a better person or set you on a path that has become central to your success and well-being? Be specific and speak from your heart. Focus on the good, only the good. Help them see that their contribution is much larger than they could have imagined, because it has carried into the world through you.

PRACTICE: Cultivating Generosity

✦ Habitually acknowledge the contributions of others, even small ones.

✦ Practice uncommon appreciation on a daily basis, and make it a game: How many people, or how many times, can I authentically appreciate or acknowledge someone today?

✦ Always exceed expectations; give or do more than is customary.

✦ When you are irritated by what someone says or does, develop a habit of making up a reason for their action that will evaporate your judgment or anger. Play with this; the reason you imagine can be goofy and utterly fantastic. What you'll find is that you will quickly build a reputation as a generous person who believes the best in others.

If you take time to cultivate gratitude and generosity, you will notice a greater desire to be of loving service. Following are two effective ways to bless others and bless the world.

PRACTICE: Discerning Your Calling

The "Thank God for Our Higher Porpoise!" section of Chapter 10 contains an exercise I've offered audiences across the theological spectrum, young and old, for more than a decade. I encourage you to do this exercise and discover your life purpose, your mission—where your great joy and the world's great needs intersect. Virtually everyone who invests the time to complete this exercise finds it to be significant, and for some it is life changing. It is one of the most effective tools I know for helping a person clarify what "God's will" is for them in outward service. This exercise isn't just for adults. There may be no questions more important to regularly ask children (at least through college) than these: "What is your cosmic task? What is your evolutionary role, your divine purpose at this time in your life?"

"That's Your Cosmic Task!"

More than a half century ago, Maria Montessori encouraged teachers to help children think expansively about their lives by gently urging them, at every opportunity, to ponder what their "cosmic task" might be. What is it that they, given their innate gifts and unique way of being in the world, might contribute in service to their larger communities—even to the Universe as a whole? While guest teaching in a Montessori classroom in Minnesota, Connie was approached by a boy intent on showing her drawings he had made of all the dinosaurs he knew. One by one he turned the pages of his self-made picture book. When he was done, Connie looked him square in the eyes and said, "For 65 million years Earth had lost the memory of dinosaurs, and now, through you, Earth is once again remembering its glorious past. Good work!" She continued, "Right now, drawing pictures of dinosaurs: that's your cosmic task!"

One way to love God and neighbor, now and into the future, is by mentoring. For virtually all of human history this was how skills, values, and knowledge passed from one generation to the next. One-to-one instruction by way of example and coparticipation is still, by far, the most effective way to ensure this transmittal. Mentors can praise in just the right ways and the right times to foster a sense of possibility and self-worth in the apprentice. There are organizations devoted to pairing up mentors with mentees: Big Brothers and Big Sisters are examples. Virtually all religious communities rely on adult mentors to work with children and youth in religious education and coming-of-age programs. And there are plenty of opportunities in public and private school systems, too.

PRACTICE: Mentoring

You surely already know enough to have something to offer someone new to your path (or fresh to life). Make an inventory of your talents, your experiences, your character traits that would be beneficial to pass on to others. What would you be thrilled to teach to another or to model for another? Now, call to mind all the communities in which you are active—and those you might like to be active in. Then, go for it!

"We are truly marvelous, adaptable creatures, products of an exciting and inspiring—even though often dangerous—evolutionary story, a story to be celebrated with conviction and enthusiasm even as we move on to new challenges."
—PAUL R. LAWRENCE AND NITIN NOHRIA

CHAPTER 12

Evolving Our Most Intimate Relationships

How might couples and family members support one another in applying evolutionary integrity principles in these most intimate of relationships? For couples, *integrity is the single most powerful aphrodisiac on the planet.* Indeed, there's nothing else remotely like it. But please don't take my word for it: you deserve to experience this divine blessing for yourself. What follows are, in my experience, essential components of healthy romantic relationships and healthy family relations.

Touch and Tenderness

"Sexual activity, indeed the frenetic preoccupation with sex that characterizes Western culture, is in many cases not the expression of a sexual interest at all, but rather a search for the satisfaction of the need for contact." —ASHLEY MONTAGU

The Furry Li'l Mammal in each of us craves touch and tenderness. Without touch, a baby dies, the human heart aches, the soul withers. Touch is not only a biological need; it is a profoundly elegant and essential form of communication. As Phyllis Davis has said, "Touch is a language that can communicate more love in five seconds than five minutes of carefully chosen words."

Most of us can relate to the joyful experience of making up with a loved one after an argument, holding one another in sacred silence. Reconciling touch feels so good—the embrace, the kiss, the caress that communicates "I love you, and I'm glad to be one with you again." When love is expressed through touch and tenderness, our Furry Li'l Mammal feels great—and that means *we* feel great.

Touch and tenderness are far more important for couple and family health than most of us realize. Research across cultures has shown that we live longer and more peacefully when we are affectionately touched on a regular basis. There is no substitute for a heartfelt hug or a timely kiss. Such communication quenches a deep thirst.

For millions of years our mammalian ancestors were reassured by parents or comrades not through words but through touch. For 99.9 percent of our mammalian journey, there were no words. The need for touch begins for mammals at birth, and continues until we die. Infants need to be touched, cradled, and rocked in order for their nervous systems to develop properly, and for healthy emotional and psychological development. This is true for other animals as well. In her book *The Power of Touch*, Phyllis Davis notes,

> All mammalian young demonstrate the necessity of touch for healthy physical and behavioral development. Even baby rats prosper when handled and petted. When they are touched and held, they outweigh, outlearn, and outlive other rats.... Children from homes with loving, touching parents look and act differently from those who are rarely touched. Touched children feel better about themselves and are less hostile, more outgoing. Well-touched children almost seem to glow.

If we do not receive enough of the right kind of affection as children, the effects can be serious. Touch deprivation can cause mental and physical retardation—even death. As a society we would do well to provide emotional and economic support for mothers or wet nurses to breastfeed infants. Reporting on studies done over the last forty years, Davis notes, "Breastfed babies have fewer respiratory ailments, diarrhea, eczema, asthma, and other ailments than bottle-fed babies. Additionally, breastfed children tend to be physically and mentally superior in their development, and the longer they are breastfed, the more striking the advances."

Scientists who have studied other mammal species, and scientists who have charted our own long evolutionary journey, have compiled stunning empirical evidence that the need to be touched, licked, nuzzled, snuggled, and playfully tumbled is universal for mammalian young. Human infants, of course, need touch more than any other age group. But even older children and adolescents benefit from reassuring touches and hugs.

Beyond the need for physical affection, children also thrive when spoken to tenderly and respectfully. A child's spirit is easily bruised by harsh words, especially coming from a parent or other beloved adult. In such cases, it is important for the adult to apologize to the child at a time when the apology can genuinely be given and received. Another option that many families use with great success is to ask for a "Take Two"—that is, to be given a chance to resay or redo whatever had been said or done in a hurtful way—like what a movie director might call for if a scene needed to be redone. Forgiveness and reconciliation bind wounded spirits. More, prompt apologies and renewed expressions of unconditional love are the only way to prevent the impressionable young amygdalas within the Furry Li'l Mammals of our children from turning a molehill of a hurt into a mountain of fear or resentment—a mountain that may cast a shadow for the rest of their lives.

There is healing touch, too. Because tender touch communicates love and care, it triggers metabolic and chemical changes in the body that assist healing. Touching also stimulates the production of endorphins—natural body hormones that control pain and enhance our sense of well-being. This is why the sick and elderly should be massaged, held, caressed, or otherwise touched as often as they wish.

Respectful Communication

"Respect is the center of the circle of community."
—MANITONQUAT

It is impossible to be in the space of deep communion with another without respectful verbal and nonverbal communication. Respect is a basic human need. Remember that our Lizard Legacy requires more than a little assistance in adjusting to the demands of civilization, and our Furry Li'l Mammal never did learn to talk. More, our Monkey Mind is quite capable of conducting a seemingly cogent conversation while remaining oblivious to the grunts and cries of the frightened creature within us and within our partner. Such verbal exchanges can do more harm than good.

Respect means different things to different people. That's why it is often helpful to ask, "What does respect mean to you? If I were relating to you with deepest respect, how would you know it?" Whatever our individual differences, most people would agree that, at the very least, respect means listening with full attention, accepting that differences are to be expected, acknowledging feelings, and speaking our truth. It also means no accusations, no name-calling, no lecturing, and no sarcasm. Respectful communication is clear and responsible. It grants to the other the level of dignity and acceptance we want for ourselves.

The difficulty, of course, is that this kind of communication does not con-sistently happen instinctually or just because we want it to. More often, our speech in difficult circumstances reflects unconscious habits developed over many years of coping with life. Without even realizing it, we absorb the dys-functional patterns modeled for us by our families and peers. Disrespectful habits are not replaced by respectful habits overnight, nor easily. Transforma-tion requires practice in a safe context.

Over the years, my family and I have used a variety of tools to help us develop habits of communicating more respectfully, which include communi-cating more often and sooner when something starts to go wrong. One tool my wife emeritus, Alison, and I found particularly useful—not only with one another but also with our children—was to regularly remind ourselves of four "func-tional relationship agreements": Each of us has the right and the responsibility (1) to speak our truth, (2) to ask for what we want, (3) to say "no" or express displeasure without fear or shame, and (4) to establish personal boundaries that foster respectful communication by ourselves and others. The latter is particularly important, because one crucial condition for respectful commu-nication is timing. Ask, "Is this a good time for me to bring something up?"

For families who wish to communicate respectfully (and for Christian fam-ilies to "grow in Christ" together) few things, in my experience, are more important than regularly scheduled family meetings that use structures and processes that become familiar (and feel safe) to all. Here, adults and children alike can offer appreciation to one another, confess wrongdoings without fear of retribution, express hurt feelings, resolve conflicts, and plan ways to have fun. Practicing this kind of relating once a week, or even once a month, makes respectful communication more likely at other times as well. It also socializes children in the coevolutionary spirit of democratic participation, taking individual and collective responsibility for what is happening in their family.

"Can We Have a Heart-to-Heart Talk?"

The most important tool that Connie and I use for moving through difficulties and deepening our love is a Heart-to-Heart Talk^{SM} communication process that we learned from our dear friends, Paul and Layne Cutright, and which we adapted from their book, *Straight from the Heart*, available at paul andlayne.com/sfth.htm. Here's how we do it: We sit close, facing each other, and one of us (usually whoever requested the process) begins, "It's important for me to say…" and then speaking what's on their heart or mind—but in no more than a few sentences, one thought or feeling at a time. The other responds with, "Thank you," "Got it," or "I understand." These two-word responses are the cue that the first person may continue with another sentence or two or three of the same sort, always beginning, "It's important for me to say…" This formulaic dialogue continues until the first person feels complete; they have expressed everything they need to, at least for the moment. Then we reverse roles. Knowing that this role reversal always follows is the reason that the second person can patiently wait their turn, without feeling compelled to interrupt or otherwise comment or challenge what the first person is saying.

Now the second person gets to share until they in turn feel complete and heard, while the first person is the listener. The process continues this way, back and forth, until it comes to a natural end. Nothing more to say? Not quite! The process cannot work its magic without one more step. We conclude with a ritualistic round or two of *appreciations*. One of us begins, "I appreciate…" and the other replies, "Thank you," "Got it," or "I understand." The same person continues speaking until they've said all the appreciative things they are led to say—about the other person, the situation, the present moment...whatever. Then the roles are reversed. Crucially, appreciations cannot be forced; they must genuinely be felt. Connie's favorite way to start her segment of appreciations, especially when she's not yet ready to appreciate me authentically, is "I appreciate myself for..." Eventually she'll get around to appreciating me for something, and then for something else: one appreciation calls forth another.

It will be really obvious when this final phase is over, and hence when the Heart-to-Heart Talk[SM] officially has come to an end. Not uncommonly, one participant will become teary eyed with gratitude and love, which then spreads to the other; or someone reaches out and touches the other's hand or face. Now the Furry Li'l Mammals are getting restless; they want to communicate too—and in the ways they know best. (Be forewarned: Your sex life may improve dramatically!)

Variations on this process have also been used with great success in less intimate settings, such as within organizations. See: paulandlayne.com.

Playfulness and Humor

"Through humor, you can soften some of the worst blows that life delivers. And once you find laughter, no matter how painful your situation might be, you can survive it." —BILL COSBY

A characteristic of all mammals is that the young play and explore. With adulthood, however, this curious, creative side tends to close down. Humans are a fascinating exception. We can remain playful, curious, and inventive all our lives. What Life essentially did when it evolved the human was to take mammalian youth, stretch it out, and call it a species. Scientists have named this process neoteny.

When children are raised in an atmosphere of love, respect, and tender touch, they are curious, playful, and creative. As we grow more responsible for our own lives and then assume responsibility for the lives of others, and as we encounter life's inevitable disappointments, we may lose more than a necessary share of the spontaneity of youth. We may become rigid, predictable, morose. This is why virtually every religious tradition teaches that the way to salvation, freedom, or enlightenment is, at least in part, to recapture the mind and heart of a child.

Humor is an essential facet of creativity and a must for healthy families. As Patch Adams, M.D., has said, "Humor is an antidote to all ills. It forms the foundation of good physical and mental health. Humor is vital in healing the problems of individuals, communities, and societies. People crave laughter as if it were an essential amino acid. Humor and fun, which is humor in action, are equal partners with love as ingredients for a healthy life."

A little bit of silliness can go a long way toward relieving the stresses of everyday life. When we can laugh at ourselves, we put a little distance between self and troubles. The burdens are still there, but they are no longer who we are. We reenter the heavenly space of acceptance and trust, where healing happens.

Meaningful Songs and Rituals

> *"Sacred ritual takes us out of this narrow, artificial human world and opens us up to the vast unlimited world of nature—both outside, in our nonhuman environment, and inside, in our own deeper layers of the older brains and cellular body knowledge."* —DOLORES LACHAPELLE

Sacred songs and rituals affect our minds at a level much deeper than the rational and the verbal. They rouse intense, even indescribable, emotions. They align us with the cycles and rhythms of both inner and outer nature. When we sing together, and when we participate in meaningful group rituals, our Furry Li'l Mammal is fed, and we feel bonded to the other Furry Li'l Mammals who are coparticipants. Melody has a particularly powerful effect on this emotional part of our brain—probably having originated from the mother–infant repertoire of sounds that mammals still use for bonding and reassurance. Our Higher Porpoise, which presides over and guards our most cherished beliefs, relaxes during a familiar group ritual or sing-along, and thus is more receptive to new ideas.

When there is drumming, bass woofers, or some other form of intense beat, our Lizard Legacy is aroused. Some reptiles (earless lizards, the tuatara, and all snakes) actually "hear" through their jawbones. They will rest their jaw on the ground to pick up vibrations. The tiny bones in our own middle ear are directly descended from jaw connectors in our ancient reptilian past. How do we know? Because even in humans, those bones originate in the jaw region of the embryo, migrating to their final position as the fetus develops.

No matter how markedly the world's tribal cultures differ, a common chord is their use of song and ritual to mark seasonal and life passages. Anthropologists tell us that this has probably been true for millennia. It seems that some things can be expressed and some bonds forged most dependably through rituals (including dance) and shared singing. Thus our health, as individuals and as communities, depends on sacred ritual more than our rational Monkey Minds may comprehend. But here is the hitch: with the exception of words embedded in song, which can sneak past the meaning police within our neocortex, the rational parts of our brain will protest if we hear (or worse, are expected to recite) anything that runs counter to cherished beliefs. That is why the idea-content of rituals must be genuinely meaningful; Monkey Mind and Higher Porpoise must be willing to let it pass, and ideally to give it their stamp of approval. Rituals that were important to us as children may, however, be exceptions.

Some of the most joyous aspects of my marriage to Connie are the playful, romantic, and sometimes goofy songs and rituals we have created that nourish us on a daily basis. Many of these rituals are simple language games—standard phrases or monologues that particular situations reliably call forth. Others include our daily walk or tea time in the afternoon. And then, of course, there's the ritual of immersing ourselves in water outdoors every full moon, no matter where we happen to be. Yes, Lake Superior and the coast of Maine are both cold in November!

"Where's My Avocado?"

In 2003, while we were in the Pacific Northwest, Connie and I saw a counselor together for a couple of months. A practice we learned then has become a ritual vital to the health and well-being of our relationship. Whenever something doesn't feel right to Connie, she says "avocado!" It's just a code word. What it means is, "Something's not working for me right now." It could be anything. Perhaps I'm driving too fast, or I say something in a tone of voice that is screened by her amygdala as threatening, or I look at her funny. There is no requirement for her to explain her reaction to me; she may not even be able to explain it to herself. Rather, if there is some sort of "ugh" for her, I want to know it, and she knows that it is safe for her to communicate in this cryptic way. The genius of this ritual is that it serves two enormously useful functions in our relationship. (1) It allows Connie to not have to carry or repress a judgment or resentment. (2) It gives me immediate feedback on my behavior. So many relationships struggle precisely because of withheld thoughts and feelings. Saying "avocado" allows her to immediately express the feeling, while alerting me to the fact that something is not working for my beloved.

This ritual has been so important to us that if more than a week or two goes by without incident, I'll sometimes ask, "Hey, where's my avocado? You're not storing up a load of guacamole, are you?"

Synergy and Service

"The future belongs to those who give the next generation reason to hope." —TEILHARD DE CHARDIN

The Universe is made of nested holons: wholes that are part of larger wholes, within still larger wholes. Each has its own integrity, its own personality, and each is more than the sum of its parts. In Chapter 2 we looked at how synergy is one of the evolutionary tools at work in the biological world. Here I would like to revisit synergy in its human relational and cultural contexts.

Years ago I came upon an article by cultural visionaries Joe Dominguez and Vicki Robin entitled "The Possible Relationship." The authors described how a shared higher purpose enables individuals to achieve wondrous results collectively. The authors used the analogy of a high-power telescope. They wrote,

The most important component allowing such a telescope to bring Saturn into perfect focus is a series of lenses, carefully polished and in proper orientation to one another. However, if the lenses are out of alignment, if one is smeared with dirt, if they are not focused correctly, or if they are the wrong distance from each other—nothing happens. The power of the telescope depends on the right relationship of its component lenses. Likewise, synergy depends on the people involved being in alignment, with a shared vision and a shared purpose, with their hearts and minds open, with a willingness to share all and a commitment to stick with it till the game is over. With all that in place, energy can flow through that single instrument and truly light up the world.

Dominguez and Robin proposed that intimate relationships are not so much for the benefit of the individuals involved. Rather, they are for the world. They explain, "When relationships ignore that they are conducted in a much wider arena called life-on-Earth and do not see as their primary purpose the enrichment of this greater whole, they tend to display symptoms of dis-ease." Relationships that make room for collaborative service to higher and wider purposes are not, however, detrimental to the well-being of the partnership or the partners. Indeed, only by standing shoulder to shoulder with an outward, shared focus do intimate relationships have the best chance to deepen and persist: "That's the secret to lasting love, for energy = ecstasy = love, and service is what opens the valve."

Service is the culmination of evolutionary integrity. Service begins with our most intimate relationships of partner and family and expands outward to larger and larger communities. Service also expands in time. We serve the past by cherishing the gifts and contributions of previous generations, both human and nonhuman. We serve the present by finding ways of blessing the lives of those around us, here and now. We serve the future by attending to where our joy and the world's needs intersect, restoring ecological integrity to our home bioregion, and by doing our part to co-create just and evolutionarily viable economic and political orders, locally, nationally, and internationally.

As with the Cosmos as a whole, synergies come together in my own life in a nested way. Connie is not only my best friend and soulmate, she is also my mission partner. Our itinerant ministry and this book are the fruit of our synergistic union.

CHAPTER 13

Transformed by the
Renewal of Your Mind

"Do not be conformed to this age any longer, but be transformed by the renewal of your mind, so that you may be able to discern what is the will of God—what is good and acceptable and perfect."
—PAUL THE APOSTLE

As noted in Chapter 6, words create worlds. So do pictures. What we habitually tell ourselves and others about the nature of reality, along with the (often unconscious) inner pictures we imagine, profoundly influence the quality of our lives and relationships. Replacing outmoded ways of thinking with inspiring new concepts, and doing the same for how we picture the world, can be transformative. This chapter offers tools for growing in evolutionary integrity via recalibrating our inner GPS systems and updating our mental software.

Deep Integrity Affirmations

"Important as the struggle for existence has been and even still is, yet as far as the highest part of man's nature is concerned there are other agencies more important. For the moral qualities are advanced, either directly or indirectly, much more through the effects of habit, the reasoning powers, instruction, religion, etc., than through natural selection." —CHARLES DARWIN

The following affirmations may assist you in integrating some of the concepts in this book. As you read or, better yet, read them aloud, I invite you to pause after each and notice if anything opens up for you. Consider the statement long enough to begin to feel it. Memorized, these affirmations can influence how you habitually see the world.

1. I am fully committed to the path of deep integrity. Growing in trust, authenticity, responsibility, and service, with the support of others, is my number one priority.

2. I am an interdependent cell within the body of Life. Everything I see, hear, and feel is part of me—part of my larger Self, my true Nature.

3. I am the sum total of 14 billion years of unbroken evolutionary development now reflecting on its immense journey. I am graced by a growing awareness of the awesome implications of my larger sacred story.

4. I am one with the living face of evolution. I participate in, and as, the ongoing emergence of the Universe and planet Earth. I do so with courage, compassion, gratitude, and care-full action.

5. I feel the loving support of the Whole. I know that my breathing, seeing, hearing, and feeling are acts of communion. My senses are portals to the holy.

6. Everything that happens is perfect for my growth and learning. I sense what is uncomfortable—in myself and in my world—and seek what wants to emerge from it.

7. I regularly solicit honest feedback and use it to grow.

8. I accept myself and others completely here and now. I trust that we each tend to do the best we can, given what we have to work with at the time.

9. I have everything I need to appreciate the gift of the present moment. I take full responsibility for my experience and refuse to blame anyone or anything for my life being the way it is.

10. I am grateful for the prehuman components of my evolved mind. I love my ancient Lizard Legacy and my Furry Li'l Mammal so much that I am attentive to when each may run into trouble in the challenges of our modern world. I call upon each to help rein in the excesses of the other, and I call upon my Higher Porpoise to lead this effort.

11. I am a compassionate and understanding person. I am getting better and better at putting myself in the shoes and experience of others and imagining what they might be going through.

12. I give my Furry Li'l Mammal room to feel and express itself in a safe place and in ways that are mindful of the sensitivities of the Furry Li'l Mammals in others. I act when centered, loving, and aware of my interconnectedness, and procrastinate acting when I'm feeling otherwise.

13. I know that I may not live to see the Sun rise again. I cherish this eternal moment.

14. I trust that whatever happens on the other side of death is just fine. I can find comfort in mythic night language without clinging to any particular belief about the afterlife.

15. I listen humbly and carefully to others who are different from me. I affirm the differences of others and am open to their influence, even as I remain faithful to the truth as I perceive it.

16. *For Christians:* I am becoming more Christ-like and Christ-centered as I regularly ask myself, "How would Jesus express integrity right now if he were in my shoes?" "What would Jesus think about this, or say or do in this situation?" And then I speak and act from that place, with love and confidence.

Imagination Matters!
Upgrading Your Mental Software

"Most of our notions about the world come from a set of assumptions which we take for granted, and which, for the

most part, we don't examine or question. We bring these
assumptions to the table with us as a given. They are so much
a part of who we are that it is difficult for us to separate
ourselves from them enough to be able to talk about them. We
do not think these assumptions, we think from them."

—WERNER ERHARD

The philosopher and historian Thomas Kuhn observed, "When paradigms change, the world changes with them." We are in the midst of a great paradigm shift. These are unsettling times. Discoveries in the natural sciences have led to a new picture of reality. Our traditional ways of understanding ourselves, our world, and our sense of the divine are all shifting. We must now update our inner imagery to fit the facts—to align with "God's Word" as revealed publicly through science. The reason this is so important is that our unconscious pictures of reality affect us far more than do the abstract concepts we profess.

For example, while Christianity has long affirmed both the transcendence of God (that God is more than Creation), and the immanence of God (that God is the pulsing reality and emerging creativity within everything), most people (theists and atheists alike) have been able to imagine only transcendence when speaking of the divine. As discussed in chapters 5 through 7, this is not surprising, given that the primary metaphor for nature during the past few hundred years has been that of a complex clock. A nested emergent Cosmos changes all that. What cannot be pictured in the mind's eye, however, remains an impotent abstraction. It is one thing for evolutionary Christians to agree with the Apostle Paul when he says, "There is one God...who is over all and through all and in all" (Ephesians 4:6), or, "In Him we live and move and have our being" (Acts 17:28). It is quite another thing to imagine how this is so. Unless these teachings can be imaged, they cannot move or change us.

The mental pictures of ourselves in relation to Ultimacy/God and the rest of Creation are still largely based on an old flat-earth cosmology. These images have colored every facet of life in the West. We as individuals and as a culture are now feeling their limits. It is time to update our inner imagery to fit what is, in fact, the case. If we are to be faithful to (God's) public revelation as it is currently available to us, our mental imagery must portray humankind as the latest development of a process that has been spiritual from the beginning. It must acknowledge that *we are part of a divinely creative Cosmos in which everything is interrelated and interdependent.* The following pairs of images illustrate the differences between the old and new cosmologies:

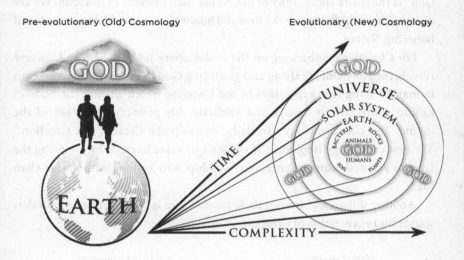

Pre-evolutionary (Old) Cosmology   Evolutionary (New) Cosmology

We have generally imagined God as transcendent to Creation (illustrated on the left.) We also imagined ourselves as separate from and above the rest of nature, as we intuitively knew that we were created in God's image. We believed that we lived *on* the Earth. What was important was our relationship to God and our relationship to other human beings. Earth was nothing but the stage upon which this drama played out.

Western economics, law, and ethics each function out of this view of reality. Morality, for example, has been concerned with human behavior toward an invisible, otherworldly God and toward one another. Our treatment of Earth was not considered a moral issue. Corporations can legally poison the air, water, and soil; forests and species can be exterminated—all because our laws are human-centered rather than life-centered.

As illustrated above right, an evolutionary cosmology enables us to think of God not only as transcendent but also intimately revealed in and through this divinely creative Cosmos. Humans are recognized as an integral part of Earth—not superior to it. The new cosmology and the old cosmology agree that of utmost importance for humans is our relationship to God. But there are two important differences in how this relationship is understood. In the new cosmology, God is acknowledged to be embodied in Creation, rather than divorced from it. And in the new cosmology the "our" in "our relationship to God" is the entire community of life, rather than humans in isolation. We are deceived if we think that we can love and honor God without loving others and honoring Nature.

For Christians embarking on the evolutionary path, the entire Universe reveals the Holy One. Praising and glorifying God has everything to do with humanity being that expression of the Universe which enables the body of Life to appreciate its beauty and celebrate this primary revelation of the divine. The old cosmology directed, "Worship the Creator, not Creation." The new cosmology implores, "If we don't REALize worship by honoring the Creator's nested omnipresence, our 'worship' will lead to death, rather than life."

Another difference between these cosmologies is the way that each understands the organization of Creation:

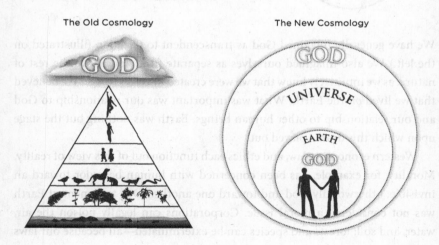

The Old Cosmology   The New Cosmology

In the old cosmology, most of us learned to see the value of things in an unchanging, patriarchal, and hierarchical way. God was Father; men were superior to women; male children were more important than female children; animals were above plants; plants were above insects and worms. (Before microscopes, our predecessors were not even aware of the protists and bacteria.) At the bottom of the pyramid were inanimate rocks and the elements of Earth. Biblical peoples believed that this was the way that God set things up in the beginning.

In contrast, the new cosmology understands that *everything* is evolving, and that we are part of a vast time-developmental Universe. As time moves on, Creation becomes more complex and more capable of realizing its spiritual potential. "God" is nothing less than a holy name for Ultimate Wholeness, the largest "nesting doll"—the one and only Reality that transcends and includes everything else. Every earthly manifestation thus reveals a face of the divine; each part is a unique expression of the Whole. In the words of Thomas Berry, "The Universe is a communion of subjects, not a collection of objects." We have no existence outside the ecological cycles of Earth, which in turn has no existence outside the solar system. The solar system has no existence outside the Milky Way Galaxy. The Milky Way has no existence outside the Universe. And the Universe has no existence outside Ultimate Reality, whatever that may be in its essence. Everything is interrelated and interdependent. The entire process oozes divinity.

The final contrast I would like to make between the old and new cosmologies has to do with how an individual human being is seen. It makes a world of difference whether we imagine a person...

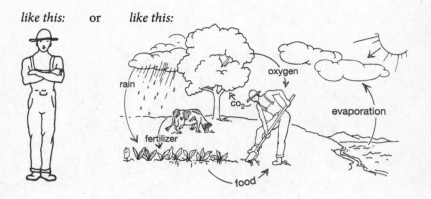

like this: or   *like this:*

The left image is of a farmer seen from the perspective of the old cosmology.

This abstraction is dangerously misleading. There are no isolated human beings except in our imagination. Any person exists only as a member of the wider community of life, air, water, and soil. As Gregory Bateson noted a half century ago, there is no such thing as an organism, only an "organism-in-environment." Or, as Elisabet Sahtouris playfully says, "There's no such thing as rabbits in habitats; only rhabitats."

The image on the right is thus a more accurate depiction of a human person. We are not stewards, or caretakers, or anything else that assumes we are separate from Nature. We have no existence apart from the living Earth. We *are* Earth. We are increasingly conscious organizations of elements composing this living planet. We are utterly dependent for our own health upon the health of the wider community of Life. Our own destiny, whether as individuals or as a species, and the destiny of Earth are identical. What we do to Earth, we do to our Self.

As we update our mental images of reality, we more easily integrate the Great Story cosmology into our daily lives. We will also experience God's loving presence and transforming power in new and richer ways.

PART V

A "God-Glorifying" Future

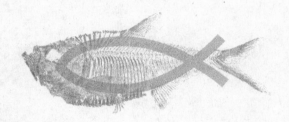

"I believe that the environmental crisis is our greatest test as a species. It may also lead to our finest moment. But we have lots of work to do. One piece of the Great Work is to learn the Great Story of our evolving Universe, to come to see it as our own sacred story and to ponder its lessons, all the while celebrating its grandeur and acting as if our very lives depended on it—because, of course, they do."

—PETER MAYER

CHAPTER 14

Collective Sin and Salvation

*"God didn't send his son into the cosmos to judge it; but that
the entire created order might be saved through him."*

—JESUS

PROPHETIC INQUIRY: What can we say about sin in a time of smart
bombs, species extinctions, global warming, and genetic engineering?
What can we say of sin in an expanding Universe of two hundred billion
galaxies?

very culture has had a way of speaking about "that which separates
or violates the integrity of the whole"—those attitudes and actions
that alienate us from life, from one another, and from ourselves. In
our culture such attitudes and actions were once widely referred to as "sin."
From the perspective of an *evolutionary* Christianity, sin is nothing that can be
believed in or not. It simply is. That which separates or violates the integrity of
the whole is an unavoidable part of human experience. The concept of sin takes
on new meaning in a nestedly emergent Universe. As discussed earlier, a fundamental truth that previous generations could not have known is the multi-
level evolutionary nature of the Universe.

*"Sin" is a religious way of speaking against action that would serve one level
at the expense of another, and that mistakenly supposes a separation in the fabric of the Whole.* Salvation is thus not merely an individual aspiration. Corporations, nations, and other organizations can taste the fruit of salvation by

operating in deep integrity and pursuing the well-being of the larger communities of which they are part, while maintaining the health of the smaller communities and individuals for which they are responsible.

Wrongdoing in a Nestedly Emergent Universe

"To put the world right in order, we must first put the nation in order; to put the nation in order, we must first put the family in order; to put the family in order, we must first cultivate our personal life; we must first set our hearts right."

—CONFUCIUS

Another truth about the nature of reality, which was first noted scientifically a century ago, is that there is a direction to cosmic creativity. Religiously, we might speak of this direction as the Universe expanding within the heart of God and becoming more complex, self-aware, and compassionate over time. An arrow of progress has been moving through Creation since the beginning. This emergent creativity sometimes works at a very slow pace, and it is not without chaos and setbacks—which are, in fact, drivers of transformation. Thus *all* is Grace, if we have but the eyes to see and the heart to trust. I like to think of this directional feature of evolution as Creation itself maturing, expressing greater cooperation, interdependence, and awareness at ever-increasing scale and evolvability. Humanity and our supportive technologies are now integral to this evolutionary process, bringing rapidly evolving consciousness and social systems into the Great Story.

As our understanding of Reality evolves, so does our understanding of "that which separates or violates the integrity of the whole." Once identified in oral cultures as the breaking of tribal codes or taboos, in literate cultures sin came to be seen as violation of the written "laws of God." In America today, while the word "sin" is not much used in political discourse, the term "illegal" does not, by itself, demarcate the full extent of our society's standards of prohibition. Among liberals and conservatives alike, there is basic agreement on which behaviors—greed, duplicity, selfishness—constitute moral transgressions beyond the prohibitions encoded in law. To a great extent, this implicit societal code has been bequeathed to us from our evolved social heritage. Yet there is much that we must now deal with as a species that our biological and social inheritance could not have prepared us to handle. Consider: What *can* we say of sin in a time of smart bombs, species extinctions, global warming, and genetic engineering? What can we say of sin in an expanding Universe of two hundred billion galaxies? Here is my response:

> *I sin when I attempt to fulfill my personal desires at the expense of the larger realities of which I am part—my community, work environment, society, bioregion, or the body of Life. I sin when I attempt to fulfill my personal desires at the expense of my peers, my brothers and sisters, my fellow human beings. I also sin when I attempt to fulfill my personal desires at the expense of the dependent realities for which I am responsible—my children or employees, the health of the cells in my body, the ideas and principles that in some way depend on me for their expression and advancement. And if "God" cannot possibly be less than a holy name for "The Wholeness of Reality, immanent and transcendent," then to sin in any of these ways is, truly, to sin against God.*

Maturity (salvation) is found, at least in part, in accepting that my real self-interest is quite distinct from my Lizard Legacy and Furry L'il Mammal interests, and knowing that this fuller expression of self-interest is served only by taking into account the well-being of the larger holons of which I am part and the smaller holons for which I am responsible.

All of us have a sense of what constitutes wrongdoing, whether or not we choose to label such actions as "sinful." My purpose here is to build on that foundation by suggesting less familiar categories of wrongdoing. The categories are less familiar only because they emerge from a nested understanding of the Universe and of the divine creativity revealed throughout the Cosmos. Indeed, not only are the categories less familiar; most of us are engaged in wrongdoing more often than we care to admit—and sometimes without our even knowing it.

Our culture lacks feedback loops to alert me to wrongdoing at holonic levels that I cannot directly perceive. For example, how would I know if the toilet I flush is connected to a septic system that adequately protects a nearby stream from unclean water? Our institutions of commerce are only beginning to alert me to which apples in the bins were grown in my own country and which were shipped halfway around the world. And why are swordfish steaks still available for purchase in any store or restaurant, given what has been discovered about the fragile state of their population and breeding stock and the importance of top predators to the entire oceanic web of life? What, too, about all the harmful things I do out of sheer ignorance? And what are my alternatives anyway?

Individual choice can and must play a role in bettering our world, but there is also a role for shifting away from business as usual. Indeed, until the right thing to do becomes the easier and cheaper thing to do, we will continue to degrade our planetary home, because precious few of us are saints. And even those who come close to qualifying for sainthood will still be harming the larger holons unless they are educated saints, up-to-date on the latest science and current events.

A nested understanding of Reality helps us see that an expanded moral calculus includes not only how our individual actions affect larger holonic levels, but also how larger social holons themselves emerge and thus can engage in right or wrong actions. As societies expand and become more complex, so too must our understandings of and referents for collective aspects of sin. *Corporate sin* includes social injustice, corporate crime, collective violence, institutional corruption, and all pursuits of group self-interest that damage or diminish the integrity and well-being of the larger holons of which the group is part, or the smaller holons of which it is composed and for which it is responsible. Crucially, corporate sin also encompasses a vast array of *crimes against Creation*. Crimes against Creation include not just those acts of pollution and degradation committed in defiance of law, but also those not yet codified by governments as wrongdoings (such as excessive production of carbon dioxide), which nevertheless damage the biological support systems of Earth or its constituent species.

When a group pursues its own self-interest at the expense of the larger or smaller spheres of sustenance, it is sinning against God. Nevertheless, in a social system for which natural self-interest and the well-being of the whole have not yet been brought into alignment, how could groups do otherwise and survive? The real problem, therefore, is one of *systemic sin*.

The fundamental immaturity of the human species at this time in history is that our systems of governance and economics not only permit but actually encourage subsets of the whole (individuals and corporations) to benefit at the expense of the whole.

Until we evolve life-centered and nestedly intelligent (synergistic) structures of global and bioregional governance, it is inevitable that morally upstanding people operating with the very best intentions will sometimes unknowingly perpetrate great evil. As Thomas Berry has said, "The assault on the natural world has been carried out by good persons for the best of purposes, the betterment of life for this generation and especially for our children. It was not bad people, it was the good people acting for good purposes within the ethical perspectives of our cultural traditions that have brought such ruin on this continent and on the entire planet."

Even good people are dangerous when they are guided by an inadequate sense of right and wrong. Thus nothing is more important, it seems to me, from a practical as well as spiritual perspective, than co-creating synergistic systems of governance at all levels—locally, regionally, nationally, and globally—that align the self-interest of individuals and groups with the self-interest of our planet as a whole. This is the heart of our great evolutionary work—our species' divine calling.

Night language can play a crucial role in bringing about this necessary shift in perspective. I have strong preferences as to how best to interpret today the religious night language that came into use millennia ago. "Thy kingdom come, thy will be done on Earth as it is in heaven" and "the second coming of Christ" are scriptural concepts that tell me in no uncertain terms that our way into a bright future for humanity and for the larger body of Life is to expand our sense of right and wrong to include actions taken by "individual" corporations and other emergent social groups that over the centuries have evolved a life of their own.

Collective Sin in an Age of Information

> "Sin is the refusal to realize one's radical interdependence with
> all that lives: it is the desire to set oneself apart from all others
> as not needing them or being needed by them. Sin is the refusal
> to be the eyes, the consciousness, of the Cosmos."
>
> —SALLIE MCFAGUE

PROPHETIC INQUIRY: What are the moral implications of the emerging knowledge of our collective impacts on the world, including other species? How can we confront evil without becoming possessed by evil in the process? How does the Christian message provide inspired guidance for working synergistically with evolution to solve social and ecological problems at a now global scale?

In an emergent Universe, utterly new life conditions and challenges regularly arise. They arise as a natural consequence of life's unfolding. When an intelligent species emerged that could kill at a distance and strategically transform its environs, a new face of sin emerged in the Universe. We humans, with our unprecedented power, brought into being the sins of ecocide and biocide—the ability to destroy entire natural systems and species. Many in the liberal religious communities have had difficulty accepting that preindustrial and tribal peoples, as well as industrial and postmodern humans, have had a large and devastating impact on the natural world, causing extinctions and massive ecosystem disruptions. But they have, and we have too.

There is a difference, however, between then and now. Before the Information Age, there could have been no moral culpability, no individual or collective guilt, for the environmental and biological harm wrought by humans. Prior to scientific awareness, massive record keeping, and cross-regional communication, there would have been no way to anticipate that one's cultural practices would lead to the extinction of prey, the salinization of soils, or the irreversible transformation of forested mountain slopes into desert scrub. But now we do collectively know the consequences of our collective actions. We have crossed a threshold, and there is no going back. In this Information Age, we have entered a new province of collective responsibility. The good news is that ecological consciousness—and a readiness to heroically face the future rather than crumble into despair—now so readily blossoms in our youth.

The Clovis culture, the mammoth hunters who first entered North America as a frontier people, may have exterminated the mammoths and unleashed an ecological cascade of extinction that resulted in the loss of two out of every three large kinds of mammals—not only the mammoths, but also the mastodons, the camels, the horses, the tapirs, the peccaries, the giant beaver, the ground sloths, the armored glyptodonts, the giant bison, and all the mythic predators who depended on them: sabertooth cats, plains lions, dire wolves, and giant bears. The

Clovis peoples, though responsible, were not, however, to blame. Lacking continental-scale communications, no group of immigrants could have discerned that the herds were vanishing on a scale from which there could be no recovery.

The ancestors of the Australian Aborigines, who first entered a land filled with marsupial beasts innocent of projectiles and fire, directly or indirectly caused nine out of every ten large species in Australia to disappear forever. But again, this early form of biocide was not their fault. They could not have known that individual actions compounded by an expanding human presence (and perhaps exacerbated by a drying climate) could result in irreversible damage. Nevertheless, this was the result.

The ancestors of New Zealand's Maori people exterminated all dozen species of moa less than twenty generations after they entered the primordial paradise of flightless, tame, and giant birds some 800 years ago. When the last moa was gone, the ancestral culture collapsed. There was no longer enough protein to support the population density that giant birds and their eggs had made possible. Thus began a terrible period of mutual extermination, perhaps even cannibalism. But even for this, the ancient peoples were not culpable. How could it have been otherwise?

In New Zealand, in Australia, and on my own continent of North America, there would have been no way for the first peoples to confirm that the moa, the giant kangaroo, or the mammoth had not just moved to some other region of the landscape, to some other people's domain. There would have been no way to predict that dwindling prey populations were not just temporary downturns. In a time before modern communications technology and detailed record keeping, no human hunter or fire maker in any of these lands could possibly sense the scale and irreversible consequences of actions that had so recently ensured an abundant life. Indeed, a little more than a century ago, the last wild passenger pigeon, which once numbered in the billions, vanished from North America—even before conservationists of the day were aware of impending doom.

All that has changed. We do have good reckoning today of the impending consequences of our interactions with all our relations in the natural world. Our eyes have been opened. Now (metaphorically speaking), for the first time since God placed us in the garden, we can recognize our nakedness and know shame. To awaken to the scale and deep roots of the harm our species has done to Creation is painful. We are horrified. It can't be true! There *must* have been a Golden Age, somewhere, sometime. But there wasn't. On this, the vast majority of scientists agree. Yet this awakening is a blessing, too. At last, we know the truth. And it is the recognition of sin that is the first step on the path to salvation.

Atoning for Collective Sin Through "Pleistocene Rewilding"

Doing all we can to prevent more extinctions and the further deterioration of waters, soils, and air are vital steps for coming into right relationship with Reality, with the Earth Community—with "God's body" as manifest on this planet. Confining our Great Work in the realm of biodiversity to the prevention of future species losses can, however, dishearten us. *Doing less bad* is important, but psychologically it is not enough. Our souls long to be engaged in efforts that replenish, that truly make amends. This is why the "ecological recovery" and "ecological restoration" movements are so exciting. By volunteering our time, energy, and money toward such endeavors, we are not so much doing penance for past sins as finding ways to feel hopeful and important about our ability to do good here and now.

Along these lines, the most audaciously inspiring proposal I know of is Pleistocene rewilding. If the first humans widely colonizing North America were, in part, the cause of mammoth and other extinctions, then perhaps it is time to reprovision this continent (in a few select and very large natural reserves) with proxies for the lost creatures—proxies who could be shaped, naturally, by evolution over the course of millennia to become truly native to this place. How do we know this is possible? Because wild horses in the American West have already repopulated the landscape. More than ten thousand years extinct on the continent of their origin, horses have returned. They are the now-feral descendents of domestic horses that the Spanish, and later, colonists brought with them from Europe.

Advocates of Pleistocene rewilding envision that elephants would be introduced into wild landscapes in North America to fill the niche that mammoths and mastodons once occupied. The African cheetah would be introduced to fill the predatory niche (chasing after the fleet American pronghorn) that the American cheetah once filled. Even the African lion could be introduced as a proxy for the giant American plains lion to keep populations of our own wild horses in check, much as lions control zebras in Africa.

For more on this idea, consult the scientific paper, "Pleistocene Rewilding: An Optimistic Agenda for 21st Century Conservation," by C. Josh Donlan et al., 2006, *American Naturalist* 168: 660–81.

Confronting Institutional Sin

"Politicians at international forums may reiterate a thousand times that the basis of the new world order must be universal respect for human rights, but it will mean nothing as long as this imperative does not derive from respect for the miracle of Being, the miracle of the universe, the miracle of nature, the miracle of our own existence." — VÁCLAV HAVEL

From an evolutionary perspective, what would *collective* recovery movements look like? What would help corporations and nation-states resist "demonic" influences that cause them to endanger the health of their communities, nature, and the world?

Those of us committed to co-creating a blessed future for our planet, and doing so to the glory of God, face a profound question: How can we confront evil without becoming possessed by evil in the process? A supremely disturbing fact of human history is that *the greatest evils have been perpetrated by those who thought they were trying to rid the world of evil.* Jesus' example and his teachings model approaches to activism that can help us resist the temptation to fight evil with evil. That is why the virtue of humility and the admonition to love one's enemies, even while resisting them, are central to the Christian path. As Christian writer Walter Wink reminds us in his book *The Powers That Be*, active nonviolent resistance—from a heart of integrity and passionate commitment to kingdom values—is Jesus' way to engage the as-yet unredeemed social structures of the "Domination System" that controls so much of the world.

In this Universe where choice is an essential part of life—where real freedoms exist—a host of evils also exist: deception, exploitation, manipulation, domination, and so forth. Central to any meaningful understanding of the Gospel, however, is that this negative realm never claims the final word. When

I look at cosmic history through sacred eyes, I see that the chaotic, destructive side of the Cosmos is consistently held within the larger arc of creative evolution. That is, God (Supreme Wholeness personified) seems to delight in taking bad news and creating good from it.

> *Crises call forth creativity.*
> *Breakdowns catalyze breakthrough.*
> *Emergencies evoke emergence.*

Is this not the central message of the cross? Is this not the prophetic claim of the Book of Revelation? Moving from the realm of day language into night language, we might say:

> *Christ will vanquish Satan (and indeed already has!).*
> *Good Fridays are followed by Easter Sundays.*
> *Crucifixion comes before Resurrection.*

By adopting the wondrous epic of evolution as my creation myth and taking the time to consider its implications, I have come to know that one of the most dependable patterns in Life's several billion year history is this: *bad news, chaos, and breakdowns regularly catalyze creativity and transformation.* Is it merely coincidence that this pattern of evolutionary history mirrors the core teaching of the Gospel? Perhaps. But a far more inspiring interpretation is that God has been revealing such truth through our traditions long before the scientific venture was birthed—and that science found another face of this truth, and dug a well from which all now can drink.

"The universe was brought into being in a less than fully formed state but was graced with the capacity to transform itself from unformed matter into a truly marvelous array of structures and life forms." —SAINT AUGUSTINE

CHAPTER 15

The Wisdom of Life's Collective Intelligence

> *"Our minds have been built by selfish genes, but they have been built to be social, trustworthy, and cooperative... Human beings come into the world equipped with predispositions to learn how to cooperate, to discriminate the trustworthy from the treacherous, to commit themselves to be trustworthy, to earn good reputations, to exchange goods and information, and to divide labor.... This instinctive cooperativeness is the very hallmark of humanity."* —MATT RIDLEY

PROPHETIC INQUIRY: From an evolutionary perspective, what can usefully be said about the Kingdom of God, or Kingdom of Heaven? What would it mean if the Kingdom of God were aspects of the evolving Universe intimately intertwined in our lives today? What if we dwell in the Kingdom of Heaven whenever we are aware of and ethically participating in that infinite and ever-evolving Reign of Reality? What would that insight imply about our purpose and responsibility for other parts of Creation—for other humans, other species, the health of our social systems and our planet?

I want to call to mind again a remarkable insight from long ago. Thomas Aquinas wrote, "A mistake about Creation results in a mistake about God." Saint Thomas wrote those words in his native Italy in the 13th century, but they have never been more relevant than here in America and right now. Truly, as our understanding of Creation matures, so too will our appreciation of the meaning and magnitude of the Kingdom of Heaven.

"The Kingdom" was central to Jesus' teaching, and for good reason. For Jesus, God's Kingdom was an inspiring vision of a religious, political, economic, and social reality, grounded in the presence and activity of the Source, Substance, Energy, and End of everything. The Kingdom had both a factual, measurable (day experience) dimension and an inspiring, nonmeasurable (night experience) dimension. For Jesus and his disciples, the Kingdom of Heaven was no far-off, unchanging place. It held meaning in each and every moment and in every situation. What is the meaning of the Kingdom of God in *our* time and place? What remains of the ancient understandings, and what enlarged meanings do we behold in our nestedly emergent Cosmos—a Universe in which God's grace is vaster and more apparent than our spiritual forebears could have known?

On Earth As It Is in Heaven

"If those who lead you say, 'Look, the kingdom is in the sky,' then the birds of the sky will precede you. If they say to you, 'It is in the sea,' then the fish will precede you. Rather, the kingdom is inside of you, and it is outside of you. When you come to know yourselves, then you will become known, and you will realize that it is you who are sons of the living Father. But if you will not know yourselves, you dwell in poverty and it is you who are that poverty."   —JESUS

Thanks to the scientific endeavor, we now know beyond any reasonable doubt that Creation, as expressed here on Earth, has become more interdependent and aware through time. It all began with a small, nonliving planet that came to life nearly four billion years ago. Interdependence and awareness ramped up as living beings evolved sense organs and as the web of life grew more intricate. Most recently, our own species (aided by science and technology) has knitted a vast and interdependent web of culture at a global scale. Humans and our supportive technologies are now an integral part of this trajectory of evolution. Thus the Lord's Prayer—"thy kingdom come, thy will be done on Earth as it is in heaven"—takes on a depth and breadth of meaning that previous generations could not have known. In light of evolutionary directionality, the Kingdom of Heaven can now be understood in a more comprehensive and this-world realistic way than ever before.

Here are contemporary ways by which this ancient biblical truth can speak to us today, while retaining its prophetic power:

✦ God's Kingdom is the realm of divine presence and activity where the 14-billion-year "Way of Life" (what the Chinese call the Tao) is celebrated. Right here on Earth is where, thanks to the human species, evolution is now consciously moving in the direction of greater cooperation, interdependence, awareness, and intimacy with the Whole—and at ever increasing speed, scale, and adaptability.

✦ When we live in deep integrity and take on a life of higher purpose, we further the work of God's Kingdom—no matter what our beliefs.

✦ The Kingdom of God is manifest at scales large and small, persistently and fleetingly, wherever and whenever the practical values of which Jesus spoke in the Sermon on the Mount and elsewhere, and the evo-integrity values he embodied (incarnated) in his own life and ministry, reign supreme.

✦ There is both an inner (psychological-spiritual) and an outer (social-political-economic) dimension to the Kingdom of God. And there is a here-and-now dimension, as well as a future dimension.

✦ Most important, Evolutionary Christianity presents "the Kingdom of Heaven" as an *eternally true* concept—true for everyone, everywhere, and at all times. The Kingdom of Heaven is not merely, or primarily, about an otherworldly place we go to when we die or after the "End Times" have opened the way. Rather, it is that realm of Reality governed by the sacred, immanent, infinite wisdom of Supreme Wholeness.

"When I Repent...I Dwell in the Kingdom of Heaven"

When I repent and ask the transforming power of divine Humility, Authenticity, Responsibility, and Service (i.e., Christ) to be the ultimate guiding priorities of my life (to be my "Lord"); *when I trust* that there is a larger reality than my ego at work in the world to which I'm accountable; *when I get real with* myself and others by owning the painful truth about my life (made bearable because I also know that God loves me anyway); *when I commit* to living in deep integrity; *when I make amends* and clean up the messes I've made from living instinctually; and *when I serve others* or bless the world—then *I dwell in the Kingdom of Heaven.* And so does everyone else who follows this path, no matter their religion or beliefs.

When I step into the shoes of my adversary (my enemy, spouse, boss, child, parent, sibling, estranged friend); when I see things from their perspective and speak from that place; when I comfort the afflicted, afflict the comfortable, and nonviolently resist the unredeemed powers that be; and when I work to right wrongs or ensure a just, healthy, and beautiful world for future generations—then *I dwell in the Kingdom of Heaven.* And so does everyone else who follows this path, no matter their religion or beliefs.

When I am engaged in creative work that will serve the larger and smaller holons of my existence, *I dwell in the Kingdom of Heaven.* And so does everyone else who follows this path, no matter their religion or beliefs.

When I appreciate the contributions of my Lizard Legacy and Furry Li'l Mammal and integrously channel those energies in service of my Higher Porpoise, *I dwell in the Kingdom of Heaven.* And so does everyone else who follows this path, no matter their religion or beliefs.

When I quiet my Monkey Mind by shifting from thinking to noticing, when I simultaneously pay attention to the sensations and sound of my breath and the beating of my heart, *I dwell in the Kingdom of Heaven.* And so does everyone else who does so, no matter their religion or beliefs.

When I am grateful for something extraordinary that comes my way or for something commonplace that might on another day have passed without notice, *I dwell in the Kingdom of Heaven.* And so does everyone else who chooses gratitude, no matter their religion or beliefs.

When I accept without resistance or resentment, and when I move actively into the future with a trust-filled "What's possible now?" mindset, *I dwell in the Kingdom of Heaven.* And so does everyone else who walks the path of faith, no matter their religion or beliefs.

Individually, dwelling in the Kingdom of Heaven is evidenced by what the early Christian scriptures speak of as "the fruit of the Holy Spirit." In addition to the list of blessed attributes that the Apostle Paul compiles in his letter to the Galatians—"love, joy, peace, patience, kindness, goodness, faithfulness, gentleness, and self-control"—we could add integrity, gratitude, courage, compassion, generosity, self-responsibility, and commitment to the well-being of the Whole over time. Those who evidence such fruit do know the Kingdom of Heaven, irrespective of their beliefs. Those who do not evidence such fruit do not know God's Kingdom. It's really no more complicated than that.

Collectively, dwelling in the Kingdom of Heaven is evidenced by furthering the evolutionary impulse in the direction of wider circles of cooperation, awareness, and compassion in our social, political, economic, educational, and religious institutions. As we shall see in later sections, this manifests when groups among us are transformed by their felt relationship to God such that they are compelled to work for the betterment of others and their surrounds—not by a sense of guilt or flight from despair, but from an in-dwelling spirit of joyful service. In such a blessed state of grace, their actions flow as effortlessly as raindrops entering a stream.

Practically, the collective embodiment of God's Kingdom goes far beyond reconciling people and organizations to the social structures that now exist. It means incarnating Christ-like values (humility, authenticity, responsibility, service) at every level of our common existence, including our social systems. The Great Work of the Kingdom in the 21st century, which is already emerging, is codesigning ways to align the self-interest of individuals, companies, and nations with the well-being of the whole of society and the whole of life.

"Thy kingdom come, thy will be done on Earth as it is in heaven" cannot possibly mean less than allowing the Holy Spirit to guide us as a species in evolving mutually enhancing relationships with one another, with the larger body of Life (Nature) and with our own creations (arts, technology, social systems, etc.), and doing so to the glory of God—that is, honoring the Wholeness of Reality. This is our destiny as a species. We stand a chance of fulfilling that destiny, however, only if the devoutly religious of the world work together, with full commitment, in serving and guiding progressive change in the decades ahead.

Any individual or institution that makes evolutionary integrity its top priority—its number one commitment—is part of the body of Christ.

Collective Deep Integrity

"Our future depends upon adapting our cultures to the realities of modern life at an unprecedented spatial and temporal scale. The idea that we can do this without a detailed knowledge of genetic and cultural evolution will appear laughable in retrospect—if we are lucky enough to persist in our current maladaptive cultures for so long. Perhaps someday we will confidently steer ourselves into the future. Until then, we will be like the Wizard of Oz in his hot air balloon, [not knowing] how it works."   —DAVID SLOAN WILSON

PROPHETIC INQUIRY: What might it mean for societies and our species to be in deep (ecological and evolutionary) integrity? What would integrous organizations and decision-making processes look like, sound like, feel like? What are proven, effective ways of discerning God's guidance for groups of different sizes and structures?

The principles of evolutionary integrity that I applied to individuals in chapters 10–12 also hold true for society. I shall be speaking in broad terms about "society," but please know that what I say here is meant to apply to any group, organization, or social system—even culture. That a society is growing in evolutionary integrity would, above all, be evidenced by its entering into partnership with the Whole and doing so in humble and loving ways. Such a society would celebrate, cherish, and creatively use the wisdom of the Whole as manifest in each and every level of existence. This includes the wisdom of individuals; of individuals aggregated into communities and organizations; of nature in all its wondrous beauty; and, indeed, the holy inspiration of Reality wherever and however it shows up.

Unity and diversity are co-arising dimensions of healthy holons. At the human scale, this pair is a natural consequence (emergent phenomenon) when people are encouraged to nurture and express their authentic selves. The more uniqueness, the more diversity.

Initially, dissonance arises from difference. Collective deep integrity involves recognizing that the dissonance and discord bear gifts. Often, these are gifts that support the evolutionary journey. Dissonance is a sign that something new is ready to emerge. This aspect of emergence can take many forms—dissent, discomfort, chaos, challenge, conflict, and crisis, to name but a few. Collective deep integrity calls on us—individually and collectively—to welcome these signs and to maintain a hopeful "What's possible now?" expectancy of the gifts we are likely to encounter just around the corner.

Evolutionary integrity for a society would thus mean creating institutions that honor diversity and dissonance. The Bill of Rights in America is a premier example. An ethos of tolerance and multiculturalism is another such manifestation. The pluralism and freedoms of democracy serve the evolutionary purpose of calling forth human diversity and dissent into the public conversation which, when handled well, generate creativity and insight that, in turn, enhance collective wisdom and capabilities.

Conversation and Creative Emergence

"All great changes begin in conversation."
—JUANITA BROWN AND DAVID ISAACS

"Insight, I believe, refers to that depth of understanding that comes by setting experiences, yours and mine, familiar and exotic, new and old, side by side, learning by letting them speak to one another." —MARY CATHERINE BATESON

To harvest the gifts of diversity and dissonance, societies would put in place institutions to gather the wisdom of creative interactions. Creative interaction almost always centers on conversation. Individuals have a chance not only to speak but also to listen and consider new possibilities. Conversations that promote understanding of everyone's gifts and limitations serve the whole.

Conversation is more than a colloquial term for informal communication. Conversation suggests authenticity, thoughtfulness, openness, spontaneity, and the possibility of something new and wondrous emerging. In these ways, conversation mimics biological and cultural evolution. Indeed, *conversation* is to be recommended as a night language description of those very processes.

Even in the social milieu of today there are promising hints of evolutionary emergence by way of conversation. Some governments have fostered collective conversations by choosing citizen members at random to consider the issues, concerns, and dreams of their community. Sometimes the participants simply reflect on how their community is doing and then report their responses. At other times the citizen recruits consider complex issues by interviewing experts from across the political spectrum, deliberating about what should be done, and reporting their findings to government officials, media, and the public at large. Such "citizen deliberative councils" are described at co-intelligence.org.

Citizen Assemblies and Citizen Juries

In the Canadian province of British Columbia, a citizens assembly consisting of 160 randomly selected individuals, was convened to study alternative electoral systems. The effort culminated in a ballot issue that proposed change in how public officials would be elected—which their fellow citizens then voted on. In Denmark, citizen consensus conferences are periodically appointed by the parliament to hold public hearings on complex technical issues, such as global warming, and then to make recommendations for governmental policy. In addition, hundreds of official and unofficial citizen juries have been held around the world to deliberate on every topic imaginable. These are all in addition to the hundreds of mediations, public dialogues, and watershed councils that every day engage thousands of diverse players in creative dialogue.

A rich fabric of such conversations would help any society solve problems, resolve conflicts, and adapt to environmental challenges. But a truly evolutionary society would achieve much more. It would regard problems and conflicts not so much as nuisances to be handled, but as opportunities for emergence, for conscious evolution into greater wholeness.

A *problem* happens because there is something we're not seeing, that we are not fully taking into account. A *conflict* occurs when two or more parts of a whole (a relationship, group, community, and so on) cannot perceive one another as valid members of the larger whole. A consciously evolving society would regard both forms of dissonance as invitations to transform. When that invitation is taken with skill, deep integrity manifests (Christ reigns). Everyone takes a step forward, and solutions emerge almost without effort. This magic begins when even a few of the participants, especially those in power, assume good intentions and cooperation. Even more, they trust that evolution is happening, no matter what happens next. Such grounding trust beyond anything that anyone can control or manipulate plays a crucial role in furthering the work.

Co-Intelligent Social Technologies

"The scarcest resource is not oil, metals, clean air, capital, labor, or technology. It is our willingness to listen to each other and learn from each other and to seek the truth, rather than seek to be right." —DONELLA MEADOWS

To bring forth the magic of emergence on a reliable basis, society would draw upon a host of sophisticated relational skills, communication systems, organizational processes, and creative decision-making. All these social technologies are expressions of collective intelligence, or co-intelligence. Their purpose is to draw out the wisdom of the larger holons. As author and process activist Peggy Holman notes, these human technologies help people to feel comfortably bonded within a larger network of shared intention. Group cohesion, in turn, dramatically increases participant openness and capacities to take difficult evolutionary steps, individually and together. Feelings of belonging and connection flourish when people feel safe to show up authentically—as unique and fully expressed individuals—whatever dissonance or coherence may result. This dynamic tension and partnership between individual and collective energies would inform all such evolutionary ways of coming together.

Ideally, co-intelligent processes would be facilitated by those who are aware that they are engaging in conscious evolution. In some cases, the facilitators or designee would begin with an invocation of one sort or another. A key realization is this: dialogue can be evolutionary and spiritual whether we call it that or not. Accordingly, powerful conversations do not depend on spiritual settings—though they may, of course, be assisted by such settings. Facilitators of conversations aimed at evoking the wisdom of the collective might do well to confer with experts in the dynamics of evolution, standing ready to call into their own processes insights from the natural world. In all these ways, collective deep integrity would manifest.

When we actually hear one another, or listen to nature, we realize that we are in the presence of a greater intelligence, a greater wisdom, a greater power. When in the felt presence of this greater wisdom, we let down our guard and call off our attack forces. Our Furry Li'l Mammal's prideful jurisdictions melt away as our awareness expands. Grace now has an opportunity to work wonders.

The Core Commons

> When the leader leads well, the people say, "We did it ourselves."
> —LAO-TZU

There is something else that happens when we hear one another's stories. We enter into what social philosopher Tom Atlee calls "the core commons"—that place within each of us where human resonates with human. The layered depths of the core commons are intimately related to our individual as well as collective evolution, and they form the bright side of what I've already referred to as our inherited proclivities, or unchosen nature.

Any of us can easily enter the core commons at the level of a group or tribe with whom we closely identify. Expanding the core commons, we work toward finding that shared resonance with strangers and wider still with those who are very much unlike us. We do this by searching out the human universals—those innate experiences, concerns, and drives that all of us share simply because we are human.

The core commons encompasses a resonance with life concerns far beyond those of the human, reaching out to other species. We readily extend the core commons to include our pets. We also naturally care about the well-being of our closest mammalian kin (chimpanzees and bonobos), as well as large, charismatic animals (whales, pandas) and the delicately beautiful (a butterfly, an endangered flower). Equally, many of us extend our concern to the well-being of a river in our community, to the wild integrity of a landscape, and all the way out to the oceans and atmosphere—to our beloved planet as a whole. We may even find commonality with all things in the Universe, as we awaken to our shared material ancestry born from the dust of stars and, before that, the Great Radiance (Big Bang) by which the Universe began. We know that every star, every galaxy, every black hole is part of God's evolving Cosmos. We thus sense, at a deep level, what some Native Americans mean when they refer to "all my relations."

What is precious here, what can inform our cultural pursuit of wisdom, is that when we open up and truly hear one another and the world around us, we naturally open to an expanded sense of kinship and expanded sense of self that can embrace the entire Universe. As Peggy Holman observes, "What is most deeply personal is universal." From that place, the boundaries melt away. We no longer are separate beings *in* a Universe. We are a mode of being, a glorious expression, *of* the Universe, companions within God's evolving Creation. We no longer identify as an isolated culture, or as a species so self-absorbed that it hoards all rights and privileges for itself.

Guided by this deepening into the wisdom of the Whole, the political, governmental, technological, economic, and other systems characterized by deep integrity would work to cleanse themselves of arrogance. They would function with respect, caution, and, ultimately, loving service. They would come from a place of caring for and valuing self, other, and the whole. A society in deep integrity would find where and how self-interest can and should intersect with the well-being of communities and nature—and it would embed those dynamics of integrity within its cultures and institutions.

The technological innovations would take into account impacts beyond their narrow purposes. Innovations that might be detrimental or dangerous—nanotechnology, bioengineering of crops, weapons of mass destruction—would move ahead only with the greatest care, regardless of potential benefits. Economic, government, and military activity would be sensitive to the wholeness of who we are as humans and to the fullness of the costs of institutional activity. "Full cost accounting" (or, "the triple bottom line") is a policy that already reckons the social and ecological (as well as financial) costs of proposed activities. Gone would be blind consumerism and profit seeking that externalize costs such that poor people, nature, and future generations pay for the shortsighted and "sinful" economic and governmental activities of today.

"How Do You Measure Sustainable Progress?"

In 1992 one hundred citizens—ranging from a corporate executive to an activist, from a priest to a teacher—formed the Sustainable Seattle Civic Panel. They wanted to build their city's long-term cultural, economic, and environmental health and vitality, with emphasis on long-term. Their work began with this question, "How does anyone know whether a community is becoming more or less sustainable, and how can sustainable progress be measured?"

They broke up into 10 topical groups: economy, education, health, environment, and so on. Each group brainstormed a long, lively list of possible measurements. The next step was to combine and winnow the contributions into a single list. After investing more than 2,500 volunteer hours in the project, the panel finally settled on 99 indicators of Seattle's sustainability. Their list included: hours of work at the median wage required to support basic needs; percentage of employment concentrated in the top 10 employers; wild salmon runs in local streams; county population and growth rate; average travel time from selected starting points to selected destinations; percentage of the population that gardens and votes in primary elections; and tons of solid waste generated and recycled per person.

Volunteers presented their list of indicators to the public in a dramatic reading interspersed with stories, readings, and poems.

Based on "Seattle Citizens Define Their Own Dow-Jones Average," a syndicated column by Donella H. Meadows

Transformed by co-intelligent methods, politics would increasingly resemble a respectful conversation. Governance would call on the wisdom of We the People to steer the ship of state. Educational and spiritual institutions would teach and explore co-intelligence and the evolutionary story that undergirds it. Those same institutions would also help each of us find our calling, as well as our practical vocation, such that our gifts could serve the world. All of society's systems would be redesigned to release the caring of its citizens—especially its youth and elders.

Cultivating Discernment Within the Whole

"When we look at the history of life on Earth, we see two complementary forces at work that, together, lead to healthy evolution. The first holds onto and preserves the learnings and contributions of the past. This is the conservative impulse. The other explores new possibilities by pushing beyond the status quo to ever wider circles of inclusion. This is the liberal impulse. Both are vital and necessary." —PAUL WEST

A society committing to evolutionary integrity would cultivate accountability and healthy feedback loops. Participants in these systems would review what was happening and notice consequences. Feedback loops would help us learn from our collective experience and detect early signs of success or failure. They would also help societies maintain balance where stability is needed and reinforce creativity where breakthroughs are needed. Information and communication flows would be vast yet accessible. Journalists would compete in presenting the full story—the whole, multiple-viewpoint story—in ways that engage the caring wisdom and response of the citizenry. Official secrets, censorship, lies, and media manipulation would be rare, and the population would be quick to counter them by rewarding whistleblowers and other risk takers on behalf of the whole. There would be a shared understanding that even painful truth is an engine of social evolution and democracy at its best.

These feedback loops are the collective manifestation of discernment. They are keys to the whole—We the People—being able to care for itself with wisdom. Thus leadership for collective deep integrity would value feedback loops that boost the self-organizing and self-evolving capacities of society.

There would be tremendous synergy between such a society and its citizens. A society growing into greater evolutionary integrity would offer healthy opportunities for all its members to satisfy their universal human needs for sustenance, safety, love, and community as well as challenge, creativity, leisure, freedom, and so on. Such a society would offer individual, cooperative, and competitive ways for people to go about satisfying those needs.

Nevertheless, "life, liberty, and the pursuit of happiness" (needs satisfaction)

would look very different in a society that evolves by way of co-intelligent social technologies. A society in deep integrity would have minimized the gross manipulation of those needs, establishing disincentives against activities that profitably addict people to "pseudosatisfiers"—activities, objects, and substances that claim to satisfy, but leave us hungry. A healthy evolutionary society would see addictive behavior, including consumerism, as a sign that people are not truly satisfying their basic needs—leaving them hungry for more, more, more—no matter how much they consume or acquire. A society in evolutionary integrity would foster feedback loops to hold accountable individuals, politicians, companies, and others who seek profit by promoting such addiction. It would become laughable for products to be sold using the tricks of sex and status. Voters would shake their heads in dismay at any politician attempting to sell a political agenda by way of raw fear and hatred. Such bizarre forms of devolution would simply not stand a chance in a society so attuned. Instead, people would be drawn in by positive visions, along with opportunities to create truly inspiring things together. These all would be deeply satisfying.

Thus we see the same manifestations of integrity at the collective level as we see at the individual level: *Trust* in the wisdom of the Whole and the humility that emanates from such trust; *Authenticity*, welcoming the fullness of whoever is present; *Responsibility*, harnessing the wisdom of the whole; and purposeful *Service* and celebration, in honor of our relationship with all existence.

From my evolutionary Christian perspective, the only hope I see for co-creating a lifegiving future (God's Kingdom "on Earth as it is in heaven") is to reincarnate such Christ-like values at all levels of society. This is our collective calling. The cultivation of such values in the sphere of collectives is surely the next evolutionary step for humankind. Tools can be accessed through such sites as co-intelligence.org and conservationeconomy.net. Among the groups we need to foster most immediately are those through which an evolving social/spiritual movement can mature.

Co-creating Our Evolutionary Spiritualities

"A wiki is a unique type of Internet site that eliminates the distance between the reader and the producer of information....

> *The main novelty of wikis derives from this total freedom of
> authorship and from the a priori trust granted to any Net
> user.... Wiki supporters contend that it is feasible to develop a
> collective intelligence that will function along the lines of a
> colony of honeybees or ants. Hence, it should be possible
> collectively to produce content or attain process excellence in
> ways we could only dream of in the past. In other words: The
> more, the smarter."*   —JÉRÔME DELACROIX

PROPHETIC INQUIRY: How does an evolutionary spirituality movement consciously evolve itself? What evolving knowledge base takes the
place of ideology and scripture, while retaining necessary levels of coherence and integrity?

The evolutionary spirituality movement has been evolving and producing its
sacred texts for decades, and indeed much longer. From the early explorations
of Julian Huxley and Pierre Teilhard de Chardin to the work of Epic of Evolution pioneers—Thomas Berry, Brian Swimme, Miriam MacGillis, Mary Evelyn
Tucker, and Eric Chaisson—to leading evolutionists such as Edward O. Wilson, David Sloan Wilson, and Ursula Goodenough to popularizers of evolutionary psychology and evolutionary brain science such as Robert Wright,
Steven Pinker, Matt Ridley, Michael Shermer, Joseph Chilton Pearce, and Paul
R. Lawrence to studious ponderings by members of the Institute on Religion in
an Age of Science to Ken Wilber's Integral Institute and Barbara Marx Hub-
bard's Foundation for Conscious Evolution to the eclectic writings and work of
Matthew Fox, Howard Bloom, Adrian Hofstetter, Duane Elgin, Elisabet Sah-
touris, Joanna Macy, Christian de Quincey, Loyal Rue, John Stewart, and
David Korten to the theological reflections of Denis Edwards, Leonardo Boff,
Ivonne Gebara, Choan-Seng Song, Chung Hyun Kyung, Vandana Shiva, Cletus Wessels, Joyce Rupp, Sallie McFague, Gene Marshall, John Cobb, Peter
Russell, Diarmuid O'Murchu, Jim Schenk, John Haught, Mary Coelho, Michael
Morwood, Marcus Borg, and John Shelby Spong to Kevin Kelly, Joël de Ros-
nay, Ray Kurzweil, and others who celebrate humanity's ever-increasing sym-
biotic relationship with technology to Don Beck and Chris Cowan's Spiral
Dynamics and Andrew Cohen's *What Is Enlightenment?* magazine to

progressive and Emerging Church leaders such as Jim Burklo, Richard Rohr, Brian McLaren, and Spencer Burke to Joel R. Primack and Nancy Ellen Abrams's *The View from the Center of the Universe* to Connie's and my itinerant Great Story ministry along with the contributions of countless others: we are collectively discovering the wonders of this emergent Cosmos, its amazing journey, and our God-given calling to contribute consciously—mindfully, heartfully—to this stupendous evolutionary adventure.

As more people join in this effort, the mix will enrich. Participation and leadership will widen, strengthening a spiritual movement that becomes as vital and evolutionary as it is unprecedented. We are diving into the creative unknown (where God lives!) and invite you to join us. The Internet is a leading edge of evolutionary emergence today. Here we can connect, converse, and collaborate with one another. Our companion website links to a wiki and other interactive portals where contributors can co-create and share songs, practices, activities, resources, and visions of this movement in and around every spiritual tradition. (See the Online Resources, p. 381.)

CHAPTER 16

Knowing the Past
Reveals Our Way Forward

"The farther backward you can look, the farther forward you are likely to see."   —WINSTON CHURCHILL

The Great Story of our immense journey contains crucial lessons for guiding humanity safely through the dangers and confusions evident today. This grand epic will propel us forward in a spirit of expectant curiosity. We will place our trust not only in the Whole but also in our own species' capacity to serve as the vessel through which the evolutionary impulse is most active at this time. In this chapter, we shall visit the highlights of our Big Picture story as publicly revealed through the sciences. We will ground ourselves in the most accurate understanding available today of where we are in time and space and who we are as children of God.

The Cosmic Century Timeline

"What is possible versus impossible depends entirely on what universe you're living in. Until you understand the universe you're living in, you cannot know what is possible."

—JOEL R. PRIMACK AND NANCY ELLEN ABRAMS

It is difficult to imagine timescales on the order of millions and billions of years. Too large to personally experience, they remain abstractions. To surmount this limitation, let us imagine the entire 14-billion-year history of the Universe—from the Great Radiance (Big Bang) to the present—as compressed into a single century, just 100 years. Our "Cosmic Century Timeline" is thus scaled in these ways:

Cosmic century	=	14 billion years
Each decade	=	1.4 billion years
Each year	=	140 million years
Each month	=	12 million years
Each day	=	400,000 years
Each hour	=	15,000 years
Each minute	=	250 years

If we place the Big Bang, or Great Radiance, on the timeline at one second after midnight on January 1, Year 0, and with today being one second before midnight on December 31 of the 99th year, then the first atomic elements (hydrogen and helium) would have formed when the Universe was just two days old. By then, this infant Cosmos had cooled just enough for subatomic particles to aggregate. By the time our Universe was ten "years" old, more than a hundred billion galaxies would have formed, and soon some would begin to merge into very large spirals indeed. Our home galaxy, the Milky Way, is one of these monstrous spiral galaxies—grown large over time by consuming smaller companions. The Milky Way is 100,000 light-years across and 16,000 light-years thick at its central bulge. A light-year is a distance measurement, defined as how far a beam of light travels in one year—at a speed of some 700 million miles per hour (11 million miles per minute).

Stars within our galaxy are in motion, collectively as well as individually. Our own star, the Sun, resides about two thirds of the way out from the center of the galaxy. For us, home is situated in a brilliant spiral structure known as the Orion Arm. The Sun, along with Earth, completes its trip around a massive black hole at the center of the Milky Way every 200 million years. That scales to about a year and a half on our Cosmic Century Timeline.

Before any spiral galaxies had assembled, there were stars. Many of these earliest stars were gigantic. By the force of their tremendous gravity, they fused primordial hydrogen into heavier elements at such a clip that they ran out of fuel and exploded as supernovas in just a few tens of millions of years. None of the first generations of stars had rocky planets around them, because there were as yet no carbon, calcium, silicon, magnesium, aluminum, and iron atoms from which to assemble such planets. Planets would emerge only following the birth and death of many suns.

Our own solar system formed from "clouds" of primordial hydrogen gas enriched by the complex atoms of exploded (recycled) stardust when the Universe was 67 "years" old on the scale of our Cosmic Century Timeline. It was this stardust that composed the bulk of the inner, rocky planets, because that close to the Sun, any residual hydrogen and helium gases would mostly have been stripped away from planetary surfaces by the force of stellar winds. The third rocky planet was blessed with an atmosphere of just enough heat-trapping gases to ensure that (eventually) water would be in liquid form at that distance from the Sun. It was large enough for a long-lived molten core to form, thus ensuring the life-replenishing flow and cycling of atoms and molecules within its fracturing and heaving crust.

When the surface temperature of our infant Earth cooled below the boiling point of water, something happened that had never happened before. It rained! That first rainfall would continue for eons. Thus, our shimmering oceans were born and periodically added to by incoming icy comets. The Universe was then 69 on the scale of our Cosmic Century Timeline. Our Sun and all its planets were just two "years" old. Earth came alive in the spring of 71, as ancient bacteria and archaea burst onto the scene—probably within Earth's crust and at hot springs on the ocean floor long before life was capable of surviving at the planet's surface.

Bacteria and archaea are unquestionably the most important expressions of planetary life. Within them, virtually all the metabolisms still in use today— all the ways of combining and disassembling and extracting energy from atoms and molecules—were tested and perfected. All other forms of life descended from these elders. More, all of us younger folk still depend on their ability to refresh the air, to turn atmospheric nitrogen into fertilizer, and to decompose and recycle any molecule that life cares to generate. Bacteria and archaea would do just fine without us; we would not last a day without them.

By the time the Universe celebrated its 72nd birthday, Earth life had tapped the energy of sunlight for the purpose of extracting hydrogen atoms from water, H_2O. This evolutionary innovation ramped up the number of carbohydrates in circulation within the web of life, and thus the mass of life itself. This was the miracle of photosynthesis. But the miracle was accompanied by release of a toxic by-product. When hydrogen was extracted from H_2O, oxygen gas was shed as waste. Poisonous oxygen gas would only much later become a gift—a gift that made possible the emergence of animals and the high-energy demands of complex animal brains.

But we are getting ahead of ourselves. Animals have not yet evolved in our story. Beginning in the year 72 of our Cosmic Century Timeline, oxygen gas combined with dissolved iron in the oceans, leaving a thick residue of rust on the ocean floors and continental shelves (which humans are now exceedingly grateful for, as these concentrated deposits are where we find industrial-grade iron). Oxygen also bubbled up into the atmosphere, accumulating there until the great pollution crisis of 87 poisoned anaerobic bacteria, thus threatening to wipe out life at Earth's surface (though not, of course, deep within the crust, where there was no sunlight to power photosynthesis). This first environmental crisis was averted when the struggling little holons of bacteria came together and learned to form cooperatives in which some members would render oxygen harmless while others continued to handle the tasks of acquiring and using energy and materials. This was the miracle of *symbiosis*.

In the early 90s, Life embarked on a number of new paths. Life discovered a new way to recombine genes (sex) that would inject just the right amount of novelty into later generations. Life became larger when the first multicellular beings evolved. And as individuals grew larger, Life figured out ways to ensure that no being could step outside the circle and escape predation. At one time, "predators" were tiny spirochete bacteria that burrowed into larger bacteria, consuming their prey from the inside out. Then amoebas came along. Even though amoebas were still single cells, they were large enough and flexible enough to wrap around a smaller being and encase it in a kind of ad hoc stomach, a vacuole into which digestive fluids could be secreted. Finally, when Life started packaging itself in multicellular forms, amoebas could no longer do the deed. It was time for teeth to evolve. Predator and prey were now engaged in a dance that would unfold novelty upon novelty—

foremost the drive to hear more, see more, and the quest to get smart. Brains (in a wormlike ancestor) emerged in July 95. Backbones appeared a year later.

Finally, in late 96, living beings came ashore, soon evolving ways to stand tall despite the pull of gravity and to prevent desiccation in the dry air. Lichens and simple plants were the first pioneers, followed soon by the arthropods (insects, millipedes, arachnids) and then small mollusks (snails), who could remain moist in their shells, hunkering down until the next rain. The first amphibians emerged four months later. Reptiles and conifer trees ventured onstage in December 97. The dinosaurs appeared in March 98, flourished, and then vanished a year later when Earth was hit by an asteroid off the coast of what is today Mexico. Mammals had begun to nurse their young before the asteroid struck, and at least a few of them managed to survive the holocaust, probably underground. Likewise, the first birds had diverged from the other dinosaurs, and they, too, survived the planetary crisis.

By the first week of April in the year 99, in Earth's 31st year—just 8 months ago on the Cosmic Century Timeline—our planet's continents (which have been in motion all along) were adorned in vibrant colors, ecstatically celebrating the diversification of flowering plants and the insect pollinators whose coevolutionary urgings had made this feat possible. Our not-so-distant ancestors, the primates, began monkeying around only a few months ago. The earliest bipedal apes (hominids) rose up on two legs and looked out across the African savanna less than two weeks ago, on December 20. The first species classified as fully human, *Homo habilis*, appeared in Africa on December 25 of the 99th year. (To use night language, God incarnated into human form on Christmas day of the 99th year on the Cosmic Century Timeline.) Our ancestors domesticated fire during the early morning hours of December 29. *Homo sapiens* emerged just 24 hours ago, at the beginning of the 365th day of the Universe's 99th year of existence.

Ours has been an amazing journey! Moreover, it was everybody's doing. It was the power of the collective, enveloped in the nurturing and sometimes challenging conditions of Earth and Cosmos, that carried the journey forward—that gave time a direction: simple to complex, same to diverse, passive to active, unconscious to conscious. Any being who ever lived made a mark, at least in some small way, on how that journey unfolded.

At no point during the past four and a half billion years (33 years on the Cosmic Century Timeline), since the origin of Earth, need anyone have entered from the outside and intentionally placed anything on this planet. Nevertheless, God was active at every moment, at every critical juncture. *God*, as I have been using the term, is no less than a holy name for Supreme Wholeness, that Ultimate Creative Reality that brought everything, step-by-step, into existence.

> *When Genesis 2:7 speaks of God forming us from the dust of the ground and breathing into us the breath of life, this is a night language way of describing the evolutionary epic that I have summarized here.*

This story has no end. Creation is ongoing. God created polar bears (from brown bear ancestors) only three hours ago. Emergence is such a blend of spontaneity and necessity that we have no evidence whatsoever that Life "knows" what comes next—at least in its luscious particulars. God has revealed this stunning fact, thanks to the public revelations made possible through the complexity sciences and the computers on which those sciences depend. This leads me to think that God enjoys surprise and open-ended adventure as much as we do. What an amazing and creative Cosmos! What an awesome and companionable God!

How do we envision and relate to the life journey yet to come? Like it or not, our technologically powerful species is responsible not only for its own fate but also for how the odyssey of evolution will continue to unfold on this planet. How will our descendants living 10,000 years hence—a half hour from now on the hundred-year time scale—how will they tell the story of *our* times? Will there even be a human expression of Earth in 10,000 years? If our lineage does survive, it will not be because everyone converted to a single belief system—nor because we all "got it" in some spiritual or ecological awakening. The crucial factor in our success will have been that we learned to align the self-interests of individuals and groups with the well-being of the entire body of Life.

Aligning Self-Interest with the Well-being of the Whole

> *"Three thousand million years ago, cooperation extended only between molecular processes that were separated by about a millionth of a meter, the scale of early cells. Now, cooperation extends between human organisms that are separated by up to twelve million meters, the scale of the planet. The same evolutionary forces that drove the expansion of cooperative organization in the past can be expected to continue to do so in the future."* —JOHN STEWART

As discussed briefly in Chapter 4, the evolution of human consciousness is driven by how information is stored and transmitted. A mutually reinforcing relationship tracks human social complexity with increasingly sophisticated "technologies of the word." The human brain, as best we can tell, has not changed structurally in any significant way since *Homo sapiens* first evolved. Yet people do not think the same today as they did a hundred generations ago. Why? Because our brains are now immersed in a swirling world of information flows and interactions that span the globe.

With each advance in data representation and communication, worldviews shift and societies reorganize. For societies at each new level of complexity and size to thrive, they must find ways to align the natural self-interest of individuals and groups of individuals with the well-being of the social whole, and to keep cheaters in check. The impact of the parts, for good or ill, must be mirrored back to the parts in congruent and consequential ways. If a part benefits the whole, the part must benefit in some way; if a part harms the whole, it must be disadvantaged in some way. These kinds of social structures, incentives, and disincentives drive the synergistic alignment of interest between part and whole. It is in this way (and only in this way) that complexity can continue along, what I like to call, "the trajectory of divine creativity." A helpful overview of how this natural process of escalating complexity is thought to have unfolded, both in the prehuman world and throughout human history, is John Stewart's *Evolution's Arrow: The Direction of Evolution and the Future of Humanity*, which is downloadable from his website (www4.tpg.com.au/users/jes999). Summarized below is how complexity evolved in the human realm.

Families and Clans. For hundreds of thousands and possibly even millions of years, families and clans were our most complex and interdependent social organizations. Prior to symbolic language—before humans could think and speak in words—cooperation was achieved at a scale of a few dozen individuals at most. Others were experienced as a threat, and they, of course, had their own isolated pods of extended relationship. Like our reptilian, mammalian, and primate ancestors before us, "eat, survive, reproduce" was, in effect, our religious creed. It wasn't, however, a dog-eat-dog world. Our genetic heritage predisposed us to care about and form rewarding emotional bonds with genetic kin and sometimes beyond kin to include those with whom we engaged in reciprocal transactions (small-scale trade), including acquisition of mates.

Tribes and Neolithic Villages. Symbolic language likely emerged within the last half million years (and possibly much more recently than that). This evolutionary innovation facilitated a new level of social complexity. For the first time, thanks to beliefs and moral codes made possible by speech, *tribes* came into existence: cooperative, interdependent organizations of several hundred human beings. Individuals, even today, are genetically and culturally predisposed to engage in reciprocal cooperation at this scale. We are also genetically and culturally predisposed to want to expel or punish those who transgress against established ways of our tribe—ways vital for social cohesion beyond the level of family and clan. Tribes in existence today evidence diverse ways of securing retributive justice (including the simplest of all: shunning) against individuals who cheat, free ride, or otherwise transgress norms perceived as important for the well-being of the whole. Tribes or ethnic groups other than one's own may be feared and demonized. Norms and inculcated morals are taught and reinforced by shared magical beliefs and collective rituals that call on invisible external powers (spirits) to assist or punish tribal members and others. The advent of horticulture led to the flowering of Neolithic villages that operated similarly.

Chiefdoms and Kingdoms. Within the last seven to ten thousand years (less than an hour ago on the Cosmic Century Timeline), a new level of social complexity emerged: chiefdoms and kingdoms. Among the innovations that drove this scale of cooperation were the further domestication of plants and animals, and the adoption of tokens and clay impressions that enable debts and favors to be recorded and tracked. This emergent social level brought with it the beginnings of external governance. A chief, king, or warlord wielded the power to reward behavior that served the chiefdom (and typically the chief, too) and to punish behavior that did not. This external governance overlaid and built upon the family bonding, tribal reciprocal cooperation, and moralistic punishment of earlier manifestations of social organization. Notably, this is the earliest form of social organization that still powerfully affects the course of world events. My own country initiated a war that it seemingly cannot win, owing in part to the hold that this scale of social organization still exerts in some regions of the Middle East.

Theocracies, Early Nations, and Empires. With the emergence of writing and mathematics (invented in several regions of the world), cooperation and complexity expanded yet again. External governance could now be more intelligently and flexibly applied. Such governance was supplemented and legitimized by myths and religious practices that predispose individuals to support their particular theocracy or nation. As with tribal and village levels of social organization, there is a stark difference between in-group and out-group moral practices and expectations. But in the case of theocracies, early nations, and empires, the in-group is vastly larger. New forms of communication and data processing enhance the spread and retention of these useful myths and the rise of national identity. The religions of nations and empires tend to be inclusive religions because they must knit together far more diversity (in heritage, even languages) than required at simpler levels of social organization. Arguably, it is the emergence of a shared holy story within each of the inclusive world religions—such as Christianity, Islam, and Buddhism—that enables disparate ethnic groups to peaceably interact in large-scale societies.

Democracies, Corporate States, and Global Markets. At this pivotal stage, ongoing in many parts of the world and commencing in the West during the 18th-century Enlightenment, myths and religions sourced solely in private revelation began to unravel. Their tenets taken literally are contradicted by empirical evidence birthed by science and worked upon by reason. External governance becomes more responsive and gains a new legitimacy through the introduction of democratic processes. The printing press, mechanistic science, and the industrial revolution (with its increasingly sophisticated communication and information technologies) make possible interdependent organizations of millions of people spanning vast geographic areas. Democratic groups benefit their members, in that they are governed by their members (self-governance). Notably, empathy moves to a new level—individuals gain the abstract cognitive capacity to "put themselves in the shoes" of individuals whom they may never have met. (Before this stage of development, empathy is concrete; it applies only in relation to peoples and individuals who are directly experienced, or well known.)

An expansion in the scope of empathy means that moral codes are internalized: individuals feel good when they help others, and they suffer remorse when they injure or fail to help others. The rise of this cognitive capacity led to the abolition of slavery, and the recognition of universal human rights. At this scale too, it would be difficult to overstate the importance of a unifying narrative to facilitate cooperation across ethnic, religious, and socioeconomic differences. Examples would include: creating holidays to celebrate glorious events in a nation's history, reciting a pledge of allegiance, singing a national anthem.

Social Democracies and the United Nations. With the emergence of this level, external governance becomes even more important, and democracy and responsiveness spread. Now there is a capacity for global empathy, thanks to electronic communications and advanced ways of accessing and spreading information and ideas. International governance is pursued mainly by consensus and negotiation, though it is not yet particularly effective at aligning corporate and national interests with the needs of the whole planet. Such limitation should not be surprising, as this level lacks a unifying sacred story, sacred songs, and shared rituals.

The Emerging Global Civilization and Planetary Governance. At this level, which is only now entering the realm of possibility within the hopes and imaginings of a tiny segment of humanity, all the nested levels of inner predispositions and external supports would be overlain by a system of global governance. This global system would be entrusted with the task of ensuring that the interests and actions of lower levels of social organization (e.g., nations and international corporations) are aligned with the interests of the global society. It would be in the self-interest of the parts to pursue only just and ecologically beneficial actions. The wise application of the principle of subsidiarity would be crucial at this scale. Tasks would be assigned to and executed by the lowest level of governance or social organization capable of doing the job. The fall of the Soviet Union provides a superb morality tale of what happens when the principle of subsidiarity is ignored. Recent acclaim (including the award of the 2006 Nobel Peace Prize to Muhammad Yunus and Grameen Bank) for the microcredit, microfinance movement is a fine example not only of the benefits of subsidiarity but also of the value of incorporating traditional ways (peer pressure) to enlist our status-conscious Furry L'il Mammals in service of institutional success.

A global scale of social organization might eventually organize or perhaps catalyze the matter, energy, and living processes of the planet (including machines and artificial life) into a mutually enhancing, symbiotic, synergistic whole. Attuned to the dynamics of self-organization (including the benefits of subsidiarity), governance at all levels would become more intelligent, more responsive, and less restrictive of freedom. The constraints applied by governance would be flexibly responsive, working in tandem with the perceived self-interests and internal dynamics of individuals and collectives. Most effective would be constraints, inducements, and accountability at the minimum scale required to effectively align component interests with the planetary system. Governance systems at all levels would evolve more rapidly, along with new means to innovate and select forms of governance, including both competitive market mechanisms and cooperative open-source sharing of new approaches and tools.

For this level of complexity and cooperation to emerge, nothing is more important than the widespread, enthusiastic adoption of a shared sacred story grounded in science as public revelation. This sacred story must validate the heart of earlier creation stories while helping all of us see that we're part of the same adventure, moving into the future together—or not at all. I agree with David Sloan Wilson, who writes, "I look forward to the day when evolutionary theory becomes part of the basic training for all people who study and run our governments and economics. This evolutionary understanding will increase our collective social intelligence so that we can manage our affairs more successfully in the future than we have done in the past." I would add, "I look forward to the day when evolutionary theory becomes part of the basic training for all people who study for the ministry and who speak from the pulpit or in any religious setting."

Who and What Are We, Really?
And Why Are We Here?

"I am the eye with which the Universe beholds itself and knows it is divine." —PERCY SHELLEY

A glorious interpretation of cosmic history provides a fresh way of identifying who and what we are and why we're here. Such an understanding must make sense both scientifically and religiously. It must inspire billions of people with disparate worldviews to join in common cause of ensuring a just and thrivingly healthy, beautiful, and lifegiving future—and doing so to the glory of God.

Who are we? The day language answer is that we are the Universe becoming aware of itself, Nature uncovering its own nature, Cosmos exploring its very essence. A night language answer, given by Meister Eckhart in the 14th century, is "We are God's sons and daughters, though we do not yet realize it." Yes! We are children of the Most High, called to be the hands and eyes and ears and heart of the Holy One for our time. That's who we are; that's who we *really* are!

What are we? "We are stardust, billion year old carbon," sang Joni Mitchell (and Crosby, Stills, Nash, and Young) in the 1960s. This is now an undisputed scientific fact, and it could not have been known until the last century. If this isn't religious knowledge, nothing is! We are also an amazing outgrowth—a fruit—of our galaxy. As Brian Swimme and Thomas Berry remind us in their book, *The Universe Story:*

> Tonight, on every continent, humans will gaze into the edge of the Milky Way, that band of stars our ancestors compared to a road, a pathway to heaven, a flowing river of milk.... Humans tonight will watch the Milky Way galaxy not only with eyes, but also with radio telescopes, satellites, computer-guided optical telescopes, with minds trained by the intricate theories of the composition, structure, and dynamic evolution of matter.... The mind

that searches for contact with the Milky Way is the very mind of the Milky Way galaxy in search of its inner depths.

Much of what has been publicly revealed over the course of the last few centuries is astonishing. More, we have learned that to understand the Universe, our galaxy, or planet Earth, we must understand the human—and vice versa. Our sense of reality is profoundly shaped by our biological and cognitive systems, our language, our culture, our place in the evolutionary story of the Universe. Conversely, the deepest truths about humanity cannot be understood apart from appreciating the nested and creative character of Cosmos and Earth, and how it is that we are blessed and burdened by the deep evolutionary roots of our quadrune brain.

Our planet was not merely created; Earth itself is creative. We humans are an expression and a glorious extension of Earth's ongoing creativity. Earth not only has life *on* it; in a very real sense ours is a living planet. Earth's physical structure—its core, mantle, crust, and continents—is the skeleton or frame for life at the largest scale that we know. The soil is a massive digestive system on which land-based ecologies depend. The oceans, waterways, and rain function as a circulatory system. They transport the blood that nourishes and purifies the body of Life. Photosynthetic bacteria, algae, and land plants serve as the planet's lungs, ever regenerating the lifegiving properties of Earth's atmosphere. Within the animal realm, a nervous system emerges in which every sensing cell and creature is a node in the net. It is a sensory system finely tuned and diversified for detecting and reacting to environmental change.

Each species is a unique expression of the collaborative creativity of Earth and Sun within the Milky Way. Each species brings its own particular gifts to the body of Life. Humanity is the vessel through which our planet now experiments with self-conscious awareness. That is, the human enables Earth to perceive and ponder its existence—and to perceive and ponder the divine Mystery out of which everything, big and small, has arisen. Humanity is a means by which Nature can appreciate its own beauty and savor its splendor. This role is not without paradox, however, as our kind is also that which threatens Earth with degradation and diminishment.

The move from seeing ourselves as separate beings placed *on* Earth ("the world was made for us") to seeing ourselves as a self-reflective expression *of* Earth ("we were made for the world") is an immense transformation in human identity.

Our Sense of Self and Our Role in the Body of Life

"What is wrong with our culture is that it offers us an inaccurate conception of the self. It depicts the personal self as existing in competition with and in opposition to nature.... We fail to realize that if we destroy our environment, we are destroying what is in fact our larger self." —FREYA MATHEWS

An important truth revealed publicly by God in recent decades, through the community of scientists, is that in a very real and measurable way one's "self" does not stop at the outermost layer of skin. When we call to mind the nested nature of reality, this truth becomes obvious. But how often do we allow this wondrous revelation to color the conduct of lives?

What we imagine as our environment is actually not "out there" separate from us. Rather, each one of us is inextricably linked to vast, ancient, and potent cosmological, geological, and biological processes. I know that my body exchanges matter, energy, and information with "the environment." The atoms that I collectively call "me" are not the same as those that made up my body a year ago. Every five days I get a new stomach lining. I get a new liver every two months. My skin is replaced every six weeks. Every year, 98 percent of my body is refurbished at the level of atoms. The particles that are continually becoming "me" come from the air I breathe, the food I eat, the water I drink. I know that these same atoms were, not long ago, supporting the identities of fish and cows, birds and trees, earthworms and algae, and my fellow humans. I give out as I take in. It thus makes little sense to overly identify with my physical self, for that is only a small part of who I am. My larger body is the body of Life itself. Earth is my larger self.

> If the Rhine, the Yellow, the Mississippi rivers are changed to poison, so too are the rivers in the trees, in the birds, and in the humans changed to poison, almost simultaneously. There is only one river on the planet Earth and it has multiple tributaries, many of which flow through the veins of sentient creatures.   —THOMAS BERRY

A living body is not a fixed thing but a flowing event, like a flame or a whirlpool: the shape alone is stable, for the substance is a stream of energy going

in at one end and out at the other. We are particular and temporarily iden-
tifiable wiggles in a stream that enters us in the form of light, heat, air,
water, milk, bread....It goes out as gas and excrement—and also as semen,
babies, talk, politics, war, poetry, and music.   —ALAN WATTS

Here is a parable: Once upon a time, a group of brain cells debated the importance of the rest of the body. Some suggested that the body was dispensable.
"After all," said one, "we are the only cells in the body that know that we know
things." "Only we can reflect on our dreams," said another, "so we must be the
only part of the body that is spiritual, right?" "Just think of the awesome
accomplishments we are capable of!" Occasionally a brain cell would protest,
arguing that the body was no less important, but to no effect. You see, the
brain cells had convinced themselves that Great Mind lived outside the body
and could be known only through their dreams. They believed that they were
destined to leave the body and dwell in a place called heaven. They also assumed
that the rest of the body was not really alive at all, that it was an inexhaustible
supply of "resources" for the benefit of the brain. . . . Sound familiar?

"We're Acting Like Cancer Cells"

Several years ago, during the Q&A period following one of my presentations, someone asked, "If we are Earth becoming conscious of itself, why
are we spoiling the air, water, and soil, as if we knew nothing?" Another
participant immediately raised a hand and asked if he could respond.

"I'm an oncologist," the man began. "I work with cancer patients every
day. From my vantage point, we are inadvertently destroying our larger
body because we lack evolutionary guidance. We're acting like cancer
cells, rather than immune cells."

He continued, "A cancer cell is a normal cell that, for one reason or
another, loses its genetic memory. Cut off from the wisdom of millions of
years of developmental guidance, it stops cooperating with the rest of the
body. It experiences itself as separate from the body, overpopulates, and
proceeds to consume the very organism that supports it."

The man paused, and then asked rhetorically, "We call our society a

consumer society, and to consume something is to eat it up, right? I believe
we are consuming the planet because, like cancer cells, we've been trying
to live without evolutionary wisdom."

What a powerful metaphor! I have been using it in my talks ever since. I
predict that when millions begin celebrating our deep-time story in religious ways, that will be the moment when we cease acting like a cancer.
We will assume an entirely new role within the body of Life—as an immune
system, searching out problem areas and protecting our communities.

Joanna Macy, a systems thinker and deep ecologist, makes a similar point. She
contends that the shift from seeing ourselves as separate beings *on* Earth to
seeing ourselves as a mode of being *of* Earth "is essential to our survival because
it can serve in lieu of morality." She explains,

> Moral exhortation doesn't work. Sermons seldom hinder us from follow-
> ing our self-interest as we conceive it. The obvious choice, then, is to extend
> our notions of self-interest. For example, it would not occur to me to plead
> with you, "Oh, don't saw off your leg. That would be an act of violence." It
> wouldn't occur to me because your leg is part of your body. Well, so are the
> trees in the Amazon rain basin. They are our external lungs. And we are
> beginning to realize that the world is our body.

Why are we here? That is, what is our role in the body of Life? What contributions can we make, individually and collectively, in the evolutionary
process?

Praise God! We now have a way of understanding the role of the human in
cosmic evolution that makes sense both scientifically and religiously. Using
traditional Christian night language, we might say:

> Our purpose, individually, is to grow in Christ and to support one another
> in staying true to God's Word and God's will. Collectively, we are here to
> create Christ-centered institutions that glorify God and embody the values
> of the Kingdom.

A day language way of saying the same thing might be:

> Our purpose, individually, is to grow in trust, authenticity, responsibility, and service to the Whole, and to support others in doing the same. Collectively, we are here to celebrate and steward what Life has been doing for billions of years and to devise systems of governance and economics that align the self-interest of individuals and groups with the well-being of the larger communities of which we are part.

We are here, as well, to love as broadly and as deeply as we possibly can— knowing that we cannot do this without the support of the entire community of Life. Our purpose is to consciously further evolution in ways that serve everyone and everything, not just ourselves. This is our calling. This is our Great Work. Indeed, this is our destiny!

> The human is the being in whom the Earth has become spiritually aware, has awakened into consciousness, has become self-aware and self-reflecting. In the human, the Earth begins to reflect on itself, its meaning, who it is, where it came from, where it's going. So in our deepest definition and its deepest subjectivity, humans are the Earth—conscious.
>
> —MIRIAM MACGILLIS

> All human activities, professions, programs, and institutions must henceforth be judged *primarily* by the extent to which they inhibit, ignore, or foster a mutually enhancing human/Earth relationship.
>
> —THOMAS BERRY

> When the Newtonian picture destroyed the comforting medieval universe and people stared out into endless space and shivered at how small they were, they felt for the first time the existential terror of cosmic insignificance. But even though the universe is overwhelmingly larger than those 17th century people imagined, we humans are not insignificant, because we are citizens of the luminous and rare; the tremendous complexity of our minds lets us do what no amount of dark matter or dark energy can ever do.
>
> —JOEL R. PRIMACK AND NANCY ELLEN ABRAMS

Evolutionary Revivals

"I really like the idea of an evolutionary revival. Revivals are American. Revivals evolved to meet a spiritual need. Why let the anti-evolutionists have all the fun?"   —RUSS GENET

The epigraph by our friend and colleague Russ Genet was one of a number of supportive responses that Connie received to an email she sent out tendering the idea of "evolutionary revivals." This happened so recently that she and I together are easily able to retrace its genesis. It all began when a friend alerted us to an interview with Harvard's esteemed biologist and Pulitzer Prize–winning author: Edward O. Wilson. Wilson's book *The Creation* had just been released. Here is an excerpt from an interview that appeared in *The Washington Post*:

> It's hard to picture, if you know him only by his scientific reputation, but E. O. Wilson confesses it freely: He loves watching preachers on television. Wilson is an internationally renowned biologist who has based his extraordinarily productive five-decade career at that great bastion of secular humanism, Harvard University. At 77, his work and his worldview are so thoroughly entwined with Darwinian theory that they're impossible to imagine without it. His reverence is for the wondrous creatures and intricate interconnections of the natural world, not for any supreme being. So what's he doing tuning in to those evangelical sermons from the megachurches?
>
> "I listen to them the way an Italian listens to opera," Wilson confesses with a lopsided grin. "I may be thinking of the texts as fiction, but I can't resist the old-time rhythm, the music, and the *superlative* performances."

When Connie read this interview, it all came together for her. Six months earlier, she and I, out of curiosity, had attended a Wednesday evening service at one of America's largest megachurches, Lakewood Church in Houston. Even at this midweek service, some 8,000 people had gathered to sing and sway and pray together. Many were "saved" that night—including Connie.

How could that have happened? Well, Connie loved the music; she sang with arms raised not unlike the most enthusiastic of those gathered. Several times throughout the service the guest preacher invited the crowd to participate in a call and response, which Connie was happy to do. After one such call and response, the speaker declared, "If you just prayed that prayer with me, then the Bible says that you've been saved!" Wide-eyed, Connie shot me a stunned glance. All I could do was smile back and think to myself, "Whoa, that was smooth." The speaker continued, "Now all of you who have made this step for the very first time, please raise your hand—raise it high, so our ushers can bring you a special gift packet." Connie cowered, and kept her hands at her side. I whispered playfully in her ear, "Saved one minute; backslid the next!"

Something about the music, the infectious enthusiasm of the crowd, the opportunity to stand and sway and let one's body express emotion with no hesitation—something about all that made a deep impression on my CREATHEISTIC wife. So when "the great E. O." (as she affectionately refers to Wilson), a renowned humanist as well as scientist, was willing to speak of his hobby of watching televangelists, Connie felt that maybe it would be okay for her to take on this hobby too. Right away, she went online and within a day or two had watched all the half dozen previous services at Lakewood that had been made available for onstream viewing. All of them featured sermons by the boyishly attractive Pastor Joel Osteen, who deftly uses scriptural passages to ground his self-empowerment message in biblical fare, while eschewing the kind of threatening material (hell and damnation) that repels so many.

Then she began to make room in her schedule to catch the live broadcasts too. She encouraged her colleagues to watch Lakewood's videocasts and asked them to report their reactions. Following are three of the comments she received. What surprised both of us was that some of our colleagues (none even close to being evangelical) had already been tuning in to these broadcasts:

✦ "I mentioned my enthusiasm for Joel Osteen to a Jewish/Atheist friend yesterday and he said that he and his wife (a former Christian/Agnostic who is intrigued by the Great Story) just love to watch Osteen."

✦ "I asked our [teenage] son, Ryan, about his perspective on what was "fun" at the PeaceJam event, which he attended last weekend (3,000 youth from all over the world listening and having conversations with Desmond Tutu, the Dalai Lama, and eight other Nobel Peace Prize winners for three days). Ryan's immediate response was, "It was when Desmond Tutu had us all stand up and wave our arms and scream, *I am a very special person!*" So, yes, there is something about having people include their bodies and voices as well as their ears that imprints the experience into each of us. It enlivens all of who we are, and our emotions also find a place in such a setting. Hooray! An Evolutionary Revival experience is just what I'm ready for."

✦ "Our church board had a recent retreat and read *The Almost Church* by Michael Durall in which he claims that [our] Unitarian Universalist churches, as well as most mainline churches, are declining and will near extinction over the next 50 years as the megachurches and fundamentalist churches grow and provide what people (especially young people) are seeking. I would hope that your approach would bring a refreshing revival/renewal to those churches that don't agree with the fundamentalist approach. We need something that will give people hope and inspiration. I actually attended a local megachurch last week to observe what they do. Wow! It was more like a rock concert, but held out the promise of a better life that people bought into. It also brought a glowing feeling of loving and belonging (again, like the audience in a rock concert). How can we compete with that?! I thought your presentation was intellectually and esthetically appealing, but can it give the inspiration we need to compete with the churches that promise salvation and eternal life? The ignorance of science and history in that movement is appalling, but can that ignorance be overcome?"

Since then, Connie has been promoting the idea of *evolutionary revivals* whenever and wherever she gets the opportunity, especially in her guest sermons at Unitarian Universalist churches. Not only has she been scouting for music that could be used for evolutionary, cross-denominational praise worship, she has coaxed her favorite singer-songwriter, Peter Mayer, to allow us to compile and sell 13 of his original songs that best express the possibilities of this movement. (*Peter Mayer Sings The Great Story* is available online through our website, ThankGodforEvolution.com.) My own wish is that Peter Mayer is the first of many. Oh for the day when Bono and U2, Van Morrison, Kenny Loggins,

Bruce Cockburn, and others might add their special talents to music-making for evolutionary revivals—and when musicians I am not even aware of, but who appeal to younger folk, will do the same for that age group. I also envision Great Story poets and rap artists, such as Drew Dellinger, performing their magic. What lively and meaningful experiences these events will be . . . praise God for the possibility of evolutionary revivals!

CHAPTER 17

Beyond Sustainability: An Inspiring Vision

> *"There is science now to construct the story of the journey we have made on this Earth, the story that connects us with all beings. Right now we need to remember that story—to harvest it and taste it. For we are in a hard time, a fearful time. And it is knowledge of the bigger story that is going to carry us through."*
> —JOANNA MACY

For two years the main workshop I offered, in both religious and non-religious settings, was "Beyond Sustainability: An Inspiring Vision of the Next 250 Years." The reason I chose 250 years was that every minute on the "Cosmic Century Timeline," which I introduced in the previous chapter, is 250 years. The question I posed at the outset was this: Given a hundred "years" of patterns and tendencies, what can we expect in *the next minute* of cosmic evolution?

In Chapter 16, I reviewed some of the milestones in life's four-billion-year journey. I also provided a crash course in human history at a very coarse scale, that is, the major advances in complexity, interdependence, and cooperation that mark the past 100,000 years—about seven hours on the Cosmic Century Timeline. During this blip of cosmic time, our ancestors invented a variety of social structures; this trend continues today. We are living in a time of freshly minted corporate states and global markets, with a dash of democracies and the United Nations stirred into the mix.

As we have seen, whenever evolution brings forth greater complexity, cooperation, and interdependence, new challenges and dangers are born, too. When a new holon emerges, it must take responsibility for the well-being of all the levels it contains. Single-celled organisms were free-floating and had to take care of their internal organelles. The cells that make up our bodies now depend on our sentient selves to provide them with food, water, and other necessities. In human culture, when empires arose, tribes and families became dependent on distant rulers to deal with them fairly: to assess them for the costs of defending the realm against intruders without depriving them of basic sustenance. We are entering a new phase of awareness and interdependence on a planetary scale, a human-made holon enveloping Earth. Catholic theologian Teilhard de Chardin used the term *noosphere* to signify this now emerging level, which works in association with the far older biosphere and the even more ancient geosphere.

There is both more opportunity and more danger (and hence the need for more responsibility) at each threshold of holonic emergence. For example, I place global democratic governance in the *good news* category. But there are those who are afraid of a One World Government—and for good reason. I envision a global governance that respects and meets the needs of all the regional and local levels, but what if it doesn't? Surely, there will be many innovations in the years to come that will brighten the prospects for the healthy step-by-step emergence of systems of planetary scale.

Following is a broad-brush introduction, gleaned from many reputable sources, as to what our species might expect during the next 250 years—during the next "minute" on the Cosmic Century Timeline. Before we get to the good news, I want to survey the *major challenges* we are likely to face as a result of our own and continuing actions. Let us also consider the *wild cards* that could interject chaos and breakdowns and that would not be the result of direct human activity. Born entirely from natural (and unpredictable) processes, these wild cards could wreak civilizational havoc, but they would not put a halt to the evolutionary process.

Why begin our imaginings with the likely and the remotely possible bad news? Because unless we take a square look at what could otherwise drive us into despair, authentic and full-bodied hope will elude us. We should pause, too, for reflecting on our tour of deep history: Let us recall that the primary driver of evolutionary creativity and transformation is chaos. Paradoxically, from an evolutionary perspective, bad news is often a good thing—a blessing in disguise.

Major Challenges in the Next 250 Years

> "Every few hundred years in Western history there occurs a sharp transformation.... Within a few short decades, society rearranges itself—its worldview, its basic values, its social and political structures, its arts, its key institutions. We are currently living through just such a transformation." —PETER DRUCKER

✦ Climate change

Well over 95 percent of the world's scientists agree that over the course of the next two and a half centuries sea levels could rise significantly. Future generations will have to cope with an increase in high-intensity hurricanes, along with more tornados, droughts, and floods—all as a result of global warming caused, in large part, by human activity. This will present a host of challenges and opportunities. Other life forms will be affected by climate change, too. In the past, species could migrate when climate warmed or cooled. But now, because of the way humans occupy much of the world, we will need to assist many animal and plant species in their move to cooler realms. (My wife, Connie, is a pioneer in this "assisted migration" movement; see torreyaguardians.org.)

✦ Continuing loss of biodiversity due to human causes

Leading biologists claim that we are in the midst of the "sixth great mass extinction" of life on this planet. As best we can tell, only five times in Earth's history have animal species gone extinct at a faster rate than what they are facing now—this time, because of human activity. Putting an end to this ravaging of the richness of life is already one of the great challenges of the 21st century.

✦ Impact of growing human population on food, energy, pollution, habitats

Population growth exacerbates many biospheric stresses for which we humans are directly or indirectly culpable. Estimates vary, of course, but most projections suggest that before leveling out, human numbers will expand from the current 6.5 billion to a peak of perhaps 8 or even 10 billion. The biospheric impacts of population growth may, of course, worsen as more of the world's peoples are lifted out of poverty. Several billion more people living at the American standard of living, and thereby using "natural resources" at our per capita rate, would require two to three more planets. Surely something has to give.

✦ Gap between the rich and poor, the haves and have-nots

In America, and elsewhere in the world, there is a growing gap between the rich and the poor, the haves and the have-nots. This trend cannot continue indefinitely; it is not sustainable politically or socially. What will catalyze a turnaround? An economic depression? Social revolution? Or perhaps something more intentional and benign?

✦ Peak oil

Sometime in the not-too-distant future, global oil production will peak and begin to decline—that is, we will be on the downhill side of the bell curve of oil production. Although there is substantial disagreement as to how "peak oil" will impact humanity and the planet as a whole, few doubt that it will be a major challenge economically and perhaps socially and politically in the coming decades.

✦ Geopolitical conflicts, including NBC (nuclear, biological, chemical) or GNR (genetic, nanotechnology, robotic) error or terror

In the next 250 years, what and where are the possibilities for geopolitical conflicts that could further stress economies, social structures, and ecologies of the world? It's anyone's guess. But as our world grows more interconnected and weapons become smarter and smaller, fewer conflicts will be contained at local or regional levels. The era when major conflicts were confined to nation-states is giving way to conflicts involving ideologies that cross geographical boundaries. As technology diminishes the age-old advantages of supporting massive human armies (and thus also diminishes the costs of waging war at small scales), world power structures will begin to shift. And who can predict what challenges NBC (nuclear, biological, chemical) or GNR (genetic, nanotechnology, robotic) error or terror will bring?

✦ *Biocomputers becoming more intelligent than human beings*

In his recent book, *The Singularity Is Near*, as well as in his earlier, *The Age of Spiritual Machines*, Ray Kurzweil contends that we are fast approaching a time when human intelligence will be exceeded by computer intelligence. When this happens, computers and machines might assume a leading role in their own evolution, and then in human evolution and the evolution of life itself. Kurzweil estimates this threshold as occurring by 2045. As technology becomes ever smarter, smaller, more powerful, and easier to obtain, will progress lead to greater peace and freedom, or greater discord and destruction? Many futurists who think seriously about the Singularity and its effects, including Kurzweil, expect the relationship to be a mutually enhancing one. But some are less optimistic. Movies like *The Matrix* trilogy, the *Terminator* series, and *I-Robot* exploit fears that naturally arise when we imagine this human–machine interface as a competitive "us against them" relationship, though even these movies incorporate examples of humans and machines working together. For example, the Oracle in the Matrix trilogy is a machine consciousness trying to facilitate an end to the human/machine war. Perhaps like Star Trek's famous android, Data, artificial intelligences will be considered sentient beings in their own right. Time will tell, of course. In the meantime, make sure that you, and your kids and grandkids, buckle your seat belts. It's going to be an adventurous ride.

✦ *Aligning self-interest with the well-being of the whole*

There may be no greater challenge and opportunity facing humanity in the coming decades than finding increasingly effective ways to align the natural self-interest of individuals and groups with the well-being of the planet as a whole. When the impact of parts on the whole are effectively mirrored back to the parts, our species will have come home.

Wild Cards

"Civilization exists by geological consent, subject to change without notice."
—ANONYMOUS

Wild cards are possible catastrophic events that may—or may not—come to pass. More than any other prospect listed in this chapter, wild cards pose hazards (and opportunities) that are difficult to predict, and "God only knows" if and when they will happen. As you begin this section, I would encourage you to keep in mind The Serenity Prayer, adapted from Reinhold Niebuhr: "God, grant me the serenity to accept the things I cannot change, the courage to change the things I can, and the wisdom to know the difference; living one day at a time; enjoying one moment at a time; accepting hardships as the pathway to peace."

Collectively, we can reduce the probability of a wild card event (for example, using telescopes to locate errant space rocks that may be headed our way). When speaking to church audiences, however, I offer individuals some light-hearted advice: First, know that *it's impossible to prepare for a wild card*; even Mormons won't be ready. Second, whatever your name for Ultimate Reality: be at peace with It. Third, for all the people in your life who you love: tell them that you love them!

With that, let us begin our survey of some of the most well known, widely accepted, and, yes, frightening, wild cards. If you like reading about such things, you can learn more about these and other potentially catastrophic events, natural and human caused, by going to armageddononline.org.

✦ *Supervolcano eruption (e.g., Yellowstone)*

Geologists have ascertained that there are a half dozen or so supervolcanos in the world and that roughly every 50,000 to 100,000 years one of them explodes. To gain a sense of what this would mean for civilization, it is important to know that a supervolcano is two or three orders of magnitude more powerful than a regular volcano: i.e., 100 to 1,000 times more destructive. The volume of exploded solids and gases is immense enough to radically alter a regional or continental landscape and to severely impact global climate for years, with a cataclysmic effect on life. While science and technology are well on their way to developing means to deter asteroid impacts, no one yet has come up with a way to deter or even delay the eruption of a supervolcano.

The last supervolcano that erupted anywhere in the world was Toba, in Indonesia, some 74,000 years ago. Geneticists tell us that less than 10 percent of the human species seems to have survived that event. Yellowstone Park is North America's supervolcano. It has erupted three times in the past two million years—roughly every 600,000 years. Here is the bad news: Yellowstone last erupted 640,000 years ago. Are we overdue for another? Perhaps. But, in truth, Yellowstone might not erupt for another 10,000 years, or longer. Of course, it could also happen next Thursday. Panic, or even any major concern, however, doesn't make sense. Why? Because if anything is in God's hands, this is. Trust (faith) is the only psychological, spiritual, or emotional option that has any practical value.

✦ Asteroid impact or extreme solar activity

Actually, there is a realistic and hopeful perspective related to the threat of asteroid impact. For the first time in our planet's 3.8 billion years of life, Earth is about to have an immune system—that is, it will soon be capable of protecting itself from incoming intruders. And it will do so thanks to us. We will have contributed something hugely beneficial to the body of Life as a whole. Remember, as I've mentioned elsewhere, we are not so much separate beings *on* Earth *in* the Universe, as we are a mode of being *of* Earth, an expression of the Universe. We are Earth becoming conscious of itself. And within the next 30 to 50 years we will have gained the knowledge of every significant object that could hit us—and the technology to deflect what would otherwise wreak havoc. Indeed, this may be one of the greatest contributions we collectively can make.

✦ Mega tsunami

A tsunami is one or a series of large waves triggered by an earthquake, volcanic eruption, or asteroid impact in or near the sea. A mega tsunami is the same thing, just bigger. An article in *The New York Times* in 2006 reported on the Holocene Impact Working Group—the scientists who have been collecting data and scanning the ocean floor and shorelines, searching for craters and enormous wedge-shaped sediment deposits (called chevrons) that are composed of material from the ocean floor. This group estimates that every five to ten thousand years a small comet or asteroid hits the ocean. The impact on world climate may be slight, but regionally the mega tsunami that is launched can devastate large stretches of coastline.

Of some concern to geologists and others today is a volcano named Cumbre Vieja on the island of La Palma in the Canary Islands, off the northwest coast of Africa. In 1949 during a volcanic eruption, a chunk of the island slid partway into the sea before halting its descent. Should another large eruption of Cumbre Vieja occur, the western side of the island is likely to collapse into the Atlantic. When this happens, the slide will cause a mega tsunami of enormous size and power. It will roll westward across the Atlantic, wiping out virtually everything along the coast of eastern North America, penetrating inland perhaps 10 or 20 miles. The skyscrapers of Boston, New York, and Miami will go down like tinkertoys.

✦ *Pole shift or magnetic field reversal*

Scientists have identified a series of past reversals in the direction of Earth's magnetic field, thought to result from natural shifts in our planet's center of gravity. Magnetic grains of iron align with the direction of the magnetic poles as lava cools. So anywhere that magma has spilled gently out of a rift, such as along the midsection of the Atlantic Ocean, Earth's record of magnetic reversals can be studied. Throughout its history, Earth's north and south magnetic poles have flipped back and forth repeatedly at intervals ranging from tens of thousands to many millions of years, with an average of 250,000 years. The last such event, called the Brunhes-Matuyama reversal, occurred some 780,000 years ago. We have no idea how a magnetic reversal might affect climate, if at all, but it surely will bear some consequences for long-distance migrators, including birds and butterflies, who have magnetometers in their brains.

✦ *Gulf Stream shutting down*

Another wild card is the possibility that the Gulf Stream and/or its extension, the North Atlantic Drift, could shut down, which has happened periodically in Earth's recent history. The last time this occurred was 12,500 years ago. A concern today is that if Greenland's glaciers continue to melt (as they are rapidly doing right now), the flush of freshwater into the North Atlantic could trigger a shutdown of the Gulf Steam or North Atlantic Drift. And, if this happens, Europe freezes over, becoming nearly uninhabitable.

On the bright side, given how interconnected humanity now is, if this wild card occurs, we can expect to see an outpouring of human compassion unlike the world has ever known.

✦ Proof of extraterrestrial life or intelligence

A gathering together of the human family might happen virtually overnight if there is unmistakable proof that we are not alone in the Universe. But even short of CNN or Fox News interviewing space aliens, undeniable evidence of bacteria on Mars or anywhere else in our solar system would itself be significant. There is nothing in the story of our Universe and its evolutionary path that would preclude such a possibility. I suspect such news would compel leaders of every religious tradition to rethink and reformulate their theologies. My wife, Connie, helped physicist Thomas Gold write his final book, *The Deep Hot Biosphere*, which hypothesizes that microbial life does, in fact, exist within (not upon) ten rocky planets and moons within our solar system.

✦ Epidemic of flu, smallpox, drug-resistant TB, emerging diseases, etc.

Few are aware that between 1918 and 1920 some 50 to 100 million people worldwide died from a particularly deadly strain of flu. An estimated 2.5 to 5 percent of Earth's human population died in that epidemic, with perhaps one out of every five people alive at that time sickened by the virus. And it struck fast, too. Half of all victims (at least 25 million) are estimated to have died in the first 25 weeks. In contrast, AIDS killed 25 million in its first 25 years. The 1918 flu quickly enveloped the world. In the United States, 28 percent of the population took ill, and 500,000 to 675,000 died. In Britain, 200,000 died; in France, more than 400,000. Entire villages perished in Alaska and southern Africa. In Australia, an estimated 10,000 people died, and in the Fiji Islands, 14 percent of the population died during just two weeks. An estimated 17 million people died in India, about 5 percent of India's population at the time.

Not only is another flu pandemic a very real possibility (many scientists say it's inevitable), but other infectious diseases, such as smallpox and tuberculosis, represent threats as well. The bacterium that causes tuberculosis (TB) currently infects (often unknowingly and without symptoms) one third of the world's population, and it kills nearly 2 million people annually. As it continues to evolve, the TB bacterium poses an ever increasing challenge, as there are already strains in existence that resist all known antibiotic treatments.

Current human activity is playing a role in how likely an epidemic might be. Our encroachment on previously isolated rainforests releases "new" micro-organisms, like ebola, which then start to mutate. Climate change accelerates the mutation and spread of germs, like malaria, and their carriers, mosquitoes, once confined to the tropics.

"Such Hope!"

So long as the divinely catalytic role of chaos goes underappreciated, we are all too easily overwhelmed when forced to think about the challenges ahead. However, when we put our faith in God's creativity, which also works through us, these very same possibilities can inspire and embolden rather than send us into fearful retreat. Already we see that, whereas older generations typically are stunned and immobilized by the now undeniable planetary consequences of human action—global warming, endocrine disruption, aquifer depletion, fisheries collapse, coral bleaching, species extinctions—the youth often are not. I shall never forget the remark of one such young person. Two weeks after *An Inconvenient Truth* (featuring Al Gore and his climate change call to action) began showing in theaters, I participated in a church discussion about the movie. Most elders were solemn, hand-wringing, distraught. But there was one 11-year-old girl among us, and it was she who chirped, "The movie gave me such hope!"

Long-term and Short-term Positive Trends

"Thinking cosmically doesn't require zipping around the Galaxy visiting aliens. It simply means integrating the new cosmic reality into our thinking whenever we try to understand what's going on in our world."

—JOEL R. PRIMACK AND NANCY ELLEN ABRAMS

✦ *Bad news, chaos, and breakdowns catalyzing creativity and transformation*

Surely, this is one of the most important trends of all. If we can count on anything, we can count on human creativity ramping up in times of severe stress. I am reminded that the most difficult and painful times in my life have also unquestionably been the most important growing times. We can expect the same, collectively.

✦ Technology further enabling and empowering human connectedness

When we look at the history of everything that has been made possible by advances in communication and transportation technologies, to take just two examples, we see individuals and groups, time and again, sharing information and sharing experience more often, more widely, and more deeply. Barring a global catastrophe, these trends will continue, and currently unimagined technologies will further the process.

✦ Circles of care, compassion, concern, and commitment widening

A fortunate side effect of our human propensity to forge ever more complex, interdependent, and cooperative relationships has been that the generations alive today feel affinity and compassion for more people (including those unlike themselves) than our ancestors typically did. Worldwide television broadcasts, the Internet, and other media now give individuals the opportunity to care about those they will never meet and those who neither speak their language nor practice their religion. This trend will continue. It will spread to more of the world's population as humankind acquires a shared cosmology that partners with traditional religious supports and is experienced and ritualized in diverse ways as holy. This will accelerate as more of the world's artists, musicians, storytellers, and television and movie producers begin to tell our common creation story in a multitude of personally and collectively meaningful ways. In these ways, the Great Story will inspire billions with different worldviews to regard one another as kin.

✦ Cooperation and interdependence expanding at multiple levels

Robert Wright's book *Nonzero: The Logic of Human Destiny* chronicles the expansion of cooperation from the beginning of life through all of human history. Simply put, a nonzero sum game is win-win/lose-lose rather than win-lose. As more and more of us are in this together, a loss for one group is a loss for all groups, and vice versa. Bill Clinton was so impressed with Wright's book that he instructed his staff to read it during the last year of his presidency. If Wright is correct, barring some catastrophe, there seems little anyone could do that would arrest this hopeful trend. (John Stewart's *Evolution's Arrow* is in this same genre.)

✦ Feedback (inner and outer) becoming more available, accurate, and helpful

Enhanced feedback is a welcome trend because life can evolve in healthy ways only when accurate and useful feedback is available. Few things contribute to and aggravate problems, breakdowns, and dysfunction more consistently than lack of feedback, coupled with scarce and mediocre ways to revise behavior based on feedback. This is why humility, authenticity, and responsibility are essential for spiritual growth and for effective service to the whole. Without these traits, it is difficult for feedback loops to function.

There is inner feedback, too—gut feeling, slumps in our spirit, intuition, our energy level, and so forth. What does your *heart* say about whatever you're contemplating? Remember, the heart is not just a pump; it's also a thinking organ. And here is the good news: even if you've never attended a 12-step meeting, never sought professional counseling, never tuned in to *Oprah* or watched Joel Osteen, and never even read a self-help book, just by being alive today you know more about your inner workings and the inner workings of our species than your great-grandparents ever thought about knowing. This trend can be trusted to continue.

Economically, too, feedback is improving. Soon we will begin to pay the true cost of products we purchase. Right now, in most circumstances, we don't. True costs—including the ecological costs of extracting raw materials and making and disposing manufactured goods—are externalized. Thus items that are costly to the Earth and human community, and that should be expensive to buy, instead are cheap. So we make decisions with our dollars that seem smart to us, but are unwise when viewed from a larger perspective. Internalizing "true cost" will be a radical and positive shift in human economics. And while it could take 25 years or longer to bring this about, I cannot imagine it will take longer than 50 years. And it's a done deal by the time the next "minute" ends (250 years from now) on our Cosmic Century Timeline.

The final thing I want to say about feedback is this: a big reason that political regimes today can't usually get away with what they could a few decades ago is because of five technologies that facilitate feedback: cell phones, digital cameras, satellites, the Internet, and global news telecasts, such as CNN.

✦ World's religions integrating evolution and ecology

Possibly within my lifetime, and certainly within the next 75 years, the majority of devout religious believers will become religious knowers. In so doing, they will not merely tolerate an evolutionary, ecological worldview; they will enthusiastically celebrate it. Why? Because the devout will come to see and experience their own core insights, their central doctrines, as larger, more meaningful, and more undeniably real than anyone could have known before. In the decades ahead, hundreds of millions of Christians, Muslims, Hindus, Jews, and others will move from flat-earth interpretations of their respective faith traditions to evolutionary interpretations. They will do so as naturally as a child moves from picture books to narrative texts.

As I hope this book demonstrates, Evolutionary Christianity is far more magnificent than is flat-earth Christianity. The same holds for every other religion. Leaders within each tradition who play catalytic, co-creative roles in helping their tradition move from its flat-earth phase to its evolutionary phase—those who help reframe the core insights and contributions of their tradition within an evolutionary context—will be doing their religion and the entire evolutionary process an invaluable service. And they will be serving God in the most practical and glorious of ways.

Likely Good News in the Next 250 Years

"The basic mood of the future might well be one of confidence in the continuing revelation that takes place in and through the Earth. If the dynamics of the Universe from the beginning shaped the course of the heavens, lighted the Sun, and formed the Earth, if this same dynamism brought forth the continents and seas and atmosphere; if it awakened life in the primordial cell and then brought into being the unnumbered variety of

> *living beings, and finally brought us into being and guided us*
> *safely through the turbulent centuries, there is reason to believe*
> *that this same guiding process is precisely what has awakened*
> *in us our present understanding of ourselves and our relation to*
> *this stupendous process. Sensitized to such guidance from the*
> *very structure and functioning of the Universe, we can have*
> *confidence in the future that awaits the human venture."*
>
> —THOMAS BERRY

✦ Human population stabilizing and then declining

Most population experts now agree that sometime in the next 50 to 100 years the increase in human numbers will come to a halt. Population will then decline toward sustainable levels. One of the most important factors driving this trend is thought to be the educational and economic empowerment of women worldwide. Whether an end to population growth occurs sooner rather than later, and whether it occurs gradually or abruptly owing to one of the challenges or wild cards already mentioned, only time will tell. But virtually no social scientist doubts that human population growth will end well before the 250-year time scale that frames this chapter's inquiry.

✦ Clean, renewable energy sources replacing toxic, nonrenewable energy sources

Here, too, scarcely anyone doubts that sometime in the next two centuries clean, renewable energy sources will replace toxic, nonrenewable ones. The only questions are *how* and *when*. If recent history is any indication (two significant breakthroughs were announced in 2006, one related to solar energy, the other, magnetic energy), the shift could occur swiftly. New forms of cheap energy will, of course, present their own challenges, as more people will gain the wherewithal to exploit more of Earth's materials and habitats. As a species, we will need to be mindful that our collective decisions and actions reflect long-term, biocentric/theocentric approaches, rather than merely shortsighted, human-centered concerns.

✦ The "Sixth Great Mass Extinction" ending

As humankind continues to grow in awareness of the interconnected web of life, and thereby comes to appreciate the moral as well as practical imperatives for maintaining biodiversity and ecological health, we can expect the preservation of species and their habitats, and the creation of corridors for their migration, to become undisputed priorities. My wife, Connie, has a vision of creating an interactive website focusing on what she calls "continental commonsense": where and how to live and where and how not to. Protecting and defending the well-being of other species and their habitats, and learning to live in ways that make sense ecologically, will become absolute priorities in the decades to come. We really are one holy community. Thomas Berry prophetically writes,

> The main task of the immediate future is to assist in activating the intercommunion of all living and nonliving components of the Earth community in the emerging Ecozoic era of Earth development. What is most needed in order to accomplish this task is the great art of intimacy and distance: the capacity of beings to be totally present to each other while further affirming and enhancing the differences and identities of each.

✦ *Biomimicry design revolution in law, medicine, governance, economics, religion, and education*

Rarely a week goes by without some announcement in the mainstream or alternative media about what we are learning from nature and how we are becoming better students of life's genius in evolutionary design. In the coming decades, as the nested emergence paradigm gains ascendancy over the mechanistic paradigm, humanity's major institutions—law, medicine, governance, education, economics, religion—can be expected to increasingly borrow ideas from the living, evolving scriptures of nature, rather than from the workings of lifeless machines. Those on the growing edge of this ecodesign, biomimicry revolution include William McDonough, Michael Braungart, Janine Benyus, Kevin Passino, Paul Hawken, Amory Lovins, L. Hunter Lovins, Sim Van der Ryn, Stuart Cowan, John Todd, and Nancy Jack Todd. Expect to see this field bear abundant fruit in the very near future.

✦ *All significant pollution problems solved*

While it is possible that doing away with pollution of a sort and scale that ecosystems themselves cannot render harmless will take a hundred years or longer, it seems likely that within the next 50 years we will have made major strides. This will require not only new technologies but renewed morality, as we humble ourselves and ever more effectively model nature's wisdom in our individual and collective business-as-usual. As I entered the final editing phase of this book, an article appeared in the March 2007 issue of *Popular Science* titled "The Prophet of Garbage." The article describes a promising new technology, plasma-gasification, that potentially could eliminate landfills worldwide and provide clean energy, to boot.

✦ *Global self-interest, personal self-interest, and corporate self-interest aligned*

I doubt that anything will transform humanity and our relationship to the air, water, soil, and life of the planet more dramatically than evolving ever more effective ways of aligning the self-interest of individuals and groups with the well-being of the whole—that is, with the larger holons of which we are part and the smaller holons for which we are responsible. This trend will unfold over many decades. A sacred understanding of cosmic history suggests that this alignment is God's will for us collectively. Many have found John Stewart's vision of "vertical markets" inspiring on this subject, which he discusses in his book, *Evolution's Arrow.*

✦ *The birth of what Joël de Rosnay has called the "cybiont": humanity, technology, and nature as one symbiotic, synergistic organism*

One of the most inspiring and realistically hopeful books of recent years is Joël de Rosnay's international bestseller *The Symbiotic Man: A New Understanding of the Organization of Life and a Vision of the Future.* A molecular biologist and futurist, de Rosnay writes of "a new form of life" that is "coming into being before our very eyes." He continues,

> In a still unconscious way we are contributing to the invention of its metabolism, its circulation, and its nervous system. We call them economies, markets, roads, communication networks, and electronic highways, but they are the organs and vital systems of an emerging superorganism that will transform the future of humanity and determine its development during the next millennium.

Cybiont is the name de Rosnay proposes for this new form of life at a planetary scale. He suggests that the cybiont "exists already, in a primitive state, as a living entity. Its birth will not occur in a single stage and the process of its evolution will never be completed." But what of the human? What will life be like for human individuals after the cybiont has emerged? Will our descendants regret the transition? Not at all, claims de Rosnay:

> From this perspective, the old question about what the people of the future will be like takes on a whole new meaning. They will be neither supermen nor biorobots. Nor will they be a supercomputer or a megamachine. They will simply be *symbiotic humanity*, living in close partnership with a social system—if they succeed in building it—that is an externalization of their own brains, senses, and muscles, a superorganism that nourishes and lives off the neurons of the Earth, neurons that we humans are in the process of becoming.

Indeed, de Rosnay envisions the human species as "living in harmony with a greater being that it helped create." I like to combine de Rosnay's vision of the cybiont with Thomas Berry's vision of the "Ecozoic Era." By whatever name, humanity will cross the threshold when it co-creates a mutually enhancing, synergistic relationship with our technology and with the body of Life of which we are part and upon which we depend.

✦ Global democratic/biocratic revolution; holistic governance

Who could doubt that in the decades and centuries to come—albeit not without setbacks—human societies worldwide will adopt more democratic forms of organization? Deeper and broader than this shift, however, will be the "biocratic" revolution. Biocracy differs from democracy in one vital way. Only in a biocracy are the health and well-being of other species, watersheds, and bioregions adequately taken into account. Only in a biocracy does governance require that the "voices" of the other-than-human world be heard and honored in all decision making. The U.S. Endangered Species Act is an important first step in this direction. Expect many more steps along this path.

This expansion of democracy into biocracy will be one facet of our increasing sophistication in making creative use of diversity of all types and bringing

"I was wondering when you'd notice there's lots more steps."

all relevant perspectives into dynamic conversation. Such holistic governance will enable the rich variety of voices to come together at altogether new scales. Wisdom of the whole will not only evoke life-serving order, but will make governance look less like government and more like conscious coevolution. As our capacities for holistic governance grow, societies will suffer less from partisan battles.

✦ *Worldwide religious revival*

This book is itself an expression of faith and a call to action for the world's religious peoples to integrate and celebrate an evolutionary, ecological worldview. By the year 2050, I envision that the majority of devout religious believers across the globe will embrace science as public revelation and cosmic history as scripture. As awareness of "the nested emergent nature of divine creativity" expands, and as more people come to know that "words create worlds" and that both day and night ways of speaking have value, God will be experienced more actively and intimately than ever before.

A multitude of sacred ways of thinking about evolution will propel a world-wide spiritual revival unlike any that have come before. By mid-century I imagine the majority of Muslims around the world will recite with conviction something like this: "There is no such thing as 'the Universe'; it is all Allah. And the more we learn about Allah, the more graciously we shall submit." Similarly, I believe that the vast majority of Christians will shed a constricted interpretation of "Christ's return" as a superhuman god-man descending from the clouds. They will regard that old way of thinking as a trivialization of a universally undeniable reality—a reality that we shall all participate in, whether or not we call ourselves Christian.

Now that we have considered some of the major challenges, wild cards, positive trends, and likely good news facing humanity, how about stepping into the future, thinking like this:

> *None of us asked to be alive at this moment in Earth's history. We did not choose to be born at this juncture in the Story. We were chosen. Each of us has been chosen by God to be alive and to participate in the most significant evolutionary transformation in 65 million years.*

This is the frame I choose to live within. It is the perspective that launched my itinerant travels six years ago, and that now has birthed this book. Would embedding your own life within such a frame of heroic participation give you, too, a sense of deep calling? Can you envision living your life as if enrolled in a mission of cosmic proportion? If you wish, take a few moments right now and allow yourself to feel your connectedness to the larger body of Life, and imagine your own great work in this emergent Universe.

✦ ✦ ✦

"*For peoples, generally, their story of the universe and the human role within the universe is their primary source of intelligibility and value. Only through this story of how the universe came to be in the beginning and how it came to be as it is does a person come to appreciate the meaning of life or to derive the psychic energy needed to deal effectively with those crisis moments that occur in the life of the individual and in the life of the society. Such a story... communicates the most sacred of mysteries.... Our story not only interprets the past, it also guides and inspires our shaping of the future.*"

—THOMAS BERRY

CHAPTER 18

Our Evolving Understanding of "God's Will"

"Any animal whatever, endowed with well-marked social instincts...would inevitably acquire a moral sense, or conscience, as soon as its intellectual powers had become as well, or nearly as well, developed, as in man....A moral being is one who is capable of reflecting on his past actions and their motives—of approving of some and disapproving of others, and the fact that man is the one being who certainly deserves this designation, is the greatest of all distinctions between him and the lower animals." —CHARLES DARWIN

A dozen years after his *Origin of Species*, Charles Darwin published a follow-up that offered evidence of evolution in the human realm: *The Descent of Man*. Darwin was by no means a "social darwinist"—a term that gained wide purchase only in the mid-20th century. That term has since been applied, post hoc, to a range of social thinkers and philosophers who regarded the human condition as inherently brutal, a struggle of each against all, or who advocated that organized brutality is the ideal form of governance. Among the most prominent social darwinists were Thomas Hobbes and Herbert Spencer, whose ideas preceded Darwin's publications.

Sadly, the horrific consequences of social darwinist thinking are still depicted in conservative religious settings as reasons to discount not only the contributions of Charles Darwin but also anyone today who uses the evolutionary worldview as a foundation for their work and ideas. Perhaps the only way that disparagement of evolutionary thinking on *moral* grounds will lose its power is if there are those willing to fight fire with fire. That is, the tide will turn only when individuals step forward to undertake the unpleasant task of showing how preevolutionary moral dictums—especially those frozen in scriptural texts—contain horrors too. Fighting fire with fire has, in fact, reached a crescendo in recent years.

Responding to Critics Who Reject Religion Because of Scripture

"We must begin speaking freely about what is really in these holy books of ours, beyond the timid heterodoxies of modernity—the gay and lesbian ministers, the Muslim clerics who have lost their taste for public amputations, or the Sunday churchgoers who have never read their Bibles quite through. A close study of these books, and of history, demonstrates that there is no act of cruelty so appalling that it cannot be justified, or even mandated, by recourse to their pages. It is only by the most acrobatic avoidance of passages whose canonicity has never been in doubt that we can escape murdering one another outright for the glory of God." —SAM HARRIS

Whew! Sam Harris is indeed fighting fire with fire. The past few years have seen an increase in virulent attacks on traditional religion from a science-based perspective. Bestselling examples are Richard Dawkins's *The God Delusion*, Christopher Hitchens's *God Is Not Great*, and Sam Harris's *The End of Faith* and *Letter to a Christian Nation*. Not as well known, but no less potent, are Michael Earl's audio programs: *Bible Stories Your Parents Never Taught You* and *The Ultimate Terrorist* (both available for free listening on his website: reasonworks.com). The impetus to challenge monotheistic "religions of the Book" is not only the "faith versus reason" divide and its debilitating consequences for social harmony, political discourse, and the teaching of science in public schools. The traditions are also faulted because believers down through the ages have justified religious hatred and violence by appealing to their scriptures.

I was first introduced to this type of critique when a man who heard me speak in Colorado Springs gave me his copy of an audiocassette program by Michael Earl. Connie and I often listen to audio books on the drives between speaking engagements. Listening first to *Bible Stories Your Parents Never Taught You,* then a few days later to *The Ultimate Terrorist,* was a painful experience. I did not want to hear what Earl was saying, yet I couldn't deny the truth of his commentary. I could no longer ignore scriptural passages—from Genesis to Revelation and in the Qur'an too—that portray God as brutal, cruel, vindictive, and genocidal. Passage after passage quoted by Earl brought images of Hitler, Pol Pot, and Stalin to mind. And yet this was supposedly *my* God, and these were passages read verbatim from the supposedly *Good* Book.

Here are just a few: Genesis 6 and 7 tell of God planning and executing the slaughter by painful asphyxiation (drowning) of billions of innocent animals and millions of children and their parents in Noah's flood. (Most people can't even let the enormity of these numbers into their heart. Can you?) Deuteronomy 3:2–6 and 7:1–2 has God commanding the ethnic cleansing of 15 to 20 million inhabitants of Canaan, including women and children. And the Book of Revelation envisions God in the future, with Jesus' assistance, brutally torturing countless animals and human beings of all ages, including children and infants.

After offering a score of equally horrifying examples—each scriptural passage read in context—Earl concludes,

> If we want to know why people kill in the name of God, and why they have been doing so for thousands of years, we must face one simple and obvious fact that almost nobody wants to confront. The fact is this: the God of Judaism, Christianity, and Islam—the God of monotheism—is a terrorist. In fact, he's the ultimate terrorist. It is an undeniable fact that the God described in the pages of the Holy Bible and Holy Koran is a blood-thirsty, ruthless, destructive terrorist.
>
> This is not mere hyperbole on my part; it is an easily verifiable fact. By every definition of the word terrorist, God qualifies. For example, the U.S. Department of Defense defines *terrorism* as *"the calculated use of violence or the threat of violence to inculcate fear; intended to coerce or to intimidate governments or societies in the pursuit of goals that are generally political, religious, or ideological."*

When we look at an event like the conquest of Canaan, the huge massacres of millions of women and children, we must not lose sight of the fact that these actions were carried out in response to orders from God. The Bible makes that absolutely clear. When we read the brutal Law of Moses, where people's brains are being bashed in with rocks for breaking the Sabbath, for having sex with the wrong people, for believing the wrong things: all of these atrocious laws can be traced back to God. And when we read in scripture about hell, about billons of unbelievers being tortured in fire for all eternity—this is God who is orchestrating all of this.

God, by any stretch of the imagination, is a terrorist. God employs the calculated use of violence or the threat of violence to inculcate fear—and he does it for religious reasons. In anybody's book, that's terrorism.

As I listened to Michael Earl's tapes, my heart ached. My mind searched frantically for a justification—any justification—and a few times I came close to finding one. But deep in my soul I knew that what he was saying was true. The Bible and Qur'an are replete with stories that portray God as anything *but* kind, loving, just, or generous. These sacred texts, in many places, portray a God who can scarcely be described in any way other than as a cosmic terrorist.

Why didn't I see this before? An even more wrenching question is, How could I have read these same passages repeatedly yet remained unfazed? I had, after all, read the Bible twice from cover to cover soon after my born-again experience. Here is another place where understanding our brain's creation story has really helped me.

Instinctively, we use our considerable human powers to rationalize choices that our deep reptilian and mammalian drives insist upon. A wandering nation in exile that finally manages to conquer the inhabitants of a rich valley would be foolish to take on the humanitarian burdens of ungrateful prisoners. And there is more: Every one of us is here today because at least some of our ancestors were the most effectively brutal people in our landscapes of origin. Some of our grandmothers, many generations back, were raped by or forcibly wedded to marauders. And that means that those marauders are our grandfathers.

To gain some perspective on how these horrific acts fit into the evolutionary psychology of humans, one can witness this same process among many social animals. Infanticide is common among social mammals when a new dominant male takes over. The new male will kill the offspring of his defeated rival so that he can mate with the now childless females. On Discovery's Animal Planet channel, the program "Meerkat Manor" tracks the day-to-day activities of Africa's little masked mongoose of the Kalahari desert. In this documentary series, family groups of meerkats are shown waging war with one another over disputed territory. In a recent episode, one family group comes upon the unguarded burrow containing the pups of the other family group and attempts to kill the pups.

Any twinges of moral doubt about attacking another people, even slaying their infants, may not shift our behavior so long as we are convinced that God is on our side. How could the ancient Hebrews have recorded their history in any other way? How could any sieging or besieged people have succeeded and survived without such "moral" support? Are we moderns any less prone to invoke God's assent, if not outright command, when we choose to engage in collective violence?

Evolution offers a much less vindictive and far more venerable understanding of God than the one portrayed in the Bible and the Qur'an. This should not be surprising, nor is it a denigration of scripture. In a divinely emergent Cosmos, how could it have been otherwise?

Imagine someone inviting you to learn about "the greatest king who ever lived." The story they then told you included more than a few instances of this king ordering genocidal ethnic cleansing: the wholesale slaughter of women and children and the extermination of entire cultures. Now imagine that when you asked questions about those particular events you were told, "Oh, don't worry about *those* incidents. Instead, concentrate on these *other* examples that show what a kind, loving, generous king he *really* was."

For me, it wouldn't matter how many good works the king had to his credit, because the stories that revealed his genocidal nature were just too gruesome to forget. Respect and adoration of this monarch would not be my natural inclination. But fear would be.

So here is where I have come to: *Of course* the Bible and the Qur'an in some places portray God as a "terrorist." Considering the time and context of scriptural origins, sacred text could not have been written otherwise. *Of course* the image of God portrayed in scripture is sometimes terrifying. Fear is a motivational force that would have contributed mightily to group cohesion at a time when disparate tribes needed to be united into one nation, or when the growth of cities meant that ethnically diverse peoples regularly commingled.

What we also know about human nature is that we are endowed evolutionarily with a superb capacity to rationalize an *is* into an *ought*, thanks to the brilliance of our Monkey Mind. Evolutionary psychology also teaches (and we all, surely, have experienced) that self-deception is a natural and emotionally calming talent. This is just as true for groups as it is for individuals. The winners will invariably explain to themselves and everyone else that God told them to do (or at least approves of) what they just did. *Of course* the Hebrew leaders, after their armies had slaughtered a village or region, would recount stories of how God commanded their actions.

The error, indeed the tragedy, is arguing that biblical portrayals of God accurately reflect the nature of Ultimate Reality. No time or culture—even our own—should be burdened post hoc with the responsibility of shaping humanity's understanding and relationship to Ultimate Reality once and for all. Each people will describe and relate to the divine as best they can for their time and their conditions. Each generation honors its ancestry by taking from the past only that which is still lifegiving. Each generation provisions posterity by remaining open to new teachings and by advising those who shall follow to do the same.

An irony of the 21st century is that for countless millions of people the Bible has become perhaps the greatest religious stumbling block of all. It holds many in bondage to ways of relating to reality, to ways of speaking of reality, that made sense in ages past, but which no longer do. Fortunately, this obstacle is likely to diminish rapidly in the coming decades because we now have a way of seeing "God's Word" in a far more realistic way than as the actual utterances of a Supernatural Being flawlessly transcribed and preserved in ancient texts.

I foresee a day in the not-too-distant future when tens of millions of religious believers—Christian, Jewish, and Muslim alike—embrace the discoveries of science as public revelation, and in so doing become religious knowers. I foresee a day when a new understanding of our scientific heritage prevails. No longer will the scientific picture of the Universe be thought to imply a cold and mechanistic rendering of cosmic processes, nor a "nature red in tooth and claw." How will this cultural shift come about? In part, because scientists themselves will applaud those who make the effort to interpret the discoveries of science in sacred ways. When God-language is used for such interpretations, that God will be seen as so much more awesome and worthy of worship than are literalist portrayals of the biblical God. God will be seen as more powerful, too, and light-years more in line with the moral stance appropriate for globally interdependent cultures and for the cross-species interdependent web of life.

Two thousand or more years after the biblical scriptures were written, humans have substantially expanded our circles of compassion beyond what is evidenced in the old texts. We see this in the way that genocide not only now has a name, but that name is invoked for the express purpose of eliciting moral outrage. Expanded circles of care are also evident in the international sanctions (global morality) that are regularly applied to motivate transgressor nations to clean up their act.

When freed from the erroneous belief that ancient holy texts reveal an accurate picture of God for all time, we can begin to appreciate how they nevertheless served as indispensable guides for many, many generations. And then, we need no longer judge unsavory scriptures harshly—even the most violent passages—or approach them with trepidation. After all, that was then; this is now.

Over the coming decades I foresee that religious believers of every tradition will embrace a far larger, more reality-based view of God than was possible even a century ago. This will be a vision of the Holy One that will draw the vast majority, regardless of religion or philosophical worldview, into a place of respect, adoration, love, and care for the larger body of which we all are part. Scripture will have become more encompassing and universally inspiring because altogether new writings will qualify as scripture. Our spirituality no longer restricted to ancient texts, we will come to know and be led by God's Word in every fact, every detail, every truth of cosmic history and of that undeniable Wholeness in which we all live and move and have our being.

Transcending Biblical Values
and Scriptural Morality

"Humans evolved as a social primate species with an ascending hierarchy of needs from self-survival of the individual (basic biological needs), to the extension of the individual through the family (the selfish gene), to a sense of bonding with the extended family (driven by kin selection of helping those most related to us), to the reciprocal altruism of the community (direct and obvious payback for good behaviors), to species altruism and bioaltruism as awareness of our membership among the species and biosphere continues to develop."

—MICHAEL SHERMER

If you're going to build a single-story house, a modest foundation will suffice. But if you plan to construct a skyscraper, a stronger foundation is required. Traditional theology and traditional understandings of morality and ethics are built on a belief-based, flat-earth cosmological foundation. It should come as no surprise, then, that a knowledge-based evolutionary cosmology offers a far more secure foundation for morality and ethics in a social milieu that is now global and boisterously multicultural.

Another way a holy evolutionary perspective can REALize religion is by expanding and deepening traditional morality and ethics along truly Christian lines. One of the challenges that we Christians, Jews, and Muslims will be wrestling with this century stems from the mismatch between today's widely accepted ethical precepts and the norms and ideals encoded in sacred texts that have remained unchanged for millennia. Because these texts are referred to as "the Word of God," however, few from *within* our traditions have been willing or able to publicly acknowledge the discordance. Michael Shermer, Michael Earl, Sam Harris, Daniel Dennett, Christopher Hitchens, and Richard Dawkins—all well-known skeptics or atheists—have made the clarion call. But who from within the traditions is so bold? Episcopal Bishop John Shelby Spong and Episcopal priest Matthew Fox do, but who else?

In America we often hear, especially in conservative settings, how the Bible is the only secure foundation for moral instruction and ethical guidance. Yet those of us who have actually read the entire Bible, and are clear-headed enough to see it as it actually is, know that it would be ludicrous (indeed, immoral) to advise a child, "Yes, dear. You should use as your own model for appropriate behavior whatever actions you read in the Bible that are attributed there to God or to what God commands us to do."

"Mixed Moral Messages"

Not long ago I was talking with a woman, the mother of two teenagers, after one of my programs. We were discussing the mixed moral messages found in ancient written scripture. The woman confided, "I wouldn't even think of encouraging my kids to apply in their own lives whatever values they found in the Bible. And most other parents I know feel the same way." She continued, "Why, then, do so many of us Christians—liberals and conservatives alike—still refer to these texts as 'God's Word'?" My response was simple: "What alternative, until now, did we have?"

Among those who are thrilled to encounter a holy view of evolution are parents yearning for inspiring ways to teach their children moral values—moral values grounded in science and commonsense, rather than based on ancient writings, which (in today's world) offer an ambiguous moral compass at best. Yes, the Bible contains hundreds of wonderful and useful passages that can help us teach our children how to become good, happy, loving, on-purpose adults. The Bible also, however, contains many grotesque and morally repugnant passages that none of us would want our children to see, much less emulate.

As Michael Earl points out in *Bible Stories Your Parents Never Taught You*, those who claim that the Ten Commandments can or should serve as an ethical foundation for us today fail to realize how far they themselves have evolved morally beyond the biblically prescribed consequences for violating these so-called laws of God. According to the Bible, "God's will" can be, and often is, brutal. Deuteronomy 13:6–10 prescribes that if someone breaks either of the first two commandments ("no gods before God" and "no idols"), they are to be put to death. Leviticus 24:13–16 and 23 instructs readers that if the third commandment ("Don't take the Lord's name in vain") is broken, the penalty is death. Numbers 15:32–36 warns that if you work on the Sabbath, thereby breaking the fourth commandment, your life will be taken from you. And according to Exodus 21:17 and Deuteronomy 21:18–21, if you curse your parents, or even if you're just a stubborn and rebellious teenager (thereby violating the fifth commandment, to honor your father and mother), God's prescribed penalty for this, too, is death. (*That'll* teach little Isaac—or, at least Isaac's younger brother, who has to watch his sibling being stoned to death for mouthing off to mom and dad.) This is not the sort of justice making or parenting practice that Americans in the main would support today.

Many Christians today advocate, or at least support the notion, that the Ten Commandments should be our moral benchmark. We are told that God is the same yesterday, today, and forevermore. But as Earl points out, today we don't kill Sabbath breakers. Nor do we stone to death our teenage daughters who lose their virginity before marriage, or our teenage sons who disobey us. And the reason we don't kill Sabbath breakers or our troublesome children is quite simple: it would be immoral. Moreover, who among us would qualify as stoner rather than stonee. Clearly, we've evolved beyond (at least some) biblical values and scriptural morality.

"When compared to the regime of Moses, the regime of the Taliban comes off looking like the ACLU."   —MICHAEL EARL

Morality One-Liners

Mosaic morality: "Obey the Lord, or die!"

Early Christian morality: "Believe in Jesus, or fry!"

Personal Evo-morality: "Live in integrity, or cry!"

Species Evo-morality: "Align self-interest and Earth-interest, or 'bye bye'!"

REALizing "Holy Scripture" and "Divine Revelation"

"'Tell me a story.' How often as children did we ask this of those who cared for us? How exciting it has been in recent years to discover that the First Book is not a collection of disconnected

> *scientific facts about a Universe that serves merely as a*
> *backdrop for our lives. Rather, it is a Story of a Universe that*
> *from within itself has unfolded stars and galaxies, mountains*
> *and oceans, plants and animals, you and me. The whole Cosmos*
> *is on a collective journey, and our individual journeys are part*
> *of that. It is a Story of the Universe that carries the meaning of*
> *what it means to be human, telling us where we are, where we*
> *come from, who we are, and what is expected of us. At our*
> *peril we ignore it. As a guide into our future, we relish it."*
>
> —JOHN SURETTE

PROPHETIC INQUIRY: What is the meaning of scripture? What if God's
primary revelation has been the Spirit-filled unfolding of evolution for the
last 14 billion years, including this very moment? How might this inform
our worship and spirituality?

Not long after writing emerged, the Bible came to be. For many in the land of
Moses and for centuries thereafter, it would have seemed a miracle to watch
someone coax words from scratches on clay tablets or from strange symbols on
papyrus or animal skins. What words would have been called forth on those
occasions? Such pronouncements would surely have included what we now call
Holy Scripture, or what Jesus' ancestors called the Torah, the first five books of the
Bible. For the Hebrew people, interpretations of the Word, even written interpretations that would become the Talmud, would be subject to question, debate,
and revision—while the Word itself stood firm. It is thus no wonder that, for
Christians, tradition places great significance on scripture as the written Word.

A much broader understanding of scripture is now emerging, however. It
includes awareness that interpretations of the Holy Word should not be teth-
ered to the meanings made manifest at any particular time. Rather, interpretations should grow commensurate with our understanding of the human
condition, the world, and indeed the Cosmos.

God's Word has always been evidenced most abundantly and faithfully on
every page of that which is fundamentally Real—the entirety of the natural
world. In contrast to such expressions of "natural revelation," the written scriptures of old have generally been referred to as "special revelation." What is the
relationship between these two modes of divine communication?

From a creatheistic perspective, the two modes are seen as complementary. Perceived conflicts between the scriptures of nature and the written scriptures most likely indicate a problem in interpreting one or the other. And who among us will not find exhilarating the invitation to do reinterpretive work in this time of religious questioning and upheaval? The Bible, taken in a literalistic, human-centered way, can sabotage rather than sustain a person's walk with God.

Scripture—be it the Word made manifest in the material Universe or that which has been revealed through human consciousness—is where we find guidance, solace, and strength. It is also where we are invited, challenged, and supported to be all that we can be, both for the present and for the future. Scripture is divine communication in any form that supports us in honoring and serving the Whole (the Holy One). For me, *scripture is everything that inspires and encourages me to grow in evolutionary integrity.* If a poem, chapter in a book, website, or movie helps me grow in Christ-like humility, authenticity, responsibility, and service to others—then for me, it is scripture. Writings and other artifacts that do not support me in this process I do not consider scripture, even if they appear on a page of the Bible.

Seen through sacred eyes, *the entire history of the Universe can now be honored as the primary revelation of God.* Written scriptures, in contrast, are derivative; they are secondary revelations. Stone, vellum, parchment, ink, and human consciousness—all expressions of nature—are prerequisites for the evolutionary emergence of written texts.

By relating to cosmic history as the primary revelation of God, science becomes empirical theology—and scientists, empirical theologians. To again quote Carl Sagan: "Science is, at least in part, informed worship." Not only does this help us see that science is a holy enterprise, a sacred endeavor, but it also provides the foundation for interpretive theology to render the gifts of science into meaningful forms: teachings, parables, liturgies, and concordances with each of the religions of the world. Thomas Berry has written, "The Universe, the solar system, and planet Earth, in themselves and in their evolutionary emergence, constitute for the human community the primary revelation of that ultimate mystery whence all things emerge into being."

With a broadened understanding of scripture, and an appreciation of the symbolic nature of human language, the Bible as Holy Writ comes alive for us in a new way. Rather than arguing over which biblical passages accurately portray history, which passages are imaginative recollections, which are poetry, and so forth, we can agree that the Holy Bible in its parts, and in its entirety, reveals truth in the only way that human language can—symbolically, and in a blend of day and night languages.

We are in the midst of a profound shift in what the Western world regards as authoritative. Two hundred years ago, if someone spoke in a public gathering, "As John says in the twelfth chapter of his gospel . . ." most who heard those words would grant that an adequate foundation had been set for taking seriously the speech or exhortation that would follow. To generate a similar positive response in a public setting today, one would appeal to mainstream science, that is, to public revelation. Indeed, for the past several hundred years, whenever a reputable source has said, "Scientists agree that . . ." few among us would quibble. We may choose to ignore confirmed, peer-reviewed scientific discoveries that strike us as irrelevant. We may demand that new scientific formulations first stand the test of time—that is, they must survive attempts within the scientific community to falsify the new ideas. Overall, we may resist those discoveries that threaten our own foundational "truths." But we risk being branded as fools or charlatans if we question the veracity of well-tested discoveries born of the scientific endeavor.

Public Revelation: "The Ever-Renewing Testament"

> *"Imagine that we could revive a well-educated Christian of the fourteenth century. The man would prove to be a total ignoramus, except on matters of faith. His beliefs about geography, astronomy, and medicine would embarrass even a child, but he would know more or less everything there is to know about God. Though he would be considered a fool to think that the Earth is the center of the cosmos, or that trepanning [boring holes in someone's skull to exorcise their demons] constitutes a wise medical intervention, his religious*

> *ideas would still be beyond reproach. There are two explanations for this: either we perfected our religious understanding of the world a millennium ago—while our knowledge on all other fronts was still hopelessly inchoate—or religion, being the mere maintenance of dogma, is one area of discourse that does not admit of progress . .*
>
> *"With each passing year, do our religious beliefs conserve more and more of the data of human experience? If religion addresses a genuine sphere of understanding and human necessity, then it should be susceptible to progress; its doctrines should become more useful, rather than less. Progress in religion, as in other fields, would have to be a matter of present inquiry, not the mere reiteration of past doctrine. Whatever is true now should be discoverable now, and describable in terms that are not an outright affront to the rest of what we know about the world."*
>
>   —SAM HARRIS

Many among us have yet to cast off the belief that God spoke clearly and was actively involved in human affairs only in the distant past. Thankfully, there is a groundswell movement among Roman Catholics, mainline Protestants, Evangelicals, Mennonites, Quakers, Pentecostals, New Thought Christians, and others, who find glad tidings in the God-glorifying ways of embracing a multibillion-year story of evolutionary emergence—a story big enough and open enough to uplift the biblical stories within its compass. Thus we arrive, with reluctance or with great expectation, but nevertheless inevitably, at a threshold:

✦ To hold that a literal interpretation of the Bible is the best or only legitimate interpretation is to foster a schizophrenic break between the religion that still guides our souls and the science that is foundational in so many aspects of our lives—including healing many of us from diseases, injuries, and birth defects that in other times would have been lethal.

✦ To continue to insist on a literal interpretation of the Bible in this age of science is to make an idol of human language, while underestimating both the extent of divine revelation and the depth of human fallibility.

We now know that, as a matter of course, it took many generations for the events described in the Bible to be recorded in written form. Yet today, by continuing to insist that ancient biblical texts are accurate records of the dictated words of an otherworldly, invisible Father, we turn millions away from the real truths available in scripture. *Adherence to literalism thus undermines the very gospel it seeks to support.* Those who think that peoples of the past would not embellish stories to their own ends, and that these departures would not magnify over the decades and in some cases centuries of oral transmittal before they were recorded in writing, do not understand human nature and the biblical portrayal of sin.

Although most Christians still call the collection of letters written two millennia ago The *New* Testament, the revolutionary idea today is that God has, for centuries, been faithfully and publicly revealing truth via facts uncovered by science. Perhaps we should call sacred interpretations of science The Ever-Renewing Testament.

There is a world of difference between a preevolutionary and an evolutionary understanding of "biblical inerrancy." With a God-glorifying understanding of deep time, one need not make an idol of human words as a carrier of God's Word. Rather, from an emergent perspective, we can see that the Bible accurately reveals how the authors and editors of the books of scripture understood themselves, their world, and the nature of Ultimate Reality two or three thousand years ago. Those understandings include many powerful insights we can use today, woven in amongst much that is of primarily historical or symbolic value, and even some components that modern sensibilities rightly find morally offensive. It is up to us to find life-serving meanings in the guidance given us by the Whole over time, no matter what the vehicles of delivery.

I pray for the day when fundamentalists, pentecostals, evangelicals, and other conservative Christians awaken to nonliteral interpretations of scripture and a sacred appreciation of deep time that offers them an even more magnificent and undeniable God, a more meaningful understanding of the Kingdom of Heaven, and a far more glorious purpose for humankind. Until then, scriptural literalists do us, and God, a great service by questioning spiritless interpretations of evolution.

REALIZING Godly Morality and Ethics

*"In this evolutionary theory of morality, asking 'Why should
we be moral?' is like asking 'Why should we be hungry?' or
'Why should we be horny?' For that matter, we could ask,
'Why should we be jealous?' or 'Why should we fall in love?'
The answer is that it is as much a part of human nature to be
moral as it is to be hungry, horny, jealous, and in love."*

—MICHAEL SHERMER

We can finally (thank God!), once again speak boldly and prophetically about right and wrong, and do so without appealing to ancient texts. From a sacred, deep-time perspective, something is right if it honors or fosters the health and well-being of the larger and smaller holons of our existence or furthers the emergence of greater cooperation and interdependence at increasing scale and evolvability. A thing is wrong if it undermines these values. Said another way, *a thing is right if it helps individuals and collectives to grow in trust, authenticity, responsibility, and service. A thing is wrong if it tends otherwise.*

In oral cultures of ancient times, the inborn moral sense would have been honed and amplified by storytelling, songs, and ceremony. When writing developed, right and wrong tended to become identified by whether or not something aligned with written laws and guidelines held sacred by the community. Judgment was also based on whether an act would promote cooperation and well-being at the level of tribe, religious group, or nation—or whether it would do the obverse. Today, thanks to print, electronics, computers, and the Internet, we've come to see that the well-being of every individual, corporation, and nation-state is integrally connected to the health and well-being of the entire body of Life. This is why right and wrong are now discerned in larger, more comprehensive ways than ever before, and why conversations to find insights and solutions that meet the needs of all parties are so central to the Great Work we are now engaged in.

People everywhere today know that love, respect, gratitude, compassion, integrity, responsibility, humility, kindness, accountability, and so on are

God's will and lead to healthy maturation, healthy relationships, and healthy communities. Similarly, we all know that hatred, pride, arrogance, self-righteousness, envy, resentment, bitterness, deception, theft, and so forth damage the human spirit and unravel social bonds. We don't need ancient writings to tell us this. It may be the case that in biblical times, the size and complexity of societies and information systems had not yet developed to a point where these moral principles were as obvious as they are to us today, just as the moral issues around war and use of fossil fuels became apparent during the 20th century. Now God's will and God's ways can be discerned throughout biological and human history, as well as in our own experience and within the quiet places of our hearts.

Wider Circles of Care, Compassion, and Commitment

"Morally laden terms such as 'good' and 'bad' have a surprisingly simple biological interpretation. Traits associated with 'good' cause groups to function well as units, while traits associated with 'evil' favor individuals at the expense of their groups. But as we all know, groups whose members are good as gold toward each other can behave toward other groups in the same way that evil individuals behave toward members of their own group.

"No one can be trusted on the basis of their job title—not scientists, politicians, priests, or self-righteous intellectuals. Trust requires accountability. Some individuals are accountable on their own, but that's not good enough at the institutional level. An effective government, religion, or scientific culture must include mechanisms that make everyone accountable."

—DAVID SLOAN WILSON

While discussing evolutionary psychology in Chapter 10, I claimed that accountability, or the lack of it, was the single best predictor of long-term integrity in individuals and groups. Integrity and accountability are essential components of God's will. So is ever-expanding love. For billions of years, evolution

(God's emergent creativity) has repeatedly produced cooperative, interdependent systems out of self-interested, even competitive, entities. It has done so by finding ways to align the well-being of the parts with the well-being of the whole. Throughout human history, the subjective manifestation of this trend has been wider circles of care, compassion, concern, and commitment.

Humans nurtured in healthy ways tend to develop and mature in a predictable sequence, beginning with the egocentrism of toddlers that then progressively transforms into expanding circles of care that may (or may not) eventually grow to embrace all of Life and all of deep time. Collectively, our species seems to be in a similar developmental process. To widen circles of care, commitment, and cooperation takes sustained societal effort and enculturation. The test of a society is not just how wide its circles of care become, but how precipitously those good feelings drop off at the borders—that is, the severity of the *in-group* versus *out-group* distinction.

We will naturally cooperate with those whom we trust or care about. Today's globally enmeshed ways of living cry out for expanded cooperation, and thus expanded circles of care and accountability—systems that build trust. Said another way, we can feel God's evolutionary unfolding urging us in those directions. For example, the appearance of international groups that go wherever they are needed, such as Doctors without Borders, the Red Cross, and the World Wildlife Fund, can be understood as the force of evolution working through us now, turning us into evolutionary agents to the glory of God.

Part of this evolutionary imperative involves absorbing the lessons of the Great Story and the ecodynamics of our living world, which so clearly demonstrate that we all are made of the same stardust and are members of the same genetic tribe. The evolutionary imperative would also place a premium on experiences through which individuals vividly sense their kinship with humanity and all of life. Such experiences may arrive unbidden, through shared crises. Other times we intentionally pursue them—for example, through participation in multicultural activities or wilderness quests. The evolutionary imperative also inclines us to envision and emplace laws and incentives that will channel human behavior toward synergy and away from harm. That same evolutionary impulse also now invites us to discover better ways to speak and listen and co-create with one another, in groups at all size scales and bridging the barriers of cultural and linguistic differences.

Realizing "Jesus as God's Way, Truth, and Life"

> "A human being is a part of the whole called the 'universe,' a
> part limited in time and space. He experiences himself, his
> thoughts and feelings, as something separated from the rest, a
> kind of optical delusion of his consciousness. This delusion is a
> kind of prison for us, restricting us to our personal desires and
> to affection for a few persons nearest to us. Our task must be
> to free ourselves from this prison by widening our circle of
> compassion to embrace all living creatures and the whole of
> nature in its beauty."
> —ALBERT EINSTEIN

PROPHETIC INQUIRY: How do we perceive Jesus' life and teachings as
providing guidance crucial for our own times? How could a science-based
religious perspective help to universalize those conclusions by making
them pertinent and accessible to peoples of all faiths, or of no faith? If Jesus
were alive today, how would he honor God in his teaching and preaching of
evolution?

As with every Christian doctrine, our understanding of the meaning and purpose of Jesus' life and message will expand as our understanding of Reality
expands. If my interpretation of Jesus as "the way, the truth, and the life" of
God is the same as that of peoples living hundreds or thousands of years ago, I
miss the magnitude and magnificence of what God has publicly revealed
through science and cultural evolution in the intervening centuries.

I cannot agree that "Jesus as God's way, truth, and life" means that only
those Christians who believe certain things about Jesus or the Bible get to go
to a special otherworldly place called heaven when they die. I used to believe
that, but I don't anymore. In hindsight, I see that my old belief cheapened,
belittled, and impoverished the universal glory of the Gospel. What Jesus' life
and ministry were actually about is far larger and more meaningful, and
offers more this-world relevance, than my old clannish, contracted "we win,
you lose" understanding. More, one need not be a Christian, nor ever have
read the Bible, in order to walk what is, effectively, the same path we Christians
aspire to—the same "one way" to a REALized, redemptive life of fulfillment and

service in this world, here and now, while simultaneously blessing future generations.

For me today, the interpretation of the Gospel that lives most vibrantly is this: "Jesus as God's way, truth, and life" means that to the extent that I live in evolutionary integrity, as Jesus lived, I am living God's way, manifesting God's truth, and bringing God's vitality and life-enhancing service into the world. *This* way of living in awareness of the Whole, in service to the Whole, as Jesus did, is not something to be merely reconciled with our vastly enlarged understanding of the Universe and of Time. Rather, it is to be enriched by it. Simultaneously, the relevance of this core Christian doctrine is universalized: any and all may benefit from its guidance without necessarily converting to Christianity as a belief system.

Many young people intuitively get this. They *get* that an understanding of evolution as holy enlarges and enlivens their faith. This may be why Evolutionary Christianity is more relevant and attractive to many young people than are flat-earth forms of the same faith.

In this book, and on the road, I have repeatedly referred to the nested emergent nature of divine creativity and grace. As the early Christian gospels portray him, Jesus expressed this form of creativity in his behavior and teaching. Moreover, he conducted his life from the stance of integrity and oneness with God. Jesus was a cell in the body who realized his relationship to the Whole and expanded his circle of care, compassion, and commitment accordingly. His compassion extended far beyond what was espoused by the religious and political leadership of the time. By example, he opened a door of possibility for the rest of us.

From a sacred evolutionary perspective, trusting in "the Lordship of Christ Jesus" is having faith that the same kind of self-giving, wisely confronting love that Jesus incarnated will, over time, prevail over sin and evil, no matter how desperate the situation may seem in the moment. For me, "faith in Jesus" means trusting the wisdom of ever-expanding *agape* love for all humans and, indeed, for all of life. It also means trusting that even catastrophes are not cosmic mistakes. Evolution uses crises to leap into new ways of being. The example of Jesus convinces me that to the degree I am able to trust and expand into loving connection, any downturn or hardship will be transforming and redemptive.

We would do well to remember that Jesus was neither conservative nor liberal, but both (or neither), which is what made him such a radical.

Jesus embodied God's will for humanity—humility, authenticity, responsibility, and loving service to the Whole, and on behalf of future generations. His ways of thinking, acting, loving, and resisting evil reveal *our* way into a God-glorifying and evolutionarily robust future. Jesus incarnated the deep integrity that is our salvation, individually and collectively. His way of living and telling the truth continue to light our path. Jesus embodied the life, the will, and the ways of Supreme Wholeness, or Ultimate Integrity (God). His expansive sense of self, his radically inclusive way of loving, and his active, nonviolent way of resisting and confronting unredeemed institutions show us the way to the Kingdom.

The more I reflect on life within and around me, the more certain I become that the only way to heaven is via the path Jesus walked—the path that might now be called evolutionary integrity. Whenever I lapse into thinking that I can know real freedom and joy without inviting Integrity Incarnate (Christ) to be the primary guiding reality of my life (my Lord), Reality has a way of demonstrating just how wrong I am. The Bible says this is so, but I know of its truth because the facts of life tell me so—because Life finds one way or another to confront me in my waywardness.

From my *evolutionary* Christian perspective, the only hope I see for ushering in God's Kingdom "on Earth as it is in Heaven" is reincarnating Christ-like values at all levels of society. Surely, the "second coming of Christ" cannot mean less than this. If we are to fulfill our species' cosmic task, Christ-like values will be a requisite.

> Speaking mythically (in night language), if Christ can be seen as the embodiment of Divine Integrity, the incarnation of "The Only Way to Wholeness," then individuals and organizations that use their power to promote fear, hatred, conformity, dishonesty, control, and irresponsibility can be said to be doing the work of the AntiChrist.

"You will know the truth, and the truth shall set you free."   —JESUS

There is little or no evidence that Jesus ever intended for us to worship him. But there's lots of evidence that he wanted us to follow him: to be in integrity like him; to trust like him; to live our truth like him; to love expansively like him; to take responsibility for our world as he did; to transform our cherished traditions like he did; to follow in his steps and do greater things than even he could accomplish, given the constraints of his time. And to ensure that our natural devotional inclination would manifest in healthy ways, Jesus assured us that "Whatever you do to the least of these (the poor, the outcast, the needy, the hungry), you do unto me."

From my createheistic perspective, I admire and honor Jesus as a unique expression of divine love, a personification of God's way, gospel truth, and eternal life. Jesus modeled ideal ways for us to live and love from a stance of deep conviction—reflecting God's will for humanity and for Creation as a whole. In his own time, it was Jesus who called for compassion, not stoning, as a communal response to sexual indiscretion. It was Jesus who consistently challenged the religious leadership of his day to be less dogmatic, less self-righteous, more generous, more integrous, and more inclusive in the love and consideration of others. In our time, we follow Jesus when we emulate the ways in which Christ-like integrity would make a real difference in the world we have inherited.

Conclusion

hat an amazing time to be alive! In 2.5 million years of human history...

✦ *Now* is when the world's great religious traditions are integrating evolution in holy and meaningful ways.

✦ *Now* is when we can see God as infinitely more real, more intimate, more understandable, and more awesome than ever before.

✦ *Now* is when we can celebrate "facts as God's native tongue" and science as our collective means of discerning God's ongoing public revelations.

✦ *Now* is when we can improve our minds, lives, relationships, and societies as we come to appreciate our brain's creation story.

✦ *And now* is when we have a shared creation myth, a cosmology, that encour-ages us to celebrate our differences, join together in common cause, and expand our circle of care to include all manifestations of life.

Thanks to our fresh understanding of the deep-time face of grace, science and religion, truly, are ushering each other into greatness. My prayer is that this book will play a role in that grand, collective journey.

Epilogue

"In order to be truthful you must embrace your whole being. A person who exhibits both positive and negative qualities, strengths and weaknesses, is not flawed, but complete."

—RUMI

I t would be impossible for me to produce a book suggesting that humility, authenticity, responsibility, and loving service are the core elements of evolutionary integrity without facing the implications of this gospel for my own life. In the course of writing, I've been forced to look honestly not only at what is real now but also at the wake I've left. Where have I been arrogant rather than humble, fearful rather than trusting, seductive rather than faithful? Where have I been deceptive rather than honest, inauthentic rather than genuine, cowardly rather than courageous? Where have I been self-righteous rather than considerate, self-absorbed rather than compassionate, irresponsible rather than accountable? And where have I been unappreciative rather than grateful, stingy rather than generous, self-centered rather than serving?

As I pondered these questions, two thoughts immediately came to mind: "Yuk!" and...

"Thank God for evolution!"

Testimonial

While on a writing retreat along the coast of Maine in autumn 2006, I felt led to do the deep integrity exercises that I was then writing about for Chapter 11—some for the first time and others for the umpteenth time. I had no idea how life changing that effort would be. Well into the process, I began to feel different, hugely different, as if I were experiencing an utterly new emotion. It was peaceful and exhilarating at the same time. I also acquired a clarity, focus, and groundedness that seemed fresh, yet familiar. The remarkable thing is that, by grace, I have not lost this state of being. An undercurrent of joy and gratitude has become my ever-present companion, even in the midst of life's persistent challenges.

What happened? By practicing what I have been preaching in this book, I entered a place of being "in Christ" that was more compelling—more REAL—than I had ever known. I trusted the process (put my faith in God), confessed everything shameful to Connie and my closest friends, made amends as responsibly and as compassionately as I could to everyone I had previously hurt (and could still locate), and passionately pursued where my great joy and the world's great needs intersect. What took me by surprise was the wave of bliss that washed over me when I let go of my last resentment, confessed my last secret, and put down the phone after talking with the last person on my amends list.

If anything this side of death qualifies as being "born again," surely this is it. No otherworldly, unnatural paradise can compare with the utterly real heaven I now experience virtually every moment of every day, free of resentment, guilt, and unfinished business. The bizarre thing is that this joy and serenity is not lessened by the sadness I still feel for those I hurt over the years, and especially for those whose wounds remain unhealed. I regularly remind myself of these people because it keeps my heart tender. Nevertheless, for the first time in my adult life I feel I could die in my sleep free of doubt, guilt, shame, resentment, fear, or anything other than love, trust, and deepest, deepest gratitude for all God's gifts, the excruciating and the exquisite.

What I didn't anticipate, not having read or heard about it anywhere, was how relatively easy it has been for me to *remain* in integrity since I came "back to Christ." I discovered a new steadiness and power in doing nothing more than living my commitment to grow in humility, authenticity, responsibility, and service, with the support of others. By genuinely appreciating my instincts—thanks to an evolutionary worldview—and creating internal and external structures of support, I now, by grace, experience an ease and freedom I've never known before regarding old habits, patterns, and temptations.

Will this pass? Will, at some point, I go back to wrestling with my Lizard Legacy or the shadow side of my Furry L'il Mammal? Perhaps. Time will tell. (Feel free to ask me about it when you see me.) But my gut, heart, mind, and support circle all suggest the more likely scenario is this: As long as I "abide in Christ," I will never lose my salvation, nor the eternal heavenly joy that comes with it. But let me slip into pride or arrogance, deception or inauthenticity, blame or resentment, or stingy, ungrateful self-centeredness, and I won't have to worry about burning in some otherworldly hell after I die. I'll be supping with Satan right here and now.

There have been times when my Lizard Legacy led me to do things that I now deeply regret. Acting irresponsibly on my instincts led me to betray the trust of both of my wives and also cost me my pastoral ministry in the mid 1990s. In all my years of recovery work, however, I don't remember ever *getting*, at least not in the way that I do now, that the key—indeed, the only key—to staying clean, sober, faithful, on-purpose, and fully empowered is integrity. Once I experienced this for myself, other insights opened as well.

On a long drive I tuned in to a Christian radio station and heard R. C. Sproul, a well-known evangelical teacher, speak about God's will. According to Sproul, the message of the Bible is clear: God's will for all of us, individually and collectively, is "righteousness," "sanctification," and "holiness." Now I don't know how many people today are able to translate these traditional religious terms into actual practices for improving their attitudes and actions—but on that drive, I thought to myself, "Sure sounds a lot like deep integrity to me!"

I also remembered hearing about an approach to addiction recovery that originated with a group of recovering alcoholics in Akron, Ohio, in the 1930s.

Rather than taking weeks or months or even years for an individual to work through the 12 steps, which is not at all uncommon, their approach was this: Once someone had a week of sobriety, that person would spend an entire weekend, with the support of a half dozen others, working Steps 1–8 as thoroughly as possible. The work thus included surrendering to Reality, owning the truth of their life, letting go of judgments and resentments, and preparing to make amends to those they had harmed along the way—all with the support of others who had experienced their own "salvation" (i.e., release from the bondage of inherited proclivities) thanks to this same transformational process. I couldn't help but think of scriptural portrayals of the first century church as "the body of Christ."

I close this testimonial with a prayer: May I continue to have the humility, strength, and peer encouragement to do what is necessary to remain in this state of grace. May I be a blessing to all those around me. May I leave a positive evolutionary legacy, in service to God. And may the light of the living Christ shine within my heart and continue to guide my steps.

Vision

I envision the day when facts are universally celebrated as God's native tongue, when evidence is honored as divine clues, and when the thought of looking to the past, rather than the community of religious knowers alive today, for our best understanding of words like "God," "sin," "salvation," "heaven," and so forth, will be unimaginable. Oh, how religion and science will then usher one another to greatness!

I long for the day when public revelation is valued above private revelation nearly everywhere, and when *day language* and *night language* thrive in their respective domains. Oh, would it come to pass that millions of people wait with eager anticipation for the next revelations from God that appear in journals like *Nature* and *Science.* May there come a time when theologians and preachers vie with one another to articulate the most inspiring meanings of such ongoing revelation!

I cherish the day when awareness of the nested emergent nature of divine creativity will be universal, and when people everywhere understand that words create worlds. What a magnificent time it will be when the question "Do you believe in God?" makes no more sense than asking "Do you believe in Life?" or "Do you believe in Reality?"

I hunger for the day when most of the world's religious believers see themselves as religious knowers; when the majority of Christians are Evolutionary Christians, the majority of Muslims are Evolutionary Muslims...

I salute the day when "the body of Christ" means all those individuals and organizations around the world who are committed to evolutionary integrity.

I anticipate a glorious day when understanding ourselves as stardust and as the Universe become conscious of itself inspires hundreds of millions of diverse people all over the world. I see, too, a time when generations live in relative peace with one another, thanks to a shared perception that death is no less sacred than life and, consequently, that this life, this moment, truly does matter.

I look forward to the day when God's active guidance will be available to all. Yes, I say, yes! There *will* be a time when young people in every tradition wonder how it was possible for their elders to favor unnatural, otherworldly interpretations of the core doctrines of their faith when natural, evolutionary interpretations of our shared journey are so much more compelling and undeniable. May there come a time, too, when billions of youth are taught in homes, churches, synagogues, mosques, schools, and through the media about the gifts and challenges of their inherited proclivities, and when healthy practices for channeling our most insistent urges are widely used and shared.

I dream of the day when aligning with the trajectory of divine creativity captures the imagination of our species; when millions of people, especially young people, are inspired to follow an evolutionary calling that serves the Whole. There will even come a time when peoples throughout the world come to regard as kin those whom their grandparents feared or hated. There will still be trying times; there will still be enormous problems to solve; but these will be regarded as evolutionary catalysts and dealt with head on in a spirit of possibility, openness, and trust.

I imagine a day when the devoutly religious give this book to their nonreligious loved ones saying, "See, God is real and faith is essential"; when scientists share it with their religious loved ones, saying, "See, evolution is divine and science is revelatory"; and when both sides read it and respond, "Oh. Got it. Thank you!"

I pray I live to see the day when billions of human beings will say, "Thank God for evolution!"

"Good and Bad Reasons for Believing"

BY RICHARD DAWKINS

"Tell a devout Christian that his wife is cheating on him, or that frozen yogurt can make a man invisible, and he is likely to require as much evidence as anyone else, and to be persuaded only to the extent that you give it. Tell him that the book he keeps by his bed was written by an invisible deity who will punish him with fire for eternity if he fails to accept its every incredible claim about the universe, and he seems to require no evidence whatsoever." —SAM HARRIS

Note: The following letter was written by Richard Dawkins, one of the world's most respected scientists. It is addressed to his daughter Juliet, who was ten years old at the time. The letter originally appeared as the last chapter of his 2003 book, *A Devil's Chaplain*. It is reprinted here by permission of the author. I include the letter for two reasons. First, it powerfully distinguishes the relative value today of, what I like to call, private and public revelation. Second, it provides an invaluable critique of traditional, flat-earth faith, while offering sound guidance for an evidential, evolutionary faith.

✦ ✦ ✦

Dear Juliet,

Now that you are ten, I want to write to you about something that is important to me. Have you ever wondered how we know the things that we know?
How do we know, for instance, that the stars, which look like tiny pinpricks
in the sky, are really huge balls of fire like the Sun and very far away? And
how do we know that the Earth is a smaller ball whirling round one of
those stars, the Sun?

The answer to these questions is "evidence." Sometimes evidence means
actually seeing (or hearing, feeling, smelling...) that something is true.
Astronauts have traveled far enough from the Earth to see with their own
eyes that it is round. Sometimes our eyes need help. The "evening star"
looks like a bright twinkle in the sky, but with a telescope you can see that
it is a beautiful ball—the planet we call Venus. Something that you learn by
direct seeing (or hearing or feeling...) is called an observation.

Often, evidence isn't just an observation on its own, but observation
always lies at the back of it. If there's been a murder, often nobody (except
the murderer and the victim!) actually observed it. But detectives can gather
together lots of other observations which may all point toward a particular
suspect. If a person's fingerprints match those found on a dagger, this is
evidence that he touched it. It doesn't prove that he did the murder, but it
can help when it's joined up with lots of other evidence. Sometimes a detec-
tive can think about a whole lot of observations and suddenly realize that
they all fall into place and make sense if so-and-so did the murder.

Scientists—the specialists in discovering what is true about the world
and the universe—often work like detectives. They make a guess (called a
hypothesis) about what might be true. They then say to themselves: *If* that
were really true, we ought to see so-and-so. This is called a prediction. For
example, if the world is really round, we can predict that a traveler, going
on and on in the same direction, should eventually find himself back where
he started. When a doctor says that you have the measles, he doesn't take
one look at you and *see* measles. His first look gives him *a hypothesis* that
you *may* have measles. Then he says to himself: If she really has measles, I
ought to see.... Then he runs through the list of predictions and tests them
with his eyes (have you got spots?), hands (is your forehead hot?), and ears
(does your chest wheeze in a measly way?). Only then does he make his

decision and say, "I diagnose that the child has measles." Sometimes doctors need to do other tests like blood tests or X-rays, which help their eyes, hands, and ears to make observations.

The way scientists use evidence to learn about the world is much cleverer and more complicated than I can say in a short letter. But now I want to move on from evidence, which is a good reason for believing something, and warn you against three bad reasons for believing anything. They are called "tradition," "authority," and "revelation."

First, tradition. A few months ago, I went on television to have a discussion with about fifty children. These children were invited because they'd been brought up in lots of different religions. Some had been brought up as Christians, others as Jews, Muslims, Hindus, or Sikhs. The man with the microphone went from child to child, asking them what they believed. What they said shows up exactly what I mean by "tradition." Their beliefs turned out to have no connection with evidence. They just trotted out the beliefs of their parents and grandparents, which, in turn, were not based upon evidence either. They said things like: "We Hindus believe so and so"; "We Muslims believe such and such"; "We Christians believe something else."

Of course, since they all believed different things, they couldn't all be right. The man with the microphone seemed to think this quite right and proper, and he didn't even try to get them to argue out their differences with each other. But that isn't the point I want to make for the moment. I simply want to ask where their beliefs come from. They came from tradition. Tradition means beliefs handed down from grandparent to parent to child, and so on. Or from books handed down through the centuries. Traditional beliefs often start from almost nothing; perhaps somebody just makes them up originally, like the stories about Thor and Zeus. But after they've been handed down over some centuries, the mere fact that they are so old makes them seem special. People believe things simply because people have believed the same thing over the centuries. That's tradition.

The trouble with tradition is that, no matter how long ago a story was made up, it is still exactly as true or untrue as the original story was. If you make up a story that isn't true, handing it down over a number of centuries doesn't make it any truer!

Most people in England have been baptized into the Church of England, but this is only one of the branches of the Christian religion. There are other

branches such as Russian Orthodox, the Roman Catholic, and the Method-
ist churches. They all believe different things. The Jewish religion and the
Muslim religion are a bit more different still; and there are different kinds
of Jews and of Muslims. People who believe even slightly different things
from each other often go to war over their disagreements. So you might
think that they must have some pretty good reasons—evidence—for believing what they believe. But actually, their different beliefs are entirely due to
different traditions.

Let's talk about one particular tradition. Roman Catholics believe that
Mary, the mother of Jesus, was so special that she didn't die but was lifted
bodily into Heaven. Other Christian traditions disagree, saying that Mary
did die like anybody else. These other religions don't talk about her much
and, unlike Roman Catholics, they don't call her the "Queen of Heaven."
The tradition that Mary's body was lifted into Heaven is not a very old one.
The Bible says nothing about how or when she died; in fact, the poor woman
is scarcely mentioned in the Bible at all. The belief that her body was lifted
into Heaven wasn't invented until about six centuries after Jesus' time. At
first it was just made up, in the same way as any story like "Snow White"
was made up. But, over the centuries, it grew into a tradition and people
started to take it seriously simply *because* the story had been handed down
over so many generations. The older the tradition became, the more people
took it seriously. It finally was written down as an official Roman Catholic
belief only very recently, in 1950, when I was the age you are now. But the
story was no more true in 1950 than it was when it was first invented six
hundred years after Mary's death.

I'll come back to tradition at the end of my letter, and look at it in
another way. But first I must deal with the two other bad reasons for believing in anything: authority and revelation.

Authority, as a reason for believing something, means believing in it
because you are told to believe it by somebody important. In the Roman
Catholic Church, the pope is the most important person, and people believe
he must be right just because he is the pope. In one branch of the Muslim
religion, the important people are the old men with beards called ayatollahs. Lots of young Muslims are prepared to commit murder, purely because
the ayatollahs in a faraway country tell them to.

When I say that it was only in 1950 that Roman Catholics were finally told that they had to believe that Mary's body shot off to Heaven, what I mean is that in 1950 the pope told people that they had to believe it. That was it. The pope said it was true, so it had to be true! Now, probably some of the things that that pope said in his life were true and some were not true. There is no good reason why, just because he was the pope, you should believe everything he said, any more than you believe everything that other people say. The present pope [1995] has ordered his followers not to limit the number of babies they have. If people follow this authority as slavishly as he would wish, the results could be terrible famines, diseases, and wars, caused by overcrowding.

Of course, even in science, sometimes we haven't seen the evidence ourselves and we have to take somebody else's word for it. I haven't, with my own eyes, seen the evidence that light travels at a speed of 186,000 miles per second. Instead, I believe books that tell me the speed of light. This looks like "authority." But actually, it is much better than authority, because the people who wrote the books have seen the evidence and anyone is free to look carefully at the evidence whenever they want. That is very comforting. But not even the priests claim that there is any evidence for their story about Mary's body zooming off to Heaven.

The third kind of bad reason for believing anything is called "revelation." If you had asked the pope in 1950 how he knew that Mary's body disappeared into Heaven, he would probably have said that it had been "revealed" to him. He shut himself in his room and prayed for guidance. He thought and thought, all by himself, and he became more and more sure inside himself. When religious people just have a feeling inside themselves that something must be true, even though there is no evidence that it is true, they call their feeling "revelation." It isn't only popes who claim to have revelations. Lots of religious people do. It is one of their main reasons for believing the things that they do believe. But is it a good reason?

Suppose I told you that your dog was dead. You'd be very upset, and you'd probably say, "Are you sure? How do you know? How did it happen?" Now suppose I answered: "I don't actually know that Pepe is dead. I have no evidence. I just have this funny feeling deep inside me that he is dead." You'd be pretty cross with me for scaring you, because you'd know that an inside "feeling" on its own is not a good reason for believing that a whippet is dead. You need evidence. We all have inside feelings from time to time, and sometimes they turn out to be right and sometimes they don't. Anyway, different people have opposite feelings, so how are we to decide whose feeling is right? The only way to be sure that a dog is dead is to see him dead, or hear that his heart has stopped, or be told by somebody who has seen or heard some real evidence that he is dead.

People sometimes say that you must believe in feelings deep inside, otherwise, you'd never be confident of things like "My wife loves me." But this is a bad argument. There can be plenty of evidence that somebody loves you. All through the day when you are with somebody who loves you, you see and hear lots of little tidbits of evidence, and they all add up. It isn't a purely inside feeling, like the feeling that priests call revelation. There are outside things to back up the inside feeling: looks in the eye, tender notes in the voice, little favors and kindnesses; this is all real evidence.

Sometimes people have a strong inside feeling that somebody loves them when it is not based on any evidence, and then they are likely to be completely wrong. There are people with a strong inside feeling that a famous film star loves them, when really the film star hasn't even met them. People like that are ill in their minds. Inside feelings must be backed up by evidence, otherwise you just can't trust them.

Inside feelings are valuable in science too, but only for giving you ideas that you later test by looking for evidence. A scientist can have a "hunch" about an idea that just "feels" right. In itself, this is not a good reason for believing something. But it can be a good reason for spending some time doing a particular experiment, or looking in a particular way for evidence. Scientists use inside feelings all the time to get ideas. But they are not worth anything until they are supported by evidence.

I promised that I'd come back to tradition, and look at it in another way. I want to try to explain why tradition is so important to us. All animals are built (by the process called evolution) to survive in the normal place in which their kind live. Lions are built to be good at surviving on the plains of Africa. Crayfish, to be good at surviving in fresh water, while lobsters are built to be good at surviving in the salt sea. People are animals, too, and we are built to be good at surviving in a world full of other people. Most of us don't hunt for our own food like lions or lobsters; we buy it from other people who have bought it from yet other people. We "swim" through a "sea of people." Just as a fish needs gills to survive in water, people need brains that make them able to deal with other people. Just as the sea is full of salt water, the sea of people is full of difficult things to learn. Like language.

You speak English, but your friend Ann-Kathrin speaks German. You each speak the language that fits you to "swim about" in your own separate "people sea." Language is passed down by tradition. There is no other way. In England, Pepe is a dog. In Germany he is *ein Hund*. Neither of these words is more correct or more true than the other. Both are simply handed down. In order to be good at "swimming about in their people sea," children have to learn the language of their own country, and lots of other things about their own people; and this means that they have to absorb, like blotting paper, an enormous amount of traditional information. (Remember that traditional information just means things that are handed down from grandparents to parents to children.) The child's brain has to be a sucker for traditional information. And the child can't be expected to sort out good and useful traditional information, like the words of a language, from bad or silly traditional information, like believing in witches and devils and ever-living virgins.

It's a pity, but it can't help being the case, that because children have to be suckers for traditional information, they are likely to believe anything the grown-ups tell them, whether true or false, right or wrong. Lots of what the grown-ups tell them is true and based on evidence, or at least sensible. But if some of it is false, silly, or even wicked, there is nothing to stop the children believing that, too. Now, when the children grow up, what do they do? Well, of course, they tell it to the next generation of children. So, once something gets itself strongly believed—even if it is completely untrue and there never was any reason to believe it in the first place—it can go on forever.

Could this be what has happened with religions? Belief that there is a god or gods, belief in Heaven, belief that Mary never died, belief that Jesus never had a human father, belief that prayers are answered, belief that wine turns into blood—not one of these beliefs is backed up by any good evidence. Yet millions of people believe them. Perhaps this is because they were told to believe them when they were young enough to believe anything.

Millions of other people believe quite different things, because they were told different things when they were children. Muslim children are told different things from Christian children, and both grow up utterly convinced that they are right and the others are wrong. Even within Christians, Roman Catholics believe different things from Church of England people or Episcopalians, Shakers or Quakers, Mormons or Holy Rollers, and all are utterly convinced that they are right and the others are wrong. They believe different things for exactly the same kind of reason as you speak English and Ann-Kathrin speaks German. Both languages are, in their own country, the right language to speak. But it can't be true that different religions are right in their own countries, because different religions claim that opposite things are true. Mary can't be alive in Catholic Southern Ireland but dead in Protestant Northern Ireland.

What can we do about all this? It is not easy for you to do anything, because you are only ten. But you could try this. Next time somebody tells you something that sounds important, think to yourself: "Is this the kind of thing that people probably know because of evidence? Or is it the kind of thing that people only believe because of tradition, authority, or revelation?" And, next time somebody tells you that something is true, why not say to them: "What kind of evidence is there for that?" And if they can't give you a good answer, I hope you'll think very carefully before you believe a word they say.

Your loving
Daddy

> "Believe nothing just because a so-called wise person said it. Believe nothing just because a belief is generally held. Believe nothing just because it is said in ancient books. Believe nothing just because it is said to be of divine origin. Believe nothing just because someone else believes it. Believe only what you yourself test and judge to be true."
>
> —SIDDHARTHA GAUTAMA (THE BUDDHA)

REALizing the Miraculous

"There are only two ways to live your life. One is as though nothing is a miracle. The other is as though everything is a miracle."
—ALBERT EINSTEIN

PROPHETIC INQUIRY: What life-serving meaning can we take from (or make of) miraculous stories that seem to exist in every religious tradition, including Christianity, but which are seldom honored or supported by any respected scientific approach?

I have relegated discussion of biblical miracle stories to this appendix for two reasons. First, the miracle stories appear to many as irrelevant, outdated, misleading, and perhaps embarrassing aspects of scripture-based religions. Second, for those readers who do find such stories inspiring and faith-enhancing, or who are curious as to how the evolutionary perspective might interpret the miraculous, I wanted to ensure that they first had a chance to read Richard Dawkins's poignant letter to his daughter on this topic.

Miracles Through the Ages

"It feels unacceptable to many people even to think of having a cosmology based on science. They misinterpret freedom of thought as requiring a refusal to believe anything. They see fanciful origin stories as spicing up the culture. The problem is, however, that spices, even in the most artful mixture, cannot compensate for the fact that there is no food—no data, no evidence; such stories are not actually about anything beyond themselves. We are not arguing to throw away the spices but to start with some food and then only use those spices that improve the food at hand. Scientific reality is the food. Aspects of many origin stories can enrich our understanding of the scientific picture, but they cannot take its place."

—JOEL R. PRIMACK AND NANCY ELLEN ABRAMS

When we look carefully at religion in the context of history and from a global perspective, a curious fact begs our attention. The further back in recorded history we peer, the more miracles we encounter and the more fantastic they generally are. Few claims of alleged supernatural occurrences of a decade or so ago have carried forward into the written records that are consulted today. Writings born of a century or more ago do contain some miraculous claims (for example, the golden tablets that launched the Mormon faith), but still, the miracles are not prolific. When we look back thousands of years, however, the written records that have survived document supernatural events and miracles galore. Indeed, the miraculous seems to have been what was often deemed worthy of passing on.

One need spend only a little time in a library or online to discover that the world's ancient religious texts are full of miraculous tales. There are hundreds if not thousands of stories, on every continent and within every tradition, of animals talking, of gods and goddesses profoundly affecting the lives of ordinary humans and the course of human history, of stars making one-time appearances and heralding world-changing events, of heroes accomplishing superhuman feats, of individuals living for hundreds of years, of angels encouraging and devils tempting, and of the blind seeing, the lame walking, and the demonically possessed made whole. There are stories of virgin births, resurrections from the dead, ascensions into heaven.

One may choose to believe or disbelieve any of these miracle stories. However, that these stories exist is beyond argument. Undeniable, too, is that miracle stories abound in many different traditions and that the older they are, the more extraordinary the claims tend to be. Not surprisingly, as a number of authors have noted, many parallels exist between the stories of Jesus recorded in the early Christian scriptures and stories that predate Christianity:

Born of a virgin: Dionysus, Horus, Tammuz, Krishna, Zarathustra, Buddha, Lao-Tzu, Attis, Heracles

Son of the Supreme God: Dionysus, Krishna, Mithras, Heracles

Death or torture by crucifixion (including bound to or embedded within a tree or stone): Dionysus, Osiris, Krishna, Prometheus

Resurrection and ascension: Osiris, Tummuz, Krishna, Mithras, Adonis

How shall we interpret similarities between the Christian story and those of other faiths and times? This was not a problem for earlier peoples when cultures did not intermingle, when there was no mass communication—no public libraries, no television, no Internet. Until recently, individuals throughout the world have mostly lived their lives unaware of the content of any creation story or religious vista other than their own. But now that we are exposed to other worldviews and religious stories (even when our parents and churches try to shield us from alternatives), this is what happens: We tend to regard the miraculous tales of our tradition as true, historical, and real—and the miraculous claims of other religions as fanciful stories. Sadly, for many of us, creation myths are "the crazy stories those people over there tell about how everything came to be. Our story is the truth!"

Perhaps another way to make sense of the similarities (though not one I recommend) is to conclude that most, if not all, of these ancient sacred stories are literally, historically true. Back then animals spoke, gods and goddesses blessed and cursed, supernova explosions were timed to coincide with important religious events, angels encouraged, devils tempted, and so forth, and that such supernatural events, for whatever reason, rarely if ever happen today.

A third interpretation—the one I find sensible and inspiring—is to regard each of these stories as an engaging and meaningful *night language* expression of something important about the nature of Reality and our relationship to It/ Him/Her, as experienced by a particular culture in one part of the world at a particular time in history. This interpretation includes claims that the origins of some miraculous stories have a basis in a scientifically verifiable event. For example, geological and archeological evidence suggests that a wall of water from the Mediterranean Sea poured through the Bosporus into the Black Sea (which was freshwater until then) about 7,000 years ago, owing to rising sea levels as Ice Age glaciers melted. Creating meaning out of a devastating event is human nature. It does not follow, though, that the event itself was punishment unleashed by a supernatural being. If we start here, we may discover that it is possible to "believe in" the miraculous stories of scripture and also to "know" that a literal interpretation of them is the least lifegiving of all. As a parishioner of mine once remarked, "I take the Bible far too seriously to interpret it literally."

From a developmental perspective grounded in deep time, nothing is lost and everything is gained by believing in the core meaning and teachings of miraculous stories, rather than in their literal truth. The Bible as sacred scripture is preserved *and* the Universe Story as sacred, updatable scripture becomes available to us, one and all. We can have both:

✦ I *believe* that God is creator and ruler of the Universe. And I *know* that this statement is metaphorical, not literal, in what it says about the nature of reality.

✦ I *know* that God has been communicating faithfully, and clearly, for hundreds of years to the entire human community through the full range of sciences. And I *believe* that this has everything to do with fulfillment of the Gospel and REALizing Christ's return.

Let us now turn to miracle stories in the Bible. In light of the distinctions between public and private revelation, and between day and night language, how can some of the central miracles of the early Christian scriptures be REALized from a creatheistic perspective?

"What can be asserted without evidence can be dismissed without evidence."

—CHRISTOPHER HITCHENS

From Born Again Believer to Born Again Knower

Connie and I visited the Creation Museum in Petersburg, Kentucky (a short drive from Cincinnati), soon after it opened in 2007. We enjoyed the visual splendor of its many exhibits, especially the life-size and animated models of humans and dinosaurs, the latter benignly coexisting with humans before the Fall, fearsome thereafter. Most impressive was the brilliant use of value-laden narrative to structure the experience: a single and memorable story-line that explained why we suffer and die—and how we can be saved. Connie and I were moved, too, by the respectful and forthright presentation of the evolutionary worldview. "Human Reason" was thus contrasted to the world-view showcased throughout the museum and labeled "God's Word."

Driving toward Chicago that same afternoon, I felt a glow from having mingled with hundreds of young families sharing the same adventure. How would the lives of those children unfold? I reflected on my own spiritual journey from traditional evangelical to evolutionary evangelical. Far from losing my faith, I had transited from born again believer to born again knower. How many years will pass before children of all faiths have a chance to encounter in exciting ways the evolutionary story—not just as science but as their creation story, a story that addresses their own biggest questions?

In my reverie on the freeway, I foresaw a time when evolutionary evan-gelicals would include a full spectrum of the devoutly religious: from those who believe in miracles, supernatural entities, and otherworldly concepts to those who believe in no such things. Uniting them all would be a shared religious knowing and experience of heavenly freedom. They would be born again knowers, incarnating Christ-like integrity and shining with joy. They would be distinguished, too, by an unshakable faith that God's Word encompasses the accumulated and ongoing public revelations delivered via scientific discovery—revelations that transcend differences of belief. Evo-lutionary evangelicals will continue to find great value in—but they will not be constrained by—the religious metaphors and understandings recorded by humans in generations past. Their faith will enable them to accept what is real, to appreciate ancestral instincts in themselves and others, while committing to practical action and support for channeling those instincts in ways that bless their communities—and thus honor God.

REALIZING "the Virgin Birth"

> *"Birth narratives tell us nothing about the birth of the person who is featured in those narratives. They do tell us a great deal, however, about the adult life of the one whose birth is being narrated. No one waits outside a hospital room for a great person to be born. This is not the way human life works. A person becomes great in his or her adult years, and the significance of that life is celebrated in tales that gather around the moment in which that powerful adult figure entered history."*   —JOHN SHELBY SPONG

PROPHETIC INQUIRY: What meaning might be made of the story of Jesus' virgin birth that would resonate with and inspire those of us grounded in a science-based evolutionary perspective?

The choice one makes to believe or not to believe, as a literal scientific fact, the miraculous stories of Jesus being born of a virgin (as reported in the gospels of Matthew and Luke) is unimportant from an evolutionary religious perspective. Some devout Christians do believe this; some do not. As well, we may or may not choose to ponder the suggestions by scholars of cultural history that many people in the ancient Mediterranean world would not have taken seriously the claims of Jesus' divinity had he *not* been born of a virgin. Consider the plethora of other such stories of virgin birth that preceded the birth story of Jesus:

Alcmene, virgin mother of Heracles

Amphictione, virgin mother of Plato

Anahita, virgin mother of Mithra (born on December 25th)

Antiope, virgin mother of Amphion and Zethus

Athena, virgin mother of Erichthnonius

Atia, virgin mother of Augustus (Roman emperor: 27 BCE – 14 CE)

Ceres, virgin mother of Proserpina

Chimalman, virgin mother of Kukulcan

Danae, virgin mother of Perseus

Devaki, virgin mother of Krishna (born on December 25th)

Ishtar, virgin mother of Tammuz (born on December 25th)

Isis, virgin mother of Horus (born on December 25th)

Juno, virgin mother of Mars

Maia, virgin mother of Hermes

Maya, virgin mother of Buddha

Mut-em-ua, virgin mother of Pharaoh Amenophis III

Myrrha, virgin mother of Adonis

Nama (Nana), virgin mother of Attis

Net (Neit, Neith), virgin mother of Ra

Olympias, virgin mother of Alexander the Great

Persephone, virgin mother of Dionysus (born on December 25th)

Rhea Silvia, virgin mother of Romulus and Remus

Semele, virgin mother of Bacchus

Semiramis, virgin mother of Nimrod

Shin-Moo, virgin mother of Somonocodom

Xochiquetzal, virgin mother of Quetzalcoatl

Scholarly and scientific quests to distinguish fact from fancy with respect to the actual life of Jesus hold little attraction for me religiously. Rather, I am compelled to ask whether spiritual guidance can be gleaned from the virgin-birth story that is deeply relevant for me as an evolutionary Christian today, regardless of the outcome of biographical research. My personal inquiry is this: *How might "the virgin birth" be REALized—that is, understood in a way that (a) validates the heart of earlier interpretations, (b) makes sense naturally and scientifically, (c) is universally, experientially true, and (d) inspires and empowers people across the theological spectrum, including non-Christians?*

Pondering the possible interpretations of Jesus' virgin birth that could be universally, experientially true, I come to this: Each and every human being who has ever brought anything of beauty, value, or importance into the world has done so only because that individual has been impregnated or in-spirited by some aspect of Beauty, Truth, Love, or other attributes of God. This divine co-creative spirit is beyond comprehension, beyond what we can call forth and direct by force of sheer will. When each of us reflects back on our own episodes of peak creativity, surely it feels as if some power greater than ourselves was at work. (The writing of this book has been such an experience for me.) There is a sense of having served, like Mary, as a vessel for something to emerge that is substantially greater than our own capacities. Truly, these peak experiences are religious moments. The story of Jesus' conception can remind us of such miracles in our own lives.

I also find it fruitful to imagine the reverse—how each of us is like a "father god" and "holy spirit" (spirit of wholeness). Who among us has not planted seeds of new life, new hope, new possibilities, within another simply by loving and cherishing them exactly as they are and exactly as they're not? Surely, this is the way most of us, most of the time, love our children. And when we stop to think about it, many of us discover there are others we have loved in this way, too. How much we have given them all! That which may have seemed to us as small gestures did in fact gestate, eventually birthing goodness in the lives we touched.

Finally, when we imagine how every human being is like the divine babe within Mary's womb, new insights emerge. From the perspective of the Whole, each one of us, especially as we are "born anew" into a life of trust, authenticity, responsibility, and service to God, becomes a divine gift, a CHRISTIAN for our world.

These are but a few REALIzed interpretations of the virgin birth that grow out of the richness of biblical stories. They are just a start. I know they will be improved upon by others who, like me, thank God for evolution.

REALizing "Christ's Resurrection and Ascension into Heaven"

"What if we were to understand the resurrection and ascension not as the bodily translation of some individuals to another world—a mythology no longer credible to us—but as the promise of God to be permanently present, 'bodily' present to us, in all places and times in our world? In what ways would we think of the relationship between God and the world were

> *we to experiment with the metaphor of the Universe as God's*
> *'body,' God's palpable presence in all space and time?"*
>
> —SALLIE MCFAGUE

PROPHETIC INQUIRY: For many non-Christians and proponents of scientific revelation, Christ's resurrection and ascent into heaven are unbelievable—and thus assumed to be irrelevant for their own spirituality and day-to-day life. Is that the end of the story? Can an evolutionary context help even nonbelievers find meaning and spiritual guidance in these stories of otherworldly events?

Given the not uncommon personification in ancient stories of the "life-death-rebirth" motif evident in the natural world, it is likely that 1st century Greeks, Romans, and Africans who became the early Christians might not otherwise have been convinced of the divinity of Jesus had stories of him excluded resurrection and ascension. Sacred stories of deities living, dying, descending to the underworld, and coming back to life abound throughout the ancient world. Mythical characters of this class include Osiris in Egyptian mythology, Adonis and Persephone in Greek mythology, Baldur and Odin in Norse mythology, Mithras in Persian mythology, and Inanna in Sumerian mythology.

Whether one interprets Jesus' resurrection and ascension as literal, historic occurrences (as many conservative Christians do), or as meaningful night language expressions of experiential insight (as many liberals do) makes little difference in the ability of these stories to transform people's lives and relationships. What Christians at both ends of the theological spectrum, and many non-Christians as well, should be able to agree on about the end of Jesus' life is what I submit is the most important lesson of all. Episcopal Bishop John Shelby Spong says it well:

> Obviously something happened after the death of Jesus that had startling and enormous power. Its power was sufficient to reconstitute a scattered and demoralized band of disciples. Its reality was profound enough to turn a denying Peter into a witnessing and martyred Peter, and to turn disciples who fled for their lives into heroes willing to die for their Lord. Easter was so intense that it created a new holy day, the first day of the week, and in

turn a new liturgical act, the breaking of bread, turning both into a weekly celebration of the presence of the living Lord in their midst. Easter was of such power that Jewish disciples taught from the time of their cradle that God alone was holy, that God alone was to be venerated, prayed to, and worshipped now could no longer conceive of God apart from Jesus of Nazareth. They could also no longer look at Jesus of Nazareth without seeing God. Whatever Easter was literally for the disciples, it meant that Jesus had been taken into God and vindicated by God. It also meant that Jesus had transcended death and was therefore ever present to the disciples as the animating Spirit. That was what the word *Easter* came to stand for in this faith community.

When I reflect on how the resurrection and ascension stories found in the Bible might be REALized today, I imagine a new Pentecost—an evolutionarily transformative revival of passion and purpose in service to Life in all its glorious diversity, beauty, and pain. Whatever else the resurrection of Jesus may mean to others, for me it means the following:

✦ Pain and suffering can be redemptive.

✦ Death is not the final word; new life for God is.

✦ Just because I am in deep integrity, have a right relationship with God, and am fulfilling my life purpose does not mean that everything is necessarily going to go well for me. And when things don't go well, I can trust God is up to something big!

✦ I can resurrect virtually any troubled relationship via the same path that Jesus incarnated: humble myself and take on the experience of the other, die to my own perspective as "the truth," take responsibility for doing the reconciling, be generous and compassionate in my communication, act with a grateful and faithful heart, and harbor no attachment that my effort should yield any particular outcome.

✦ I participate in the transformation of humankind's social structures along just and sustainable lines when I follow in Jesus' footsteps: knowing I'm a child of God, honoring my past, befriending the marginalized, loving my neighbor as self, courageously speaking my truth, and being the change I wish to see in the world.

Devout religious believers will surely continue to find inspiration and truth in diverse interpretations of the miracle stories of old. Nonetheless, we are each in a position to choose how we regard the stories unfolding in our own times and even in our own neighborhoods. Is the blossoming of a tiny flower in the crack of a sidewalk an ordinary event, or is it too a miracle? What about the way that the low-angle sun of autumn transforms brown leaves into chips of shimmering gold? Might the intense gaze of a toddler be a miracle too, as is the voice of a beloved that bounces off a satellite and then is channeled to our ear? I can send an entire book full of pictures and text through the air, using broadband wireless access, to friends and colleagues around the world, and at the same time. Now if that's not a miracle, nothing is!

What would it mean for me if, from time to time, I were to look at everything around me afresh, through childlike eyes of wonder, awe, gratitude, and curiosity? What would it mean if I *knew* and felt in my bones that everything is a miracle?

> When I was in Sunday school
> We learned about the times
> Moses split the sea in two
> Jesus made the water wine
> And I remember feeling sad
> That miracles don't happen still
> Now I can't keep track
> Because everything's a miracle
> Everything, everything's a miracle
>
> —PETER MAYER, *Holy Now*

✦ ✦ ✦

"The new cosmic story emerging into human awareness overwhelms all previous conceptions of the universe for the simple reason that it draws them all into its comprehensive fullness. Who can learn what this means and remain calm?"
                                                —BRIAN SWIMME

Invitation

I opened *Thank God for Evolution* with ten promises and am ending with a fivefold invitation...

1. Testify and share the good news.

If the sacred evolutionary perspective offered here has been transformative in some way, please share your testimonial with us. My publisher and I have discussed the possibility of a follow-up volume consisting mostly of people's stories of transformation. You will find suggestions for writing up your experience on ThankGodForEvolution.com, one of the two companion websites to this book. If the ideas or practices in *Thank God for Evolution* have made a difference in your life, work, or relationships, we want to hear from you.

2. Come together online.

ThankGodforEvolution.net, a second companion website, is a meeting ground where people of all religious traditions, philosophies, and spiritual paths who resonate with the perspectives offered in this book can come together, participate in an interactive course based on *TGFE*, and receive training to teach a *TGFE* curriculum in their local schools, places of worship, and communities. From Baptists to Buddhists and beyond, all are encouraged to engage in fresh conversations about matters of ultimate concern and learn how to apply this work to their own lives.

3. Expand your learning.

TheGreatStory.org is an affiliated educational website that holds the content and curricula that we have developed and used in our own evolutionary ministry, along with links to other online materials. Here you will find curricula for children, dramatic scripts for teaching the evolutionary story and values in playful ways, and much more.

4. Introduce others to this perspective.

Initiate a discussion group among friends, colleagues, your religious community, or your book club. You can download a *free study guide* for this book (and its companion DVD) at ThankGodForEvolution.com. DVDs of Connie's and my earlier programs are also available through the website.

5. Come meet us.

Connie and I would love to share these ideas with you in person. You can check our posted schedule online to see when we will be speaking in your town or region. We welcome speaking invitations from religious institutions of all traditions as well as from secular organizations. Please visit TheGreatStory .org, the website that directly supports our itinerant, evolution-celebrating ministry.

Acknowledgments

"In the end, everyone will know that everyone did it."
—ANONYMOUS

I did not write this book. God did. To claim it as mine would be the height of arrogance. Giving God the credit is the most responsible acknowledgment I can make—and yes, it is unquestionably *night language*. "God did it" is a poetic way of saying that this bouncing baby you hold in your hands was birthed through me, but it took the entire body of life to conceive and write it.

Painters, poets, and songwriters know this feeling. So do scientists and mathematicians, whose ideas and solutions often seem to come from nowhere and all of a piece. Any of us who have ever spontaneously composed a lullaby or a love letter, or have been amazed by the wise and eloquent speakers we become when excited by conversation or a supportive audience—we, too, share this experience. Understanding our brain's creation story can, of course, help us comprehend some of the mystery here, but the mystery (and the gratitude) remain. More, we feel tendrils of connection to the minds and hearts all around us and to those who came before. Call it God, call it collective intelligence, call it creative evolution: all of us are enmeshed within and thus potentially conduits for expression of emergent novelties far greater than our own felt capacities.

That said, some manifestations of God's grace are easy to identify, beginning with my own family. I thank my parents and siblings, my first beloved Alison and our children, Sheena, Shane, and Miriam—through whom I have received life's greatest gifts and grandest lessons. I also thank all the individuals and institutions that have nurtured, challenged, and chastened me over the years.

I am grateful to those who read the manuscript and offered suggestions, especially Tom Atlee, Cathy Russell, Michael Patterson, Jon Host, Jason John, Larry Edwards, Terri Anderson, Tom Buxton, John Stewart, Diarmuid O'Murchu, Matthew Fox, Christian de Quincey, Leslie Chance, Roger Meyers, and Gail Koelin, and to all who graciously and generously provided endorsements. What a blessing!

A special thanks to Benjamin De Pauw and his team at Ursa Minor for creating a state-of-the-art companion website for this book and the movement and to Paul West, my communications consultant, whose enthusiasm for this message is surpassed only by his competence and creativity. Tom Atlee, a pioneer in the world of collective intelligence, and Peggy Holman, a leader in the process arts, provided many of the ideas and some of the wording in Chapter 15. They also volunteered their efforts in shaping the prophetic inquiries. Tom and Peggy, and integral futurist Susan Cannon, have been especially supportive and dear friends and colleagues over the past two years. I also acknowledge the vital personal support given by my companions in recovery and by the men in my Evolutionary Integrity group.

I am indebted to Terri Anderson, Mia Van Meter, and Clare Hallward, without whom this book might not have been written. Thanks also to all who have taken us into their home and made us feel like family, to those who financially contributed to our ministry or volunteered their time, expertise, or wise counsel, and to those who gave us retreat time and space.

I am especially grateful to the trustees of the Foundation for Global Community for their generous donation that supported development of the book's companion website and video book trailer.

Grace is also evident in the poignancy and power of the epigraphic and embedded quotations drawn from scores of mentors and colleagues: some living, some dead, some famous, some not.

While I have been pregnant with this book for a long time, its birth would not have been possible without Laura Wood, my editor during the first phase of this project at Council Oak Books. Indeed, it was Laura who proposed that the timing was perfect for a book like this. She countered my initial hesitation by suggesting that we begin with transcripts of my oral presentations. In less than a year, the little book she envisioned became a major work; her editorial guidance was crucial at every step.

Special thanks to my amazing literary agent, Jillian Manus, the only person I know with more energy and passion for life than I have, and to Carolyn Carlson, my editor, and Clare Ferraro, president of Viking Penguin, for their visionary leadership and enthusiastic support of this project.

Foremost, I acknowledge the divine immanence expressed through my beloved mission partner, spouse, and best friend, Connie. I relied on my science-writer wife for the first drafts of the science components of this book. It was she who contributed the short course on evolution that became Chapter 2 (drawing on material from two of her previous books: *Evolution Extended* and *Green Space Green Time*). Connie as well drafted the sections on the Hubble Telescope, stardust, and death in Chapter 5. Although we co-created the ideas and terminology in Part III (on evolutionary brain science and evolutionary psychology), when I asked her to sketch out a few paragraphs, she went into a trance for two days and produced more than forty pages: much of chapters 9 and 10. "I was channeling Loren Eiseley," she laughed.

It was Connie's idea—her insistence—to ensure that this book teemed with anecdotes drawn from our experiences on the road. She keeps a journal of our travels and has a gift for organizing bits of data and memories into retrievable form. "This book requires a single voice—yours," she told me when I asked her why she didn't want to be listed as coauthor. And so, as she has done professionally in her career, Connie played the role of my holy ghost for those parts, too.

Finally, I acknowledge the contributions of generation upon generation of scientists and evolutionary thinkers. I'm mostly an evangelist, a popularizer, a storyteller. The real saints in this movement are those who labor in their labs or in the field, often without the public recognition they so rightfully deserve.

Resources

The Big Picture

Barlow, Connie. *Green Space Green Time* • *From Gaia to Selfish Genes*

Berry, Thomas. *Dream of the Earth* • *Great Work* • *Evening Thoughts*

Bryson, Bill. *A Short History of Nearly Everything*

Christian, David. *Maps of Time: An Introduction to Big History*

Combs, Allen. *The Radiance of Being*

Dawkins, Richard. *The Ancestor's Tale* • *The Blind Watchmaker*

Diamond, Jared. *Collapse* • *Guns, Germs, and Steel*

Flannery, Tim. *The Eternal Frontier*

Fortey, Richard. *Life: An Unauthorized Biography*

Genet, Russ. *Humanity: The Chimpanzees Who Would Be Ants*

Goodenough, Ursula. *The Sacred Depths of Nature*

Loye, David. *Darwin's Lost Theory*

McIntosh, Steve. *Integral Consciousness and the Future of Evolution*

Margulis, Lynn. *Symbiotic Planet*

Matthews, Clifford, et al., eds. *When Worlds Converge*

Primack, Joel R. and Nancy Ellen Abrams. *View from the Center of the Universe*

Rogin, Neal. *Awakening the Dreamer* (DVD)

Rue, Loyal. *Everybody's Story: Wising Up to the Epic of Evolution*

Sagan, Carl. *Cosmos* (DVDs or videos)

Swimme, Brian. *Canticle to the Cosmos* (DVD) • *The Universe Is a Green Dragon*

———and Thomas Berry. *The Universe Story*

Volk, Tyler. *Gaia's Body* • *Metapatterns* • *What Is Death?*

Wilber, Ken. *A Theory of Everything* • *Integral Spirituality*

Wilson, David Sloan. *Evolution for Everyone*

Wilson, Edward O. *Consilience*

The Trajectory of Divine/Cosmic Creativity and Human History

Barlow, Connie, ed. *Evolution Extended*

Beck, Don and Chris Cowan. *Spiral Dynamics*

Bloom, Howard. *The Lucifer Principle* • *Global Brain*

Carroll, Sean B. *Endless Forms Most Beautiful* • *The Making of the Fittest*

Chaisson, Eric. *Epic of Evolution: Seven Ages of the Cosmos*

Corning, Peter. *Nature's Magic* • *Holistic Darwinism*
Dawkins, Richard. *Climbing Mount Improbable* • *The Selfish Gene*
Eiseley, Loren. *The Immense Journey* • *Starthrower*
Elgin, Duane. *Awakening Earth*
Hubbard, Barbara Marx. *Conscious Evolution*
Huxley, Julian. *Religion Without Revelation*
Liebes, Sidney, et al. *A Walk Through Time*
Logan, Robert K. *The Sixth Language*
Margulis, Lynn and Dorion Sagan. *Microcosmos* • *Dazzle Gradually*
Morowitz, Harold. *Emergence of Everything: How the World Became Complex*
Morris, Simon Conway. *Life's Solutions*
Ong, Walter. *Orality and Literacy* • *The Presence of the Word*
Richerson, Peter and Robert Boyd. *Not by Genes Alone*
Russell, Peter. *Waking Up in Time* • *The Global Brain*
Sahtouris, Elisabet. *EarthDance: Living Systems in Evolution*
Stewart, John. *Evolution's Arrow*
Teilhard de Chardin. *The Human Phenomenon* • *The Divine Milieu*
Wright, Robert. *Nonzero: The Logic of Human Destiny*

Living in Deep Integrity

Blanton, Brad. *The Truthtellers* • *Practicing Radical Honesty*
Butler-Bowdon, Tom. *50 Self-Help Classics*
Campbell, Susan. *Getting Real* • *Saying What's Real*
Canfield, Jack. *The Success Principles*
Cutright, Layne and Paul Cutright. *Straight from the Heart*
Eisler, Riane. *The Power of Partnership*
Finney, Charles G. *Experiencing Revival*
Hendricks, Gay and Kathlyn Hendricks. *Conscious Living* • *Conscious Loving*
LaChance, Albert J. *Cultural Addiction* • *Architecture of the Soul*
Landmark Education Corporation. "Curriculum for Living"
McCarthy, Kevin. *The On-Purpose Person* • *The On-Purpose Business*
Morler, Edward. *The Leadership Integrity Challenge*
Murray, Andrew. *Humility: The Journey Toward Holiness*
Osteen, Joel. *Your Best Life Now* • *Become a Better You*
Schuller, Robert Anthony. *Walking in Your Own Shoes*
Schuller, Robert H. *Don't Throw Away Tomorrow* • *Self Esteem*
Selby, John. *Seven Masters, One Path*
Tozer, A. W. *The Best of A. W. Tozer, Books One and Two*
Warren, Rick. *The Purpose Driven Life*

Parents and Children

Brotman, Charlene. *The Kids' Book of Awesome Stuff*
Hart, Sura and Victoria Hodson. *Respectful Parents, Respectful Kids*
Louv, Richard. *Last Child in the Woods*

Martignacco, Carole. *The Everything Seed*
Morgan, Jennifer. *Born with a Bang* • *Lava to Life* • *Mammals Who Morph*

Humanity, Technology, and the Future

de Rosnay, Joël. *The Symbiotic Man*
Garreau, Joel. *Radical Evolution*
Johnson, Steven. *Emergence*
Kelly, Kevin. *Out of Control* • *New Rules for the New Economy*
Kurzweil, Ray. *The Singularity Is Near* • *The Age of Spiritual Machines*

Evolutionary Psychology, Brain Science, and Ethics

Brizendine, Louann. *The Female Brain*
Bromberg, S. E. *The Evolution of Ethics*
Burnham, Terry and Jay Phelan. *Mean Genes*
de Waal, Frans. *Our Inner Age. Primates and Philosophers: How Morality Evolved*
Diamond, Jared. *The Third Chimpanzee*
Haidt, Jonathan. *The Happiness Hypothesis*
La Cerra, Peggy, and Roger Bingham. *The Origins of Minds*
LaChapelle, Dolores. *Sacred Land Sacred Sex*
Lawrence, Paul R. *Being Human*
———and Nitin Nohria. *Driven*
Legato, Marianne. *Why Men Never Remember and Women Never Forget*
Leopold, Aldo. *A Sand County Almanac*
Lewis, Thomas, et al. *A General Theory of Love*
Midgley, Mary. *The Ethical Primate* • *Heart and Mind*
Mithin, Steven. *The Singing Neanderthals*
Pearce, Joseph Chilton. *The Biology of Transcendence*
Pinker, Steven. *The Blank Slate*
Ridley, Matt. *The Origins of Virtue* • *The Agile Gene*
Roughgarden, Joan. *Evolution's Rainbow*
Sapolski, Robert. *Monkeyluv: Essays on Our Lives as Animals*
Shermer, Michael. *The Science of Good and Evil*
Wilson, Edward O. *On Human Nature*
Wright, Robert. *The Moral Animal*

Visions and Tools for an Evolutionarily Sustainable Future

Atlee, Tom. *The Tao of Democracy*
Alexander, Christopher. *A Pattern Language*
Benyus, Janine. *Biomimicry: Innovation Inspired by Nature*
Brown, Juanita, and David Isaacs. *The World Café*
Brown, Lester. *Plan B 3.0: Mobilizing to Save Civilization, 3d ed.*
Chiras, Dan and Dave Wann. *Superbia*
Dixon, Thomas Homer. *The Upside of Down*

Eisler, Riane. *The Real Wealth of Nations*
Elgin, Duane. *Promise Ahead: A Vision of Hope and Action*
Hawken, Paul, et al. *Natural Capitalism* • *Blessed Unrest*
Holman, Peggy, ed. *The Change Handbook*
Korten, David. *The Great Turning*
McDonough, Bill and Michael Braungart. *Cradle to Cradle*
McKibben, Bill. *Deep Economy* • *Hope, Human and Wild*
Mollison, Bill. *The Global Gardener* (DVD) • *Permaculture*
Suzuki, David. *The Sacred Balance* • *Good News for a Change*
Thayer, Robert L. *LifePlace: Bioregional Thought and Practice*
Tucker, Mary Evelyn. *Worldly Wonder* • *Worldviews and Ecology*
Van Der Ryn, Sim and Stuart Cohen. *Ecological Design*
Williamson, Marianne, ed. *Imagine*

The Limitations of Flat-Earth Faith

Ayala, Francisco. *Darwin's Gift*
Dawkins, Richard. *A Devil's Chaplain* • *The God Delusion*
Dennett, Daniel. *Breaking the Spell*
Earl, Michael. *Bible Stories Your Parents Never Taught You*
Harris, Sam. *The End of Faith* • *Letter to a Christian Nation*
Hitchens, Christopher. *God Is Not Great*
Kitcher, Philip. *Living with Darwin* • *The Advancement of Science*
Miller, Kenneth R. *Finding Darwin's God*
Murdock, D. M. *Who Was Jesus?*
Rue, Loyal. *Amythia* • *Religion Is Not About God*
Scott, Eugenie C. *Evolution vs. Creationism: An Introduction*
Edis, Taner. *An Illusion: Science and Religion in Islam* • *Science and Nonbelief*

Toward an Evolutionary Faith

Boff, Leonardo. *Cry of the Earth* • *Cry of the Poor*
Borg, Marcus. *The Heart of Christianity* • *Jesus*
Burke, Spencer. *A Heretic's Guide to Eternity*
Burklo, Jim. *Open Christianity: Home by Another Name*
Cannato, Judy. *Radical Amazement* • *Quantum Grace*
Carter, Jimmy. *Our Endangered Values: America's Moral Crisis*
Cleary, William. *Prayers to an Evolutionary God*
Coelho, Mary. *Awakening Universe, Emerging Personhood*
Collins, Francis S. *The Language of God*
Csikszentmihalyi, Mihaly. *The Evolving Self* • *Flow* • *Creativity*
Dalai Lama. *The Universe in a Single Atom*
Edwards, Denis. *The God of Evolution*
Fabel, Arthur and Donald St. John, eds. *Teilhard in the 21st Century*
Fox, Matthew. *Original Blessing* • *Creation Spirituality*
Funk, Robert, et al. *The Five Gospels*

Gebara, Ivonne. *Longing for Running Water*
Gingerich, Owen. *God's Universe*
Haught, John. *Is Nature Enough?* • *God After Darwin*
Hofstetter, Adrian M. *Earth-Friendly: Revisioning Science and Spirituality*
Hubbard, Barbara Marx. *The Revelation* • *Emergence*
Hyun Kyung, Chung. *Struggle to Be the Sun Again*
Kimbell, Dan. *The Emerging Church*
LaChance, Albert J. *The Modern Christian Mystic* • *Jonah*
Lacroix-Hopson, Elaine. *Creation, Evolution, and Eternity*
Lerner, Michael. *The Left Hand of God*
Loye, David. *The River and the Star*
McFague, Sallie. *The Body of God* • *Super, Natural Christians*
McLaren, Brian. *A New Kind of Christian* • *A Generous Orthodoxy*
Macy, Joanna. *Coming Back to Life* • *World as Lover, World as Self*
Manji, Irshad. *The Trouble with Islam Today*
Marshall, Gene. *The Call of the Awe* • *Christianity in Change*
Midgley, Mary. *Evolution as a Religion*
Mitchell, Stephen. *The Gospel According to Jesus*
Moorwood, Michael. *Tomorrow's Catholic* • *Praying a New Story*
Murray, William. *Reason and Reverence*
Noll, Mark. *The Scandal of the Evangelical Mind*
O'Murchu, Diarmuid. *Catching Up with Jesus* • *Evolutionary Faith*
Redfield, James, et al. *God and the Evolving Universe*
Rohr, Richard. *From Wild Man to Wise Man* • *Contemplation in Action*
Roughgarden, Joan. *Evolution and Christian Faith*
Rupp, Joyce (art by Mary Southard). *The Cosmic Dance*
Schenk, Jim. *What Does God Look Like in an Expanding Universe?*
Sheldrake, Rupert and Matthew Fox. *Natural Grace* • *The Physics of Angels*
Shiva, Vandana. *Earth Democracy: Justice, Sustainability, and Peace*
Sider, Ron. *Scandal of the Evangelical Conscience*
Slifkin, Natan. *The Challenge of Creation: Judaism's Encounter with Science*
Song, Choan-Seng. *And Their Eyes Are Opened*
Spong, John Shelby. *Jesus for the Nonreligious*
Thich Nhat Hanh. *Living Buddha, Living Christ* • *Going Home*
Wallis, Jim. *God's Politics* • *Faith Works* • *The Call to Conversion*
Wessels, Cletus. *The Holy Web* • *Jesus in the New Universe Story*
Wink, Walter. *The Powers That Be*
Wright, N. T. *Simply Christian*

Online Resources

ThankGodForEvolution.com

The website directly supporting this book, including free study guides, Michael's blog, and speaking schedule, a wealth of audio and video clips, online sales of T-shirts and caps sporting the *TGFE* fish logos, and dynamically updated and annotated links to the best online resources for exploring more deeply the themes and topics discussed in *TGFE*.

ThankGodforEvolution.net

Enroll in a free course based on the book, learn how to teach the Great Story to others, and find others also interested in a sacred understanding of evolution. You will have the opportunity to post your ideas, questions, and projects to the forum board, post your own content and stories to the community discussion, and become part of a dynamic community of inspired individuals and organizations committed to a sacred and meaningful future.

TheGreatStory.org

The main educational website in the Great Story/sacred evolution movement. Here you will find a condensed timeline of the fourteen-billion-year epic of evolution, along with ready-to-use items for ceremony, teaching, and reflection. These include science-based parables in narrative and script formats, instructions for assembling your own set of Great Story beads (aka, cosmic rosaries), downloadable songs and litanies, and reports and photos of "sacred sites" where particular episodes of our immense journey can be celebrated. You will also be able to access curricula for teaching our common creation story and values to children, and you will encounter a wealth of examples of how enthusiasts have transformed sacred science into meaningful ceremony.

Who's Who

and Sources of Quotations

Abrams, Nancy Ellen—lawyer, writer, and Fulbright scholar; coauthor with her husband, Joel R. Primack, of *The View from the Center of the Universe.*

Ackerman, Diane—author, poet, and naturalist; best known for her book *A Natural History of the Senses;* p. 31: *A Natural History of Love* (1994), p. 260.

Adams, Patch—medical doctor, social activist, professional clown, author, and founder of the Gesundheit! Institute; his life was portrayed in film in 1998.

Agent Smith—fictional character and primary antagonist in *The Matrix* film series.

Aquinas, Thomas—13th-century Italian Catholic priest, philosopher, and theologian; the foremost classical proponent of natural theology.

Atlee, Tom—founder of the Co-Intelligence Institute and author of *The Tao of Democracy: Using Co-Intelligence to Create a World That Works for All.*

Augustine of Hippo, or Saint Augustine (354–430)—a key figure in the development of Western Christianity; author of *The Confessions.*

Bache, Christopher—leader in transformative learning; author of *Dark Night, Early Dawn: Steps to a Deep Ecology of Mind.*

Banathy, Bela H. (1919–2003)—Hungarian linguist and systems scientist; author of *Guided Evolution of Society.*

Barlow, Connie—science writer and evolutionary evangelist; author of *The Ghosts of Evolution;* pp. 28, 142: *Green Space Green Time* (1997), pp. 14, 227.

Bateson, Gregory (1904–1980)—anthropologist, social scientist, linguist, and cyberneticist; pp. 127, 248: *Steps to an Ecology of Mind* (1972), p. 491.

Bateson, Mary Catherine—writer and cultural anthropologist; author of *Composing a Life;* p. 267: *Peripheral Visions* (1994), p. 14.

Beck, Don—contributor to developmental views of human psychology and culture; coauthor of *Spiral Dynamics: Mastering Values, Leadership, and Change.*

Benyus, Janine—science writer and lecturer on environmental matters; author of *Biomimicry: Innovation Inspired by Nature.*

Berry, Thomas—cultural historian and Catholic geologian; pp. 62, 136, 310–311, 317: *Dream of the Earth* (1988), pp. 137, 17–18, 137, xi; pp. 134, 247: *Evening Thoughts* (2006), pp. 86, 96; p. 8: (in) *Deep Ecology for the 21st Century,* ed. Sessions (1999), p. 18; p. 72: "The New Story," *Teilhard Studies* 1 (1978); p. 134: *Dawn Over the Earth* (video; undated); p. 192: "Mystique of the Earth" interview in *Caduceus* No. 59 (2003); p. 255: "Ethics and Ecology," paper delivered at 1996

Harvard seminar on environmental values (Internet publ.); p. 293: (quoted in) *Critical Social Issues in American Education* (2004), p. 399; p. 329: (quoted in) "12 Principles" at ajlachance .com/12principles.htm; p. 112: (in) Dowd essay (in) *Radical Grace* newsletter (ca. 1991); p. 312: (in) "The Big Picture" (by M. Dowd, 1992); pp. 106, 290, 329: personal communication.

Bertalanffy, Ludwig von (1901–1972)—Austrian-born biologist and one of the founders of general systems theory; p. 93: (in) *Unity Through Diversity: A Festschrift for L Bertalanffy* (1973), p. 1059.

BIBLE PASSAGES

pp. 96, 241: Romans 5:12, 12:1–2; p. 133: Genesis 2:7; p. 184: Matthew 22:37–40; pp. 191, 251, 339: John 13:34–35, 3:17, 8:32; pp. 214, 244: Acts 2:8, 17:28; p. 262: [Gospel of Thomas 3:1–5]; p. 265: Galatians 5:22–23; p. 339: John 8:32.

Black Elk (1863–1950)—medicine man of the Oglala Lakota Sioux; p. 56: *Black Elk Speaks.*

Bloom, Howard—cross-disciplinary thinker; author of *The Lucifer Principle*, *Global Brain*, and *How I Accidentally Started the Sixties.*

Boff, Leonardo—Brazilian theologian, philosopher, and writer known for his active support of the rights of the poor and excluded.

Borg, Marcus—well-known progressive Christian writer and a leader of the Jesus Seminar; author of *The Heart of Christianity, Jesus,* and *Reading the Bible Again.*

Braungart, Michael—German chemist and noted green industrial designer who advocates "upcycling," not recycling; coauthor of *Cradle to Cradle.*

Brewer, John—longtime epic of evolution enthusiast, writing from the perspective of Unitarian Universalism; p. 154: personal communication, email to C. Barlow (2004).

Brown, Juanita—co-originator with David Isaacs of the World Café, innovative dialogue; author of *The World Café.*

Burke, Spencer—former megachurch pastor and founder of TheOOZE.com, a progressive evangelical website; author of *A Heretic's Guide to Eternity.*

Burklo, Jim—Christian pastor and activist with The Center for Progressive Christianity; author of *Open Christianity: Home by Another Road.*

Campbell, Neil (1946–2004)—scientist best known for his *Biology* textbook, coauthored by Jane Reece, now in its seventh edition and used worldwide; p. 79: *Biology* 2nd edition (1990), p. 434.

Canfield, Jack—motivational speaker and author; co-creator of *Chicken Soup for the Soul* book series; pp. 60, 216: *The Success Principles* (2005), pp. 3, 157.

Carlson, Eric—physicist at Wake Forest University and contributor to meaningful dialogue between scientists and religionists; p. 69: (in) *Cosmic Beginnings and Human Ends*, Matthews, ed. (1994), p. 88.

Carroll, Sean B.—developmental biologist and highly regarded popularizer of development biology; pp. 24, 79: *Endless Forms Most Beautiful* (2005), pp. 296, 294; pp. 40, 82–83: *Making of the Fittest* (2007), preface, preface.

Carter, Rita—a British medical writer; p. 153: *Mapping the Mind* (1999), p. 180.

Chaisson, Eric—astrophysicist and science educator; author of *Epic of Evolution* and *Cosmic Evolution.*

Churchill, Winston (1874–1965)—British statesman, orator, and prime minister of the United Kingdom from 1940 to 1945 and from 1951 to 1955.

Cobb, John B.—process theologian who integrates Alfred North Whitehead's metaphysics into Christianity and applies it to social justice issues.

Coelho, Mary—Quaker epic of evolution enthusiast; p. 109: *Awakening Universe, Emerging Personhood* (2002), p. 184.

Cohen, Andrew—spiritual teacher who promotes evolutionary enlightenment through EnlightenNext and *What Is Enlightenment?* magazine.

Collins, Ed—naturalist and leader in prairie ecological restoration; practitioner of Earth-honoring spirituality; p. 180: personal communication, email to C. Barlow (2003).

Corning, Peter—systems researcher; author of *Holistic Darwinism*; p. 42: *The Synergism Hypothesis* (1983), p. 103.

Cowan, Chris—contributor to developmental views of psychology and culture; coauthor of *Spiral Dynamics: Mastering Values, Leadership, and Change.*

Cowan, Stuart—coauthor of *Ecological Design.*

Darwin, Charles (1809–1882)—eminent English naturalist and author of *On the Origin of Species* and *The Descent of Man*; p. 40: *On the Origin of Species* 2nd edition (1860), final paragraph; pp. 251, 318: *The Descent of Man*; (1871), chapter 21, chapters 3 and 21.

Davis, Phyllis K.—counseling professional and leader in somatic psychology; p. 232: *The Power of Touch* (1999), pp. 31, 104, 231, and 232.

Dawkins, Richard—evolutionary biologist, renowned writer; author of *The Selfish Gene*; pp. 39–40: *Ancestor's Tale* (2004), pp. 588 and 599; p. 41: *The Blind Watchmaker* (1986), p. 184.

Deacon, Terrence—neuroscientist and biological anthropologist; author of *The Symbolic Species: The Coevolution of Language and the Brain.*

Delacroix, Jérôme—leader in cooperative Internet technologies; author (in French) of *Les Wikis*; pp. 274–275: www.m21editions.com/us/wikis.shtml.

Dellinger, Drew—spoken-word poet, activist, and Great Story enthusiast; founder of Poets for Global Justice and author of *Love Letters to the Milky Way.*

Dennett, Daniel—leader in the philosophy of mind, science, and biology; author of *Consciousness Explained* and *Breaking the Spell*; p. 29: *Darwin's Dangerous Idea* (1995), p. 73.

de Quincey, Christian—leader in consciousness studies; author of *Radical Nature: Rediscovering the Soul of Matter.*

de Rosnay, Joël—French futurist, molecular biologist, and science writer; pp. 313, 314: *The Symbiotic Man* (2000), pp. xi–xiii, xvi.

Dick, Philip K. (1928–1982)—writer known mostly for his works of science fiction; p. 121: *How to Build a Universe That Doesn't Fall Apart* (1978).

Dobb, Edwin—journalist and writer; contributor to *Harper's* magazine; p. 103: "Without Earth There Is No Heaven" in *Harper's*, February 1994, p. 33.

Dobzhansky, Theodosius (1900–1975)—geneticist, evolutionary biologist, and a leader in the dialogue between science and religion; pp. 78–79: "Nothing in Biology Makes Sense Except…" in *American Biology Teacher*, March 1973, pp. 125–29; p. 134: *The Biology of Ultimate Concern* (1967), p. 115.

Dominguez, Joe (1938–1997)—teacher of financial independence and voluntary simplicity; coauthor of *Your Money or Your Life*; p. 240: in *Friends and Lovers* magazine, Summer 1985, p. 55 (reprinted 1997 *In Context* magazine).

Drucker, Peter (1909–2005)—a leader and prolific writer of innovative approaches to corporate leadership and business management; p. 300: *Post-Capitalist Society* (1993), p. 1.

Earl, Michael—author of *Bible Stories Your Parents Never Taught You* audio programs; pp. 320–321: *The Ultimate Terrorist*, audio essay (2002 ReasonWorks).

Eckhart, Meister (1260–1328)—German theologian, philosopher, and mystic.

Edwards, Denis—Australian theologian; author of *The God of Evolution*.

Einstein, Albert (1879–1955)—theoretical physicist of world renown for his theory of relativity; p. 117: in *The Universe and Einstein* by Barnett (1948), p. 109; p. 336: letter by Einstein of March 4, 1950; p. 357: public domain.

Eiseley, Loren (1907–1977)—anthropologist noted for his pioneering blend of science writing with personal reflection; author of *The Immense Journey*; p. 147: "Starthrower" in *The Unexpected Universe* (1972), p. 76.

Elgin, Duane—a leading advocate of approaches for cultural transformation that blend spirituality and ecology; author of *Promise Ahead* and *Awakening Earth*.

Eliade, Mircea (1907–1986)—historian of religion, celebrated for his cross-cultural examinations of myth and ritual; author of *The Eternal Return*.

Emerson, Ralph Waldo (1803–1882)—a writer, philosopher, and leader of the transcendentalism movement in America; p. 32: *Ralph Waldo Emerson Journals*, volume 1831.

Erhard, Werner—creator of transformational models and applications for individuals and organizations; founder of "the est training"; pp. 243–244: quoted in *Beyond Our Consent*, by Robert Harris (2004), p. xlvii.

Ferris, Timothy—astronomer and renowned writer of popular astronomy books; author of *Coming of Age in the Milky Way* and *The Mind's Sky*; p. 36: in *Cosmic Beginnings and Human Ends*, Matthews, ed. (1994), p. 88.

Fox, Matthew—theologian and leading exponent of creation spirituality; author of *Original Blessing*, *The Coming of the Cosmic Christ*, and *Creativity*; p. 105: *Creation Spirituality* (1991), p. 74.

Gebara, Ivonne—Brazilian feminist theologian; author of *Longing for Running Water* and *Out of the Depths*.

Genet, Russ—astronomer and epic of evolution enthusiast; author of *Humanity: The Chimpanzees Who Would Be Ants*; p. 294: personal communication, email to C. Barlow (2006).

Goldberg, Elkhonon—neurologist; p. 152: *The Executive Brain* (2001), p. 2.

Goodenough, Ursula—cell biologist and leader in the dialogue between science and religion; author of *The Sacred Depths of Nature*.

Goodfield, June—British scientist and writer; author of *An Imagined World*.

Gore, Al—45th vice president of the United States, environmentalist, and author of *Earth in the Balance*, *An Inconvenient Truth*, and *The Assault on Reason*.

Gould, Stephen Jay (1941–2002)—evolutionary biologist and historian of science; prolific writer of science essays and books for popular audiences; p. 46: (in) *The Logic of Life*, Boy and Noble eds. (1993), pp. 15–42; p. 79: *Hen's Teeth and Horse's Toes* (1994), pp. 254–55.

Grassie, William ("Billy")—religious scholar and founder of the Metanexus Institute on Religion and Science; p. 139: "Science as Epic" in *Science and Spirit*, Vol. 9, issue 1 (1998).

Gump, Forrest—fictional protagonist and narrator in both the novel and movie *Forrest Gump*, played by Tom Hanks, who won an Academy Award for the role.

Harjo, Joy—poet and musician, member of the Muscogee (Creek) Nation of Oklahoma; author of *She Had Some Horses* and *How We Became Human;* p. 113: *Secrets from the Center of the World* (1989), p. 56.

Harris, Sam—critic of scriptural literalism; author of *Letter to a Christian Nation;* pp. 319, 330–331, 349: *The End of Faith* (2004), pp. 78, 21–22, 19.

Haught, John—renowned Roman Catholic theologian; author of *God After Darwin* and *101 Questions on God and Evolution;* p. 1: paper delivered at Epic of Evolution Conference, Chicago (1997); p. 24: *Deeper Than Darwin* (2003), p. 162.

Havel, Václav—Czech writer and world advocate of human rights; first president of the post-Soviet Czech Republic; author of *Art of the Impossible;* p. 259: (in) *Cosmic Beginnings and Human Ends*, Matthews, ed. (1993), p. 88.

Hefner, Philip—theologian and leader in the dialogue between science and religion; author of *The Human Factor: Evolution, Culture, and Religion;* p. 63: paper delivered at Epic of Evolution Conference, Chicago (1997); p. 108: "How Science Is a Resource" (2002) in *Zygon* 37(1): pp. 55–62.

Hitchens, Christopher—British-American author, journalist, and literary critic; p. 360: *God Is Not Great* (2007), p. 150.

Hobbes, Thomas (1588–1679)—English philosopher whose 1651 book, *Leviathan*, set the agenda for nearly all subsequent Western political philosophy.

Hofstetter, Adrian—Dominican nun and author of *Earth-Friendly: Re-Visioning Science and Spirituality Through Aristotle, Thomas Aquinas, and Rudolf Steiner.*

Holman, Peggy—process consultant; author of *The Change Handbook: The Definitive Resource on Today's Best Methods for Engaging Whole Systems.*

Hubbard, Barbara Marx—futurist and advocate for conscious evolution; author of *Conscious Evolution* and *Emergence: The Shift from Ego to Essence;* p. 47: (in) www.evolve.org (undated); p. 177: interview of Hubbard in *What Is Enlightenment?*, wie.org/j30/hubbard.asp; p. 207: *Emergence* (2001), p. 86.

Huxley, Julian (1887–1975)—British evolutionary biologist, humanist, and award-winning writer; author of *Religion Without Revelation;* pp. 33, 133: *New Bottles for New Wine* (1957), pp. 13, 13.

Hyun Kyung, Chung—Korean Christian theologian; author of *Struggle to Be the Sun Again*, an introduction to Asian women's theology.

James, William (1842–1910)—noted psychologist and philosopher; author of *The Varieties of Religious Experience;* pp. 48–49: *Pluralistic Universe* (1901), p. 25.

Johnson, Elizabeth—a Roman Catholic sister of St. Joseph, theologian, and leader in Christian feminism; author of *She Who Is* and *Consider Jesus;* p. 133: adapted from essay in *Earthlight*, Vol. 25, Spring 1997, with permission of the author.

Kant, Immanuel (1724–1804)—German philosopher, widely regarded as one of the most influential thinkers of modern Europe; author of *Critique of Pure Reason.*

Kauffman, Stuart—theoretical biologist and complex systems researcher; author of *At Home in the Universe;* p. 46: *The Origins of Order* (1993), pp. 6, xii, 173.

Keller, Helen (1880–1968)—deaf-blind author, activist, and inspirational force.

Kelly, Kevin—cross-disciplinary thinker, Internet innovator, and futurist; author of *Out of Control* and *New Rules for the New Economy;* p. 65: "Evolution of the Scientific Method" (2004) at kk.org/thetechnium; p. 71: "Major Transitions in Technology" (2006) at kk.org/thetechnium.

Koestler, Arthur (1905–1983)—cross-disciplinary thinker and celebrated writer; author of novels, social philosophy, and science-related books; p. 85: *Janus: A Summing Up* (1978).

Korten, David—a leading voice in the resistance to corporate globalization; author of *When Corporations Rule the World* and *The Great Turnings*.

Kuhn, Thomas (1922–1996)—philosopher and historian of science; p. 244: *The Structure of Scientific Revolutions* (1962), p. 111.

Kurzweil, Ray—inventor, futurist, transhumanist; author of *The Age of Spiritual Machines* and *The Singularity Is Near.*

LaChance, Albert—cultural therapist and Great Story teacher; author of *Cultural Addiction*, *The Modern Mystic*, and *Architecture of the Soul.*

LaChapelle, Dolores (1925–2007)—deep ecology scholar and writer; p. 237: *Sacred Land, Sacred Sex* (1992), p. 128.

Lakoff, George—cognitive linguist, author (with Mark Johnson) of *Metaphors We Live By*; p. 105: *Metaphors We Live By* (1980), p. 239.

Lavanhar, Marlin—liberal thinker and activist; senior minister of All Soul's Unitarian Universalist Church in Tulsa, Oklahoma; p. 128: unpublished sermon, spring 2006.

Lawrence, Paul—cross-disciplinary scholar whose work spans from organizational management to evolutionary psychology; coauthor of *Driven: How Human Nature Shapes Our Choices*; p. 230: *Driven* (2002), p. 284.

Leopold, Aldo (1887–1948)—ecologist and inspirational force in the environmental movement via his book *A Sand County Almanac.*

Lerner, Michael—rabbi, political activist, and editor of *Tikkun*, a progressive Jewish and interfaith magazine; author of *The Left Hand of God.*

Lovins, Amory—leader in advancing environmentally healthy technologies and ideas; author of *Winning the Oil Endgame.*

Lovins, L. Hunter—champion of ecologically sustainable practices; coauthor of *Natural Capitalism.*

MacDonald, Gordon—a noted evangelical Christian and writer, editor of leadership journal at *Christianity Today*; p. 160: 2006 blog, "Leaders Insight: When Leaders Implode" posted at www.christianitytoday.com/leaders/newsletter/2006/cln61106.html.

McDonough, William—a leading ecodesigner and architect of sustainable buildings and transforming industrial processes; coauthor of *Cradle to Cradle.*

McFague, Sallie—renowned feminist ecotheologian; author of *Models of God*, *The Body of God*, and *Super, Natural Christians.*

MacGillis, Miriam—Dominican nun, farmer, and leader in the bioregional movement; influential popularizer of the ideas of Thomas Berry and the Great Story; p. 293: "The Fate of the Earth" in *Soul in Nature*, Michael Tobias, ed.

McLaren, Brian—a leading voice in the Emerging Church movement; author of *A New Kind of Christian*, *A Generous Orthodoxy*, and *The Secret Message of Jesus.*

McMenamin, Mark—paleontologist and evolutionary thinker; author of *The Garden of Ediacara* and *Hypersea: Life on Land.*

Macy, Joanna—scholar and activist of Buddhism, general systems theory, and deep ecology; author of *World as Lover, World as Self* and *Coming Back to Life*; p. 298: in *Thinking Like a Mountain* (1988), John Seed et al., eds.

Manitonquat (Medicine Story)—a Wampanoag elder working with American Indians in and out of prison; author of *Ending Violent Crime.*

Margulis, Lynn—biologist known for her work on symbiosis as a primary driver of evolutionary emergence; coauthor of *Symbiotic Planet* and *Microcosmos;* p. 42: "Marriage of Convenience," in *The Sciences*, September 1990.

Marshall, Gene—Christian thinker and writer; author of *The Call of the Awe, Christianity in Change,* and *The Reign of Reality.*

Mathews, Freya—Australian ecophilosopher; author of *The Ecological Self, For Love of Matter,* and *Reinhabiting Reality.*

Mayer, Peter—folksinger-songwriter and a leader in expressing ideas and values of the Great Story via song.

Maynard Smith, John (1920–2004)—British evolutionary biologist instrumental in applying game theory to evolution; author of *The Major Transitions of Evolution.*

Meadows, Donella "Dana" (1941–2001)—pioneering environmental scientist, teacher, and activist; lead author of *Limits to Growth;* p. 269: "Seeing the Population as a Whole," *The Economist*, June 1993.

Meeker, Tobias—Buddhist-Catholic hospital chaplain, administrator, and a leader in the field of medical ethics.

Montagu, Ashley (1905–1999)—British anthropologist, humanist, and author of many books that brought race and gender issues to public awareness; p. 231: *Touching* (1972), p. 192.

Morwood, Michael—Australian ecotheologian; author of *Tomorrow's Catholic, Praying a New Story, God Is Near,* and *Is Jesus God?*

Morris, Simon Conway—British paleontologist; author of *The Crucible of Creation;* pp. 38–39 (paragraph 1): *Life's Solution* (2003), p. 283; p. 39 (paragraph 2): "We Were Meant...," *New Scientist*, November 16, 2002, p. 26.

Mumford, Lewis (1895–1990)—historian of technology and science, urban futurist, and prolific author; p. 341: *The Transformations of Man* (1956), p. 179.

Murray, W. H. (1913–1996)—Scottish mountain climber; p. 181: *The Scottish Himalayan Expedition* (1951).

Niebuhr, Reinhold (1892–1971)—liberal Protestant theologian, best known for his writings on how the Christian faith intersects with modern politics.

Neo— protagonist in *The Matrix* science fiction film trilogy.

Nohria, Nitin—cross-disciplinary scholar; coauthor of *Driven: How Human Nature Shapes Our Choices.*

Oliver, Mary—contemporary American poet with an Earth-centered spiritual focus; p. 185: "The Summer Day" in *New and Selected Poems* (1992).

O'Murchu, Diarmuid—Irish priest, social psychologist, and evolutionary theologian; author of *Quantum Theology, Evolutionary Faith,* and *Transformation of Desire.*

Osteen, Joel—senior pastor of Lakewood Church in Houston, Texas, North America's largest and fastest-growing church; author of *Your Best Life Now.*

Pearce, Joseph Chilton—cross-disciplinary thinker; author of *Crack in the Cosmic Egg, Magical Child, Evolution's End,* and *The Biology of Transcendence.*

Pinker, Steven—cognitive scientist grounded in an evolutionary worldview; author of *How the Mind Works* and *The Blank Slate.*

Popper, Karl (1902–1994)—Austrian-born British philosopher of science; author of *Conjectures and Refutations, The Open Universe, In Search of a Better World*; p. 151: *Of Clouds and Clocks* (1966), p. 27.

Pratney, Winkie—New Zealand Christian evangelist whose worldwide ministry focuses on youth; author of *Youth Aflame: Manual on Discipleship.*

Primack, Joel R.—astrophysicist; coauthor of the *Theory of Cold Dark Matter* and *The View from the Center of the Universe;* pp. 13, 19, 22, 85–86, 86–87, 93, 107, 118, 126, 134, 137, 277, 293, 307, 358: *View from the Center* (2006), pp. 276, 84, 151, 281, 119, 279, 253, 276, 274–75, 262–63, 267, 267, 120, 240, 85.

Rhodes, Thomas—senior minister at Unitarian Universalist Fellowship of Raleigh; p. 101: liturgy composed for April 1, 2007, church service.

Ridley, Matt—British science writer; author of *The Red Queen* and *Nature Via Nurture;* p. 261: *The Origins of Virtue* (1996), p. 249.

Robin, Vicki—a leader for social transformation and founder of Conversation Café and Let's Talk America; coauthor of *Your Money or Your Life.*

Rohr, Richard—Franciscan priest known for his progressive ideas; author of *Everything Belongs: The Gift of Contemplative Prayer* and *The Enneagram.*

Rue, Loyal—philosopher and contributor to the dialogue between science and religion; author of *Amythia* and *Religion Is Not About God;* p. 67: *By the Grace of Guile* (1994), p. 289; p. 84: *Everybody's Story* (1999), p. xii.

Rumi—13th-century Persian poet and theologian of Sufi Muslim faith.

Sagan, Carl (1934–1996)—astronomer and celebrity popularizer of meaning-filled science; author of the 1980 *Cosmos* series and other award-winning books; p. 7: *The Varieties of Scientific Experience* (2007), p. 31; p. 89, 107: *Cosmos* (1980/2002 edition), pp. 343, 218.

Sahtouris, Elisabet—a Greek American evolutionary biologist and futurist; author of *Earth-Dance* and *A Walk Through Time: From Stardust to Us;* p. 248: "The Evolving Story of Our Evolving Earth," paper presented at "How Evolution Works" (1999), published at: http://www.ratical.org/LifeWeb/Articles/H3Kevolv.html.

Sandburg, Carl (1878–1967)—one of America's most celebrated poets of the 20th century pp. 169, 171, 174, 176, 181: "Wilderness" in *Cornhuskers* (1918).

Shelley, Percy (1792–1822)—English Romantic poet, considered one of the finest lyric poets of the English language; p. 288: "Song of Apollo" (poem).

Shenk, Jim—ecospiritual activist and epic of evolution enthusiast; editor of *What Does God Look Like in an Expanding Universe?*

Shermer, Michael—science historian and editor of *Skeptic* magazine; author of *Why Darwin Matters;* pp. 325, 333: *The Science of Good and Evil* (2004), pp. 20, 57.

Shiva, Vandana—Indian physicist, ecofeminist, and environmental activist of world renown; author of *Earth Democracy* and *Alternatives to Economic Globalization.*

Song, Choan-Seng—Taiwanese scholar of Christian and Asian thought; author of *Third Eye Theology* and *Jesus in the Power of the Spirit.*

Southard, Mary—sister of St. Joseph, cofounder of *Spiritearth*, and ecoevolutionary painter and sculptor, p. 75: in *Spiritearth*, 1994, Vol. 4, No. 4.

Spencer, Herbert (1820–1903)—English philosopher who coined the term "survival of the fittest" and whose works provided the foundation for social Darwinism.

Spong, John Shelby—retired Episcopal bishop and liberal theologian; author of *Jesus for the Non-Religious*; pp. 362, 365–366: *Rescuing the Bible from Fundamentalism*; (1991), pp. 215, 223.

Sproul, R. C.—Calvinist evangelical theologian and popular radio evangelist; author of *Defending Your Faith: An Introduction to Apologetics*.

Stewart, John—Australian cross-disciplinary thinker; pp. 10, 283: *Evolution's Arrow: The Direction of Evolution and the Future of Humanity* (2000), pp. 9, 11.

Sunderland, Jabez (1842–1936)—a Unitarian minister and reformer whose 1902 book, *The Spark in the Clod*, depicted evolution as enriching Christianity; p. 132: sermon, 1925.

Surette, John—Jesuit priest and epic of evolution enthusiast, cofounder of the Spiritearth network for Earth-enhancing spirituality; pp. 327–328: in *Spiritearth*, 1995, Vol. 5, No. 4.

Swimme, Brian—leading Great Story popularizer; pp. 6, 288–289: *The Universe Story* (1992), pp. 229, 44–45; pp. 21, 367: *The Universe Is a Green Dragon* (1984), pp. 67, 162; p. 102: *The Hidden Heart of the Cosmos* (1996), p. 40; p. 124: "Where Does Faith Fit?" interview in *U.S. Catholic* (1997), www.thegreatstory.org/SwimmeUSC.pdf; p. 133: "Comprehensive Compassion," interview in *What Is Enlightenment?* (2003), issue No. 19.

Teilhard de Chardin, Pierre (1881–1955)—French Jesuit priest and paleontologist; author of *The Divine Milieu*; p. 29: *The Phenomenon of Man* (1959), p. 219.

Todd, John—leader in the fields of ecodesign and ecotechnology; author of *From Eco-Cities to Living Machines*.

Todd, Nancy Jack—cross-disciplinary leader in sustainability and ecospirituality; author of *A Safe and Sustainable World: The Promise of Ecological Design*.

Toulmin, Stephen—British philosopher and educator; author of *Human Understanding*, *The Return to Cosmology*, and *Return to Reason*; p. 143: *The Discovery of Time* (1982), p. 17.

Tucker, Mary Evelyn—cofounder with John Grim of the Forum on Religion and Ecology; author of *Worldly Wonder: Religions Enter Their Ecological Phase*.

Van der Ryn, Sim—a leader in the field of sustainable architecture; author of *Design for Life* and coauthor of *Ecological Design*.

Volk, Tyler—Earth systems scientist and cross-disciplinary scholar; author of *Gaia's Body*, *What Is Death?*, and *Metapatterns*; p. 27: quoted in *Green Space Green Time* by C. Barlow (1997), p. 291.

Watts, Alan (1915–1973)—philosopher best known as a popularizer of Asian philosophies for a Western audience; author of *The Way of Zen*; pp. 290–291: *Does It Matter?* (1970).

Wessels, Cletus—Dominican priest and theologian; author of *The Holy Web* and *Jesus in the New Universe Story*.

West, Paul—market activist who has helped motivate institutions to adopt evolutionary new policies on environmental protection and human rights; pp. 123, 200, 273: personal communication to M. Dowd, 2007.

Whitehead, Alfred North (1861–1947)—founder of process philosophy; pp. 80, 105: *Science and the Modern World* (1929), p. 270.

Wilber, Ken—integral thinker and author working outside the academic mainstream; author of *Integral Spirituality* and *A Theory of Everything*.

Wilson, David Sloan—evolutionary biologist and a pioneer in the evolution of religion; author of *Darwin's Cathedral*; pp. 12, 13, 113, 141, 144, 162, 205, 209, 266, 287–288, 334: *Evolution for Everyone* (2007), pp. 304, 281, 67, 70, 294, 73, 267, 315–16, 232, 295, 125.

390 *Who's Who*

Wilson, Edward O.—biologist pioneering an evolutionary perspective on human nature; celebrated and prolific author; pp. 24, 25, 34: *On Human Nature* (1978), pp. 201, 201, 169; p. 294: quoted in "Science and Salvation," *Washington Post*, September 20, 2006, p. C1.

Wink, Walter—biblical scholar and leader in progressive Christianity; author of *The Powers That Be*, *The Human Being*, and *Jesus and Nonviolence*.

Wright, Robert—cross-disciplinary, evolutionary thinker, and popularizer of evolutionary psychology; author of *The Moral Animal* and *Nonzero*; pp. 146, 158, 166, 185: *The Moral Animal* (1994), pp. 368, 368, 211–12, 151.

Yunus, Muhammad—a Bangladeshi banker and economist, awarded the 2006 Nobel Peace Prize for Third World microcredit financing.

Index

Page numbers in *italics* refer to illustrations.

Abrams, Nancy Ellen:
  as quoted, 13, 19, 22, 85–87, 93, 107, 118, 126,
    134, 137, 277, 293, 307, 358
  role in evolutionary spirituality, 276
accountability:
  for holding urges in check, 161–62, 166, 175,
    186–89, 214–15
  at level of societies, 273–74
Ackerman, Diane, as quoted, 31
Adams, Patch, as quoted, 237
adaptation, as term, 34
addictions:
  evolutionary insight into, 167–68, 171, 172
  evolutionary understanding of for healing,
    15, 123–24, 150, 156
  role of reptilian brain in, 148, 150
  societal role in curbing, 1, 274
  and twelve-step programs, 157, 185–86,
    309, 346
adolescence:
  appeal of evolutionary brain science to, 92,
    163, 203
  cultural traditions for sexual urges of, 173
  death as biblical punishment for rebellion
    of, 327
  delay in maturation of prefrontal cortex, 173
  higher purpose as helpful for, 153
affirmations, for evolutionary empowerment,
    61–62, 241–43, 263–64
afterlife:
  honoring night language referents of,
    116–17
  how evolution affects views of, 97, 100
agency as evolutionary driver, *see* initiative
Agent Smith, as quoted, 45
aggression:
  role of reptilian brain in, 149
  scriptural depictions of God engaging in,
    319–23
  territorial forms of, 322

amends, psychological importance of making,
    223–25, 233, 264, 344
amygdala, 148, 233, 239
amythic, 66
ancestors, ancient stars as, 85, 89–90, 91
anger, evolutionary insight into, 167
animal life, evolutionary sequence of, 280–81
Answers in Genesis, 96–97
anthropocentrism, as result of flat-earth
    faith, 74
Anti-Christ, 338
apatheism, 132
Apostle Paul:
  mention of, 85, 197
  as quoted, 96, 241, 244, 265
appreciation of instincts, 235–36, 264, 344, 361
Aquinas, Thomas, as quoted, 109, 200, 261
archaea, as early life, 279
artificial intelligence, futurist concerns about,
    302
assisted migration, 300
asteroid impacts, as unknowable to ancient
    peoples, 142
atheism, 129, 130, 131
  origin of, 111
  prominent exponents of, 325
  relationship to prayer, 124
Atlee, Tom:
  as proponent of "core commons," 270
  as quoted, 187–88
atoms:
  as created in stars, 53, 85–86, 89–91, 119
  ecological cycling of, 290
Augustine:
  as introducing concept of Original Sin, 158
  as quoted, 260
authenticity:
  as aspect of deep integrity, 59–60, 186–89,
    274, 309
  practices for growing in, 221–23

authority, as insufficient reason for belief, 352
autopoiesis, 121
awareness, as increasing through time, 51

bacteria:
    as early life, 279
    role in creating free oxygen, 53–54, 279
    sophistication of, 68
bad news, *see* challenges for humanity
Bakker, Jim, sex scandal of, 159
Banathy, Bela H. as quoted, 47
baptism:
    Evolutionary Christianity form of, 189
    in the Holy Spirit, 1, 213
Barlow, Connie:
    mention of, 3–6, 26, 33, 44, 66, 87, 88, 89,
        91–92, 93, 98–100, 127–28, 130, 163, 176,
        183, 203, 214, 222–23, 229, 235–36, 238–39,
        240
    as quoted, 28, 142
Barlow, Halsey, as quoted, 163–64
Bateson, Gregory, as quoted, 127, 248
Bateson, Mary Catherine:
    mention of, 30
    as quoted, 267
Beck, Don, role in evolutionary spirituality, 275
belief:
    as contrasted to scientific knowledge, 65, 107,
        124, 204, 351–56
    as dependent on language, 104, 105, 107
    as opposite of faith, 196, 243
    see also religious belief vs. knowledge
belief and disbelief, as outcome of private
        revelation, 65–66
Benyus, Janine, as contributor to ecological
        improvement, 312
Berry, Thomas:
    as quoted, 8, 25, 62, 72, 75, 106, 112, 134, 136,
        192, 226, 247, 255, 288–89, 290, 293, 310,
        312, 317
    role in evolutionary spirituality, 275
Bertalanffy, Ludwig von, as quoted, 93
Bethe, Hans, 95
Bible:
    as a blend of day and night language, 330
    as enlivened by evolutionary interpretation,
        330
    faulty science in, 11–12
    how to interpret miracle stories within,
        357–76
    morally offensive passages in, 325, 326
    portrayal of God as cruel in, 320–22
    as quoted, 63, 96, 97, 133, 184, 191, 213, 241,
        244, 251, 262, 265, 282, 339
    as recorded by fallible people, 331–32

    as sabotaged by literalist interpretations,
        329–32
biblical literalism, *see* scriptural literalism
Big Bang, 29, 89, 90, *90*, 133, 278
big picture, *see* cosmology
biocide, 256–57
biocracy, 314–15
biodiversity:
    preservation of as Great Work, 258, 300, 311–12
    sacred value of, 289
biomimicry, 312
Black Death, as leading to schism between
        religion and science, 55–56, 109
Black Elk, as quoted, 56
black hole, 89
blank slate, theory of psychological flexibility,
        144, 155
Bloom, Howard, role in evolutionary
        spirituality, 275
Boff, Leonardo, role in evolutionary spirituality,
        275
bonobo chimps, social uses of sex by, 161
Borg, Marcus, role in evolutionary spirituality,
        275
born again, naturalized, REALized
        understanding of, 344, 364
born anew, as evolutionary version of born
        again, 364
boundaries, importance of personal, 234
brain, practical value of evolutionary
        understanding of, 144–54
    see also evolutionary brain science
Braungart, Michael, as contributor to ecological
        improvement, 312
breastfeeding, importance of, 232
Brewer, John, as quoted, 154
Broad, William J., as quoted, 23
Broadway, Bill, as quoted, 23
Brown, Greg, songwriter as quoted, 170
Brown, Juanita, as quoted, 267
Buddha, mention of, 91, 197
Buddha, as quoted, 356
Buddhism:
    literalist interpretations of, 73
    outlook for evolutionary form of, 10
    response to Tsunami of 2004, 23
Burke, Spencer, role in bringing evolution to
        Evangelicals, 276
Burklo, Jim, role in bringing evolution to
        Progressive Christians, 276

calling, discerning one's, 229–30, 272, 316
    see also Higher Porpoise
Campbell, Neil A., as quoted, 79
cancer, as prevented by natural cell death, 94

Canfield, Jack:
  as proponent of feedback practices, 216
  as quoted, 60
carbon, as atom crucial for life and made within
    stars, 101
Carlson, Eric, as quoted, 69
Carroll, Sean B., as quoted, 24, 40, 79, 82
Carter Rita, as quoted, 153
catastrophic events:
  as evolutionary drives, 52–58, 62, 195, 199,
    204, 249, 252, 260, 299–300, 307, 337
  meaning-making of, 22–23, 90–101
  see also wild cards
Catholic Church, sex scandals within, 161–62,
    175
celebration, psychological importance of, 226, 249
Chaisson, Eric, role in evolutionary spirituality,
    275
challenges for humanity:
  biodiversity loss, 300
  climate change, 300
  disparities of wealth, 301
  evolution of governance, 302
  geopolitical conflicts and terror, 301
  Peak Oil, 301
  population growth, 300–301
chance and necessity, 121–22, 134, 282
cheating, societal effects of, 283
cheetah:
  human relatedness to, 33
  as once native of North America, 258
chemical elements, origin of, 89–101
children:
  importance of play for, 236
  importance of touch and tenderness for, 232–33
choice, 151, 259
Christianity:
  causes for resistance to evolution of, 7–8, 10,
    12, 13–14, 31, 66, 96, 109–10, 130
  flat-earth forms of, 72–74
  key metaphors of, 106
  literalist interpretations of, 73; *see also*
    scriptural literalism
  see also Bible; Evolutionary Christianity
Christ-like evolutionary integrity, 191–95, 265,
    329, 337, 343
Churchill, Winston, as quoted, 277
Circle of Life, 102
civilizations, historical collapse of, 113
climate change, 155, 253, 300, 305, 307
Clinton, Bill, sex scandal of, 162
Cobb, John, role in evolutionary spirituality, 275
codependence:
  role of twelve-step programs in healing,
    185–86

  as understood via evolutionary brain science,
    175–76
Coelho, Mary:
  as quoted, 109
  role in evolutionary spirituality, 275
Cohen, Andrew, role in evolutionary
    spirituality, 275
co-intelligence, 267–74
collective intelligence, 281–82, 287–88, 313–15
Collins, Ed, as quoted, 180
communication:
  evolution of, 8–9, 283
  practices to enhance, 233–36, 238–39
compassion:
  as aspect of deep integrity, 186
  expansion of human ability for, 194, 270–71,
    306, 324, 335
  practices for growing into, 223–25
  as quality of God, 106, 107, 252
competition, as evolutionary driver, 40–41, 43,
    44
complexity:
  evolution of, 43–44, 46, 50–53, 68–69, 79–82,
    86, 93; *see also* directionality in evolution
  evolution of cultural forms of, 28, 43, 283, 299
  as growth in information transmittal, 68–69
  role of death in generating, 94–96
  science of, 121–22
confession, psychological value of, 222, 235, 344
conflict, as driver of evolutionary emergence,
    267
conflict religion vs. science:
  basis of, 7–8, 10, 12, 13–14, 31, 65–74, 96,
    109–10, 350–56
  examples of, 1–2, 5, 128, 319–27
  as failure to distinguish day vs. night
    language, 115
  at Grand Canyon, 66
  harmful effects of, 67, 331
  historical origin of, 108–11
  reconciliation as inadequate, 140, 142, 337
  resolving, 2–3, 7–13, 25–26, 75–77, 100,
    112–13, 115, 118–20, 133–34, 136
  resolving by distinguishing day and night
    language, 140, 316
  resolving by evolving the faiths, 139–41,
    315–16, 341
  resolving by universalizing religious
    doctrines, 141
  resolving via Createism, 129–30, 329
  resolving via the Great Story, 2–6, 118–22,
    310, 316, 324
  in schools, 24, 66
  as ushering each other to greatness, 341, 346
  young people leading in resolving, 140

Confucianism, literalist interpretations of, 73
Confucius:
    mention of, 91, 197
    as quoted, 252
conscious evolution, 47, 269, 275, 293
consciousness:
    role of prefrontal cortex in, 153
    whether continues after death, 97
consumerism:
    as addictive behavior, 274
    damage caused by, 135
    as sin, 271
continental common sense, 312
convergent evolution, 36–40
conversation:
    as night language for evolutionary processes,
        267
    as propelling human evolution, 267–69
cooperation:
    aligning interests as crucial for future,
        283–88, 313, 335
    as evolutionary driver, 10, 42–45, 69
    evolution of, 9–10, 28, 149, 283–85
    expansion of, 10, 69, 252, 265, 271, 283–88,
        299, 308–9, 335
    reciprocal forms of, 149, 284
Copernican Revolution, 80, 112, 128
Copernicus, Nicolaus, 73, 78
core commons, 270–72
Corning, Peter A., as quoted, 42, 43
Cosby, Bill, as quoted, 236
Cosmic Century Timeline, as way to sense deep
    time, 277–82, 298
Cosmic Christ, 193
cosmic evolution, 85–86, 89–93
cosmic history, as new form of scripture, 316
cosmic task, 101, 229
    see also Great Work
Cosmic Uroboros, as metaphor for nestedness of
    Universe, 85
cosmology:
    as affecting environmental ethics, 247–48
    as answering biggest questions, 21, 34, 81–82,
        103, 288
    Big Bang, 29
    as Big Picture view of reality, 13, 19, 24, 34,
        81–82, 133, 277, 361
    as core of human culture, 19, 21–22, 24, 317
    as necessarily faulty in pre-scientific times,
        142
    nestedness of stories in, 22
    new vs. old, 244–48, 245, 246, 247–48
    as publicly revealed through science, 277
    social and psychological importance of, 13,
        21, 24, 93, 126, 277, 288–89

transmittal of in pre-literate cultures, 28
    see also worldview
Cowan, Chris, role in evolutionary spirituality,
    275
Cowan, Stuart, as contributor to ecological
    improvement, 312
Createheism:
    as beyond belief, 129, 198
    for bridging theism and atheism, 129, 295
    definition of, 131
    as open to divergent beliefs, 129, 329
    perspective of on Jesus, 130
    relationship of to Christianity, 194
    as religious outgrowth of evolutionary view,
        128–31
creation, as ongoing, 72
creationism:
    resistance to evolutionary view of death,
        96–97
    Young Earth, 1, 11, 66, 96
    see also conflict religion vs. science
creation science museums, 66, 361
creation stories:
    cultural isolation as having contributed to,
        359
    as depicting evolution of human brain,
        162–66
    Great Story as modern example of, 25–26, *90*,
        141–42
    Great Story as wondrous version of, 142, 328,
        341
    human need for, 11–12, 28, 33, 285, 317
    need for a new story, 72, 287–88
    as shapers of worldview, 24, 361
    see also cosmology; Great Story
creativity:
    as divorced from nature in mechanistic
        worldview, 111
    as propelled by challenges, 52–58, 61,
        198, 260
    as quality of Universe, 52, 55, 57, 84–85,
        119–20, 121–22, 129
    see also nested emergent creativity of
        Universe
Creator vs. Creation, 245–46
Crick, Francis, 73
crises, role in creativity, 260
    see also catastrophic events
crocodile, as nurturing its young, 149
cruelty, examples of among animals, 172
cultural conservatism, support for via
    evolutionary science, 173
cultural evolution:
    ancient peoples as unaware of, 143
    drivers of, 55–56, 68, 69–72

history of, 9–10, 43, 55–56, 69
increasing complexity in, 28
cultural history, as part of The Great Story, 25
cultural traditions, wisdom of, 156, 189
culture, transmission of information in, 28,
68–69
Cutright, Layne and Paul, 235
cybiont, as term, 313–14

dark matter and energy, 142
Darwin, Charles:
mention of, 32, 38, 46, 73, 78, 79, 80, 94, 95,
192, 318–19
as quoted, 40, 41, 241, 318
Davis, Phyllis, as quoted, 231, 232
Dawkins, Richard:
as author of Appendix A, 349–56
as critic of religion, 18
as prominent scientist, 73
as proponent of selfish gene theory, 172
as quoted, 39–40, 41
as skeptic of scriptural morality, 325
day vs. night experience:
definition of, 113–14
of Kingdom of Heaven, 262
day vs. night language, 113–17
correlation with public vs. private revelation,
113–14
as helpful for talking about death, 116–17
for making religion REAL, 140, 360
for resolving religious conflict, 316
for talking about Demonic temptation, 162
for talking about faith, 199
for talking about God, 121–23, 132
for talking about integrity, 195
for talking about nonmeasurables, 125
for talking about resurrection and the
Rapture, 260
for understanding Holy Spirit, 214
for understanding Original Sin, 167
as used in the Bible, 330
Deacon, Terrence, his theory of language, 104
death:
biblical explanation of, 96
as crucial for evolution of life, 92–101, 279
evolutionary explanation of, 45, 90–101
as impetus for religious belief, 93–94
importance of higher purpose for easing, 177
litany of, 98–99
as natural and generative, 93–94, 279
night language interpretations of, 116–17, 366
as no less sacred than life, 94, 347
as portal to evolutionary worldview, 94, 97–100
teaching children about, 98, 101
trust in vs. beliefs about, 243

deception:
evolutionary advantages of self-deception,
156–57, 322
evolutionary emergence of, 151–52, 221
in humans, 167
practices for growing past, 221–23
social mammals, 161
as urge to lie, 169
deep hot biosphere theory, 306
deep integrity:
accountability as crucial for, 175, 215
affirmations to support, 241–43
as aphrodisiac, 231
Christ-like values of, 191–95, 265, 329, 337,
366
as culminating in service, 240
definition of, 17, 183
as demarcating the body of Christ, 266
ecumenical definition of salvation through,
198
Evolutionary Christian version of, 135, 183,
191–95, 197
four characteristics of, 59–60, 186–89, 274,
341, 363
for furthering God's Kingdom, 263
as God's will, 338, 345
horizontal and vertical forms of, 210
how to grow in, 57–61, 163, 210–30, 241–48
importance of, 123–24
as manifestation of enlightenment, 185
nested manifestation of, 265
as self-reinforcing, 344–45
at species and societal levels, 266–67, 270–74,
293
as valuing of past and concern for future, 135
ways to tame Lizard Legacy, 214–15
as way to experience rapture, 204, 338, 344
deep time:
Cosmic Century Timeline as way to sense,
277–82
effect on values, 183
Jesus' core message as discerned by, 202
as opposite of Young Earth Creationism, 96–97
sacred understanding of, 332
as term, 33, 65, 183
understanding of as dependent on
technology, 143
value of for understanding humanity, 142
as worldview, 75
deism, 129, 131
Delacroix, Jérôme, as quoted, 274–75
Dellinger, Drew, as poet and rap artist of the
Great Story, 297
democracy, as serving an evolutionary
purpose, 267

demonic temptation, evolutionary
understanding of, 162
Dennett, Daniel:
as philosopher of evolution, 29
as skeptic of scriptural morality, 325
DePaolo, Donald J., as quoted, 23
de Quincey, Christian, role in evolutionary
spirituality, 275
de Rosnay, Joël:
as proponent of cybiont future, 313–14
role in evolutionary spirituality, 275
Devil:
as name for powerful forbidden urges, 161
naturalized usage of term, 169
Dick, Philip K., as quoted, 121
dinosaurs, extinction of, 36–37, 54–55, 142,
229, 281
directionality in evolution, 8, 24–25, 36–41, 43,
51–52, 69–71, 84–85, 252, 262–63, 281–82,
298–99
directionality in social evolution, 283
disease, risk of major epidemics, 306–7
diversity:
importance of cultural, 266–67
as increasing through time, 51
Dobb, Edwin, as quoted, 103
Dobzhansky, Theodosius, as quoted, 78–79,
122, 134
Dominguez, Joe, as quoted, 239–40
Downey, Bella, 26
dreaming, as vital to mammals, 148
Drucker, Peter, as quoted, 300
Durall, Michael, as author, 296

Earl, Michael:
as quoted, 320–21, 327
as skeptic of scriptural morality, 319–20,
325, 326
Earth:
as habitable planet, 23, 262, 279–80, 289
history of, 279–81
as our larger Self, 247–48, 290
sacred understanding of, 289
earthquakes, 23, 304
Eckhart, Meister, as quoted, 226, 288
ecocide, 256–57
ecological ethics:
as blossoming in youth, 256
as enhanced by evolutionary worldview, 75,
312
see also environmental ethics
ecological restoration, 258
ecology:
cycling of atoms, 290
death as structuring web of life, 94–96

how science expands sense of self, 290
top predators for stabilizing, 253–54
economics:
as benefiting from feedback of true costs, 309
as needing to evolve, 254, 271–72, 293, 309,
312, 313
practical benefits of understanding evolution
for, 287
vulnerability to geopolitical and other
challenges, 301
Ecozoic era, as term, 312, 314
Edwards, Denis, role in evolutionary
spirituality, 275
Einstein, Albert:
his view of God, 117
mention of, 47, 73, 80, 91
as quoted, 117, 336, 357
Eiseley, Loren, as quoted, 25, 33, 147
elephants, as surrogates for mammoths, 258
Elgin, Duane, role in evolutionary spirituality, 275
Eliade, Mircea, as quoted, 34
embryology, death as generative in, 94
emergent evolution, 10, 33, 35, 42–44, 52, 78,
84–85, 119, 121–22, 133–34, 266, 282
in cultural context, 192, 265
human brain as example of, 147, 164
new opportunities and dangers at each stage
of, 256, 299
role of conversation in, 267–69
as unknown to ancient peoples, 143
Emerson, Ralph Waldo, as quoted, 32
emotions:
as emerging from brain components, 148–50,
153
natural value of, 165–66
see also evolutionary brain science; paleo-
mammalian brain
empathy:
evolution of, 286
for understanding evildoers, 322
Endangered Species Act, example of biocracy,
314
endorphins, as stimulated by touch, 233
energy, shift to renewable, 311
enlightenment, evolutionary version of, 61
environment:
caring for as a way to honor God, 106, 112
as our source, 120
environmental ethics:
as affected by portrayal of God, 127, 245–48
collective action as enriching, 254
as enhanced by evolutionary view, 133–36,
245, 249, 288
as expanding circles of care, 270–71, 333–35,
341

as facet of our relationship to God, 191
hopeful trends toward, 309, 310–16
as primary ethics, 191
science as crucial for, 254
as shaped by our sense of self, 290–93
Epic of Evolution:
as foundation for all meaning, 63
as the greatest religious story, 53, 84, 93
mention of, 29, 30, 33–35, 36, 44, 45, 50, 56, 89
as revitalizing the Christian gospel, 2–4, 200–202
ways to interpret, 139
Erhard, Werner, as quoted, 243–44
ethnic cleansing, as ordered by the God of scripture, 320–22
eukaryotic cell, evolution of, 42–43
Ever-Renewing Testament, sacred interpretations of science as, 330–32
evidence, as proper basis for belief, 350–56, 358
evil:
as consequence of fighting evil, 256, 259
as difficult term for liberals, 146
how evolution assists understanding of, 322
how ignorance can lead to, 59, 199, 254–55
human capacity to recognize, 167
in-group vs. out-group forms of, 335
as less powerful than love, 337
psychological value of personalizing such forces, 169
scriptural acts of God now viewed as, 319–21
survival value of, 321
Evo-devo, as new science-supporting fact of evolution, 78, 121
evolution:
causes of, 29–30, 31–32, 40–46, 69–71, 78–79
of chemical elements, 53, 85, 119
creative role of death in, 93–101
directionality in, 8, 24–25, 36–41, 51–52, 69–71, 84–85, 252, 262–63, 281–82
as enriching religion, 10, 13, 75–77, 141–43, 316, 337; *see also* Evolutionary Christianity
as fact, 8, 78–82
future of, 282
importance of teaching in churches, 66, 341
major transitions in, 69–72
as meaningful, 8, 24, 29–30; *see also* meaning-making
metaphors of, 29, 41, 85, 86, 133
as ongoing, 25, 89–90, 91, 133, 134, 282
religious resistance to, 1–3, 7–8, 10, 78, 128; *see also* conflict religion vs. science
as sacred process, 8, 75
as worldview, 10, 24, 29, 47, 63, 73–74, 84–100, 121–22, 127–32, 195, 198, 244, *245, 246,* 290–93, 315–16, 347–48

evolutionary arms race, 41, 44, 280–81
evolutionary brain science:
benefits for adolescents, 203, 347
as key understanding for evolutionary evangelicals, 361
practical value of, 74, 144–54, 158–68, 194
see also quadrune brain
Evolutionary Christianity:
affirmations to support, 243
atheist responses to, 123
Christ-like evolutionary integrity, 135, 182–84, 191–95, 265, 329, 337, 366
core aspects of, 10, 123–24
definition of, 75
how to evolve toward, 10, 75–76, 100, 140, 209–10, 331–32
implications for morality, 166–68, 193–95, 325–27, 333–35
importance of working toward, 140–41, 310, 315–16
interpretation of miracle stories, 360, 363–64
naturalizing the Gospel, 200–205, 210, 260, 337, 344
naturalizing the kingdom of Heaven, 226, 262–66, 274, 338
need for ongoing reinterpretations of core tenets, 200
relationship of to Universe, 244–47, 293
relationship to the Bible, 331–32
scriptural literalist responses to, 122–23
understanding of collective sin, 251–60
understanding of salvation, 182–85, 195–200, 210, 237, 344–46
view of Jesus, 263, 336–39
view of purpose of humanity, 293
see also REALizing
evolutionary convergence, 36–40
evolutionary cranes vs. skyhooks, 29
evolutionary emergence, *see* emergent evolution
evolutionary enlightenment, deep integrity as central to, 185
evolutionary epic, *see* Epic of Evolution
evolutionary ethics:
core features of, 123, 265
effect on environmental ethics, 133–34
as enriched by biblical values, 166
need for, 74, 76, 97, 100–101
nestedness as core feature of, 299
see also deep integrity
evolutionary evangelicals, 2
as term, 361
evolutionary faith, as grounded in religious traditions, 139–40, 189, 316
evolutionary integrity, *see* deep integrity
evolutionary parables, 26, 101, 189

evolutionary psychology:
    for advancing compassion, 194
    benefits for adolescents, 203
    cruelty as understood via, 321–22
    for developing witness capacity, 164, 169, 171
    for economic success, 287
    emphasis on human universals, 154
    as explaining cultural norms, 156, 355
    as explaining misplaced priorities of media, 155
    insights into gender differences, 155–56
    mismatch theory of, 146, 150, 155, 168, 171–72, 194
    opposition to blank state theory, 155
    for self-help, 164, 194, 347
    for transcending scriptural literalism, 323
    for understanding the human condition, 74, 146, 154–58, 223
evolutionary revivals, 294–97
evolutionary spirituality:
    benefits of, 94, 97
    contributors to, 275–76
    examples of, 34, 316
    interactive website for, 18, 141, 276
exaptation, definition of, 34
existential view, as not required by science, 93, 126
extinction:
    as challenge to religious views, 94
    discovered by science, 94, 257
    history of, 300
    during human prehistory, 256–57
    human role in, 253, 256–58
    mass, 54–55
    sixth major mass, 300, 311–12
extraterrestrial life, possibility of, 306
eyes, evolution of, 38, 39

facts as God's native tongue, 15, 77–81, 197, 324, 341, 346
fact vs. theory, 78–80
faith:
    day vs. night language expressions of, 198–200
    flat-earth vs. evolutionary, 73–75, 94, 139–40, 349
    in God, 58–59, 68, 196–200
    growing evolutionary forms of, 139–41, 199
    as opposite of belief, 196, 243
    practical value of, 195
    as trusting the Universe, 68, 195–200
    as trust in imperfections serving a purpose, 198–200
    as way to endure suffering, 58
faithfulness to God, 134–36

Fall (The), as interpreted through evolutionary worldview, 158–62
feedback:
    as expanding through time, 309–10
    for perceiving wrongdoing, 253
    at societal level, 273, 283, 309
    for spiritual growth, 215–16, 239
feminism, as prerequisite for benign expressions of evolutionary psychology, 156
Ferris, Timothy, as quoted, 36
flat-earth thinking, 11, 66, 73–75, 94, 95, 167, 244–46, 310, 325, 349
flow, as mental state, 150
food addictions, evolutionary explanation of, 171, 172
forgiveness:
    as assisted by evolutionary view, 147, 168, 233
    as distinguished from amends, 225
Foundation for Conscious Evolution, 275
four quadrant model, 114
Fox, Matthew:
    as advocate of "original blessing," 144
    as quoted, 105
    role in evolving Christian faith, 275, 325
free will:
    evolutionary brain science as demonstrating, 151
    openness of Universe to, 259
frontal lobes (of brain), *see* prefrontal cortex
Furry Li'l Mammal:
    ability to override Lizard Legacy, 174–76
    affirmation for, 243
    as craving touch and tenderness, 231–32
    definition of, 16
    description of, 148–50, *149,* 164
    as distinct from real self-interest, 253
    for healing addictions, 157
    melody and ritual effect on, 237
    for microfinance success, 287
    role in codependence, 175–76
    role in depression, 152
    as seat of emotions, 148–50
    self-help exercise for, 190
    selfish concerns of, 184, 345
    status-seeking of, 149, 151, 153, 157, 161–63, 165, 167, 169, 174–76, 190, 287
    vulnerability to addictions, 150
    see also paleo-mammalian brain

galaxies:
    increasing knowledge of, 49–50, 87–88
    mergers of, 95
    origin of, 278
Galileo, 73

Gebara, Ivonne, role in evolutionary spirituality, 275

gender conflicts, as understood via evolution, 163, 172

gender differences:
cultural practices that track, 155–56
psychological expression of, 155–56

generosity, ways to cultivate, 228

genes:
regulator, 32
role in evolution, 32, 164

genes-eye view of nature, 172, 174, 261, 325

Genet, Russ, as quoted, 294

genetic engineering, 252

genocide:
Bible stories of, 322
modern outrage against, 324

geology, death as generative in, 95

Germann, Doug, as quoted, 106

global warming, 155, 253, 305, 307

glossolalia, evolutionary understanding of, 212–14

God:
belief in vs. experience of, 105–7, 111–12, 124–27
as beyond belief, 120, 129–30, 198
biblical portrayal of as cruel and vindictive, 67, 320–21
compassionate quality of, 106, 107, 252
creatheistic portrayal of, 129–31, 198
as diminished in a clockwork Universe, 110
as experienced in mystery, 107
as experienced in natural processes, 84, 110, 125–27, 316
faith in, 58–59, 198–200
getting right with, 59, 198
history of shifting metaphors for, 110–12, 245–46
how became solely transcendent, 111, 245
how humans sin against, 253
how metaphors limit understanding of, 106
how notions of affect environmental ethics, 127, 245–48
immanence and omnipresence of, 106, 107, 110, 119–22, 130, 245–46
importance of personal metaphors for, 130
incarnated as the Christ, 193
as like a warrior or king, 110
loss of immanence of, 111, 127, 130, *245, 246*
masculine metaphors for, 110
moral character of, 67
as name for Ultimate Reality, 60, 85, 86, 119–23, 125, 127, 135, 182, 253, 282
naturalized understanding of, 119–22, 125–32, 245–47, 282, 347–48

in nature, 106, 109, 131, 245–46
as personal, 123–26, 130, 248
as portrayed in different "isms," 130–32
relationship to suffering, 106
as residing outside the Universe, 87, 110–11, 135, *245, 246*
science as enlarging our concept of, 2, 119–20, 125–27, 201, 245–46, 332, 341
as still active in the world, 203, *245, 246,* 282, 331
as still communicating, 7, 8, 68–72, 87, 135, 196, 209, 324, 331, 360
supernatural characterization of, 87, 130, 135, 244–45, 324
Supreme Wholeness as name for, 123, 125, 127, 131, 135, 163, 179, 260, 263, 282, 338
transcendent nature of, 120, 131
trivialized or diminished notion of, 87, 119, 127, 244, 324
as the ultimate terrorist, 320–21
as undeniable, 119, 125–27
as understood through day or night language, 119–23

God's love, interpreted through deep-time view, 106, 201, 245–46

God's will:
as alignment with higher purpose, 180
beliefs about, 135, 333
as expanding circles of care, 335
as manifest in deep integrity, 186, 338, 345
as revealed via evolutionary understanding, 163, 184, 201, 205, 209, 229, 313

God's Word:
as basis of authority for creation science, 361
historical understanding of, 328
as revealed through science, 244, 323–24, 361
as still emerging, 328, 361

Gold, Thomas, as proponent of alien microbial life, 306

Goldberg, Elkhonon, as quoted, 152

Golden Age, absence of, 257, 321

Goodenough, Ursula, role in evolutionary spirituality, 275

Goodfield, June, 143

good news:
biocracy and holistic governance, 314–15
biodiversity crisis ending, 312
biomimicry design advances, 312
catastrophes as catalyzing creativity, 308; *see also* catastrophic events, as evolutionary drivers
cooperation expanding, 308–9; *see also* cooperation
deep-time perspective on, 307–16
evolutionary revitalization of, 202–3

good news *(cont.)*
  feedback increasing, 309–10
  global governance, 299
  pollution decreasing, 312–13
  population stabilizing, 311
  shift to renewable energy, 311
  synergetic evolution of humans, 313–14
  technology enabling connectedness, 308
  worldwide religious revival, 315–16
Gore, Al, in global warming film, 307
Gospel (The):
  according to evolution, 49, 57–59, 137
  naturalizing the good news of, 200–205,
    337, 344
Gospel of Thomas, as quoted, 262
Gould, Stephen Jay, as quoted, 34, 37, 46, 79
governance:
  evolutionary view of, 255, 283–88
  need for a shared sacred story, 287
  recommendations for future of, 265, 273–74,
    283–88, 293, 299, 302, 312–15
Grameen Bank, as recipient of Nobel Peace
    Prize, 287
Grand Canyon, creationist interpretations of, 66
grand narrative, history of Universe as, 25
Grassie, William, as quoted, 139
gratitude:
  spiritual importance of, 226, 264
  ways to cultivate, 58, 226–28
gravity, role in creation of elements, 90, 279
Great Radiance, as name for Big Bang, 34, 89,
    90, 91, 270, 278
Great Story:
  atheist response to, 123
  as bridging science and religion, 24–26,
    119–20, 141–42, 215
  as catalyst for spiritual transformation, 2–6,
    141, 203, 331
  as containing lessons for humanity, 277
  as crucial for governance and economics, 287
  definition of, 24–25, 141
  as encompassing religious diversity, 61
  as enriched by evolutionary brain science,
    144–45
  envisioning God through, 119–20, 331
  human role in, 25, 29–30, 132
  importance of for social evolution, 287
  as marriage of science and religion, 25–26
  as our common creation story, 25, 50, 203,
    249, 287, 341
  parables, 26, 101, 189
  as product of public revelation, 142
  scriptural literalist response to, 122–23
  as self-correcting, 26, 27
  six core attributes of, 25–26, 287

Great Work:
  humanity, 134, 255, 258, 265, 293
  of individuals, 180, 197, 249, 316; *see also*
    Higher Porpoise
grief, evolutionary understanding for healing,
    97–100
Gump, Forrest, 29

Haggard, Ted, sex scandal of, 159–60
Ham, Ken, as quoted, 96
Harjo, Joy, as quoted, 113
Harris, Sam:
  as quoted, 319, 330–31, 349
  as skeptic of scriptural morality, 319, 325
Haught, John:
  as quoted, 24, 154
  role in evolutionary spirituality, 275
Havel, Václav, as quoted, 259
Hawken, Paul, as contributor to ecological
    improvement, 312
Hawking, Stephen, 73
hearing, as evidence of evolution, 238
heart:
  guidance from, 124, 309
  metaphors of Ultimacy that engage with, 130
Heart-to-Heart talk, 235–36
Heaven:
  day vs. night referents to, 262
  naturalized view of, 60, 123, 198, 202, 226,
    261–66, 291, 336–39, 344
Hebrew prophets, 197
Hefner, Philip, as quoted, 63, 108
helium:
  as end-product of Sun's power process, 95
  formation of during birth of Universe, 278
  as power source of Red Giant stars, 90–91
Hell:
  as experienced in this world, 185
  as a matter of belief or disbelief, 66
  naturalizing, 185, 202, 345
  scriptural view of, 321
hierarchy, biblical form of ranking, 246–47
Higher Porpoise:
  affirmation to support, 243
  definition of, 16
  delayed development of in teens, 174
  description of, 149, 152–54, 164
  examples of, 152–53, 177
  and God's will, 180, 185
  how to find it, 178–81, 229–30
  importance of during crisis, 176–77
  as leaving positive evolutionary legacy,
    215, 346
  as requiring emotional drives, 165–66
  response of to sacred ceremonies, 237

role in trumping other drives, 157, 167,
176–77, 180, 243
self-help exercises for, 190, 243
struggle with Lizard Legacy, 162
as valuing the future, 154
see also prefrontal cortex
Higher Power, 58, 124, 187
higher purpose, 104, 177, 185, 193, 263
see also Higher Porpoise
Hinduism, outlook for evolutionary form of,
10, 76, 141
history:
of human social systems, 283–88
of life on Earth, 280–81
pre-biblical, 11
timeline of cosmic, 277–82
timeline of Earth, 279–81
Hitchens, Christopher:
as quoted, 360
as skeptic of scriptural morality, 319, 325
Hobbes, Thomas, 318
Hofstetter, Adrian, role in evolutionary
spirituality, 275
holarchy, as term, 86
Holman, Peggy:
as developer of co-intelligent processes, 269
as quoted, 271
Holocaust, 135
holons:
of complexifying social systems, 283–88
definition of, 85
as helpful for understanding sin, 253
of human relationship, 162
nested character of, 119, 132, 239
as providing a nested view of morality, 299,
313, 333
understanding of for environmental ethics,
253–54, 313
Holy Spirit:
feeling the presence of, 214
as guidance for our species, 265
homosexuality:
Haggard portrayal of as repulsive, 159–60
as natural, 158–60
in primates, 160
hope:
as arising through crises, 307
for ending religious conflict, 324–27
as enhanced by evolutionary worldview, 62,
76, 143, 202, 239, 296, 307–16
as enhanced through ecological restoration, 258
as requiring a square look at the bad news, 299
see also good news
horses, evolutionary understanding of native
range, 258

hospice movement, 97
Hubbard, Barbara Marx:
as quoted, 47, 177, 207
role in conscious evolution, 275
Hubble, Edwin, 73
Hubble Space Telescope:
mention of, 49, 86–88
photos of as spirituality rich, 87–88
human:
as moral animal, 161, 318
propensity to tell stories, 114–15
relationship to environment as influenced by
views of God, 127
relationship to God, 133–36, 245–48
relationship to God through prayer, 124
relationship to Universe, 56–57, 90–91, 92,
112–13, 124, 131, *245, 246, 247,* 347
humanism, 1, 111, 130, 295
humanity:
close genetic relationship to other animals,
103
as collectively immature, 254
evolutionary history of, 281
as like a cancer cell, 291–92
purpose of, 25, 29–30, 33, 52, 53, 56–57, 63,
129, 131; 132–33, 134, 203, 293; *see also*
human role
as responsible for future evolution, 282
symbolic language as distinguishing, 104
human nature, benefits of evolutionary
understanding of, 15–16
human role:
as celebrants of the Universe Story, 103, 132,
136, 226
Christian expression of, 292
as contributor to the body of Life, 133
as evolution becomes conscious of itself, 33,
52, 53, 131, 133, 252, 288–91, 304, 347
as expression of Earth's creativity, 289, 292,
304
as expression of Universe, 60–61, 85, 86–87,
102, 133, 136, 244–45, 271, 288–89, 292, 304
as immune system of Earth, 304
as integral part of evolution, 262, 277
as participants in evolution, 262
in prehistoric extinctions, 256–58
as recovered memory of Earth, 229
as serving God, 310
human species:
futuristic view of, 312–14, *315*
past evolution of, 41, 43, 47, 85, 141
human universals:
cultural responses to gender differences, 156
definition of, 147, 270
examples of, 156, 164

humility:
  as component of deep integrity, 59, 184,
    186–89, 259, 309
  practices for enhancing, 216–17, 243
  psychological benefit of, 58
humor, importance of, 236–37
Huxley, Julian:
  as interested in convergent evolution, 38
  as interpreter of evolution, 29
  as proponent of evolutionary spirituality,
    275
  as quoted, 25, 33, 133
hydrogen:
  as fuel in stars, 90, *90*, 279
  origin at beginning of Universe, 278
  as required element for photosynthesis, 280
hypothesis, definition of, 350
Hyun Kyung, Chung, role in evolutionary
    spirituality, 275

id, reptilian brain as correlated with, 161
idolatry, scriptural literalism as, 78, 82, 135
imagination, as expressed through worldview
    metaphors, 243–46
immense journey, as synonym for Epic of
    Evolution, 25, 28, 33, 100, 226, 242, 277
infanticide, among animals, 322
information flow, role of in evolution, 68–72,
    80, 283
in-group vs. out-group dynamics, 9–10, 67, 74,
    135, 335
inherited proclivities:
  as our unchosen nature, 15–16, 146, 157, 164,
    169, 347
  troublesome aspects of, 158, 168, 174, 346
  see also evolutionary psychology; quadrune
    brain
initiative, role of in evolution, 45–47
Inquisition (The), 135
instincts:
  evolutionary appreciation of, 144–48, 162–63,
    166, 184, 203, 344, 361
  evolutionary roots of, 168
  see also evolutionary brain science
Institute on Religion in on Age of Science, 275
Integral Consciousness, as model of
    development, 153
Integral Institute, 275
integrity, *see* deep integrity
intelligent design:
  as assuming a mechanistic Universe, 121
  bridging to, 121
  mention of, 8, 12, 143
  refutation of via evolutionary brain science,
    37, 147, 174

Internet:
  as expanding cooperation, 308, 333
  role of in cultural evolution, 70–72, 128, 202,
    274–75
intimacy, evolutionary understanding as
    enhancing, 231–39
intuition, seeking guidance from, 124
Isaacs, David, as quoted, 267
Islam:
  literalist interpretations of, 11, 68, 73, 74
  outlook for evolutionary form of, 10, 76, 141,
    316, 346
  response to Tsunami of 2004, 23
  scriptural portrayal of God as sometimes
    cruel, 320–22

James, William, as quoted, 48–49
Jesus:
  as central for Evolutionary Christianity, 76,
    201, 336–39
  as the Christ, 124, 193–94, 366
  Christ-like integrity of, 182, 191–95, 243, 259,
    337
  deep-time understanding of, 202
  his view of Kingdom of Heaven, 262–63, 338
  as Lord, 263
  as quoted, 191, 251, 262, 339
  see also REALizing
Johnson, Elizabeth, as quoted, 133
Johnson, Mark, as quoted, 105
Judaism:
  evolutionary form of, 10, 76
  literalist interpretations of, 11, 73
  response to Tsunami of 2004, 23
justice:
  biblical guidance as inadequate for, 326–27
  evolution of social manifestations of, 284

Kant, Immanuel, 104
Kauffman, Stuart, as quoted, 46
Kaufman, Gordon, as quoted, 30
Keller, Helen, as quoted, 195
Kelly, Kevin:
  as quoted, 65, 70–72
  role in evolutionary spirituality, 275
Kingdom of God, *see* REALizing the Kingdom
kin selection, role of in evolution, 325
kinship, evolutionary expansion of, 270
Koestler, Arthur, as quoted, 85
Korten, David, role in evolutionary spirituality, 275
Kosmos, as name for Universe, 126
Kuhn, Thomas, as quoted, 79, 244
Kurzweil, Ray:
  as futurist of technology, 302
  role in evolutionary spirituality, 275

LaChance, Albert, mention of, 2
LaChapelle, Dolores, as quoted, 237
Lakewood Church (Houston), 294–95
Lakoff, George, as quoted, 105
Landmark Education Corporation, 114, 225
language:
    as a cause of stress, 152
    challenges of for faith in God, 196
    as complicating factor in sexuality, 174
    evolution of, 103–4, 284, 328
    idolatry of through scriptural literalism, 78,
        82, 135
    metaphors as central to, 104, 130
    reverential, 117
    role in social evolution, 103, 284
    role of neocortex in, 150
    speaking in tongues, 212–14
    transition from orality to literacy, 328
    see also day vs. night language
Lao-Tzu, as quoted, 270
Lavanhar, Marlin, as quoted, 128
Lawrence, Paul R.:
    as proponent of evolutionary psychology,
        151, 275
    as quoted, 230
Lemurs, as beneficiaries of continental
        break-up, 95
Leopold, Aldo, as quoted, 25
Lerner, Michael, as quoted, 23
life:
    cybiont future of, 313–14
    as dependent on death, 92–101
    evolutionary history of, 85, 141, 143, 280–81,
        299–300
    see also directionality in evolution
light-year, definition of, 278
limbic system, *see* paleo-mammalian brain
literalism, *see* scriptural literalism
Lizard Legacy:
    affirmation for, 243
    blessings of, 174, 243
    as celebrated in song, 170
    challenges of, 171–74, 184, 345
    as distinct from real self-interest, 253
    as held in check by Furry Li'l Mammal, 157,
        174–76
    instinctual drives of, 148, 164
    role in addictions, 148
    role in sex scandals, 160–62
    self-help exercises for, 189, 214–15
    as stimulated by rhythm, 238
    as term, 16, 170
    ways to tame, 214–15
    see also reptilian brain
Lord's Prayer, as quoted, 262

love:
    as central teaching of Jesus, 191, 194
    as expressed through service, 240
    of God, 226, 229
    for one's enemies, 184, 217–21, 259
    as quality of God, 199, 201, 203
Lovins, Amory, as contributor to ecological
        improvement, 312

MacDonald, Gordon, as quoted, 160, 170
McDonough, William, as contributor to
        ecological improvement, 312
McFague, Sallie:
    as quoted, 255
    role in evolutionary spirituality, 275,
        364–65
MacGillis, Miriam:
    as quoted, 293
    role in evolutionary spirituality, 275
McLaren, Brian, role in bringing evolution to
        Evangelicals, 276
MacLean, Paul, his theory of triune brain, 147
McMenamin, Mark:
    and convergent evolution, 38
    as quoted, 42
macroevolution, 32
Macy, Joanna:
    as quoted, 292, 298
    role in evolutionary spirituality, 275
Magellan, Ferdinand, circumnavigation of
        globe, 95
Malthus, Thomas, his understanding of death,
        96
mammals:
    distinguishing features of, 148–49
    evolution of, 55
    need for touch, 231–33
mammoths and mastodons, Pleistocene
        Rewilding as response to loss of, 258
Manitonquat, as quoted, 233
Margulis, Lynn, as quoted, 42
marital relations:
    how a shared higher purpose can enhance,
        177; *see also* Higher Porpoise
    how evolutionary view can enhance, 231–39
marriage, as understood through evolutionary
        sciences, 174
Marshall, Gene:
    as quoted, 125
    role in evolutionary spirituality, 275
Mary, ascension of, 352–53
Mathews, Freya, as quoted, 290
Mayer, Peter:
    as quoted, 249, 367
    as singer-songwriter of the Great Story, 296

Maynard Smith, John:
  and convergent evolution, 38
  as evolutionary theorist, 38, 69
Meadows, Donella, as quoted, 269
meaning-making:
  of catastrophes, 22–23, 360
  as contrasted to existential view, 93, 126
  Epic of Evolution as ground for, 63, 93
  as expressed through ritual and song, 237–39
  of the fact of death, 93–102, 177
  as gift to science, 13
  healthy spiritual practices for, 217–21
  importance of, 21, 33
  inescapability of, 114
  poetic examples of in science, 79, 85–86, 91,
    98–99, 101–2
  role of creation stories in, 24, 33
  science as ground for, 29, 68, 93
  see also day vs. night language; Epic of
    Evolution
meditation, as understood via evolutionary
    brain science, 150, 211–12
Meeker, Tobias, mention of, 2
meerkats, morality and immorality as viewed
    through, 322
megachurches, 294–95, 296
mentoring, as spiritual practice, 229–30
metaphors:
  in biblical story of creation, 133
  as central to religions, 105–7, 123
  as central to symbolic language, 104, 114,
    126–27
  as crucial for night language, 114
  of evolution, 29, 41, 85–86, 133
  ongoing shift in for understanding God and
    Universe, 244–47
  prayer described through, 123
  for Universe, 110–12, 123, 126–27
metareligious, attribute of the Great Story, 26, 129
meteor impact, as cause of dinosaur extinction,
    54–55
microfinance movement, 287
Milky Way Galaxy:
  our sacred relationship to, 288–89
  size and origin, 278
miracle stories:
  controversy over, 357–76
  see also REALizing miracles
mismatch theory, of evolutionary psychology,
    146, 150, 155, 168, 171–72
Mitchell, Joni, as composer of stardust lyrics, 288
Mohammed, 67, 85, 91, 197
Monkey Mind:
  ability to rationalize wrongdoing, 322
  as Buddhist name, 150

definition of, 16
  description of, *149*, 150–52, 164
  as internal generator of stress, 152, 234
  meditation as means to discipline, 150, 152
  practices for taming, 211–14
  practices to engage in helpful ways, 157
  resistance of to irrational ritual, 238
  role in taming Lizard Legacy, 175
  self-help exercise for, 190
  see also neo-mammalian brain
monotheism:
  basis of, 110
  scriptural literalism as diminishing,
    111–12
monotreme mammals:
  absence of dreaming in, 148–49
  as transitional life forms, 148–49
Montagu, Ashley, as quoted, 231
Montessori, Maria, as quoted, 229
morality:
  cultural universals of, 252
  the dark side of scriptural literalism, 319–27
  Darwin's view of as evolving, 318
  ecological, 75; *see also* evolutionary ethics
  evolutionary parables for teaching, 101
  evolution of, 333–34
  evolution of human forms of, 284–86, 318
  as expanding in an Age of Information,
    255–59
  inconsistencies of scripture-based, 67, 76
  in-group vs. out-group dynamics, 9–10, 67,
    74, 135, 335
  as linked to cultural cosmologies, 73
  nested nature of, 333
  role of prefrontal cortex in discerning, 167
  sexual in human, 161
  social, 283–88
  see also sin
Mormonism, as based in miracle story, 358
Morris, Simon Conway, as quoted, 38–39
Morwood, Michael, role in evolutionary
    spirituality, 275
Moses:
  Law of, 321
  mention of, 11, 67, 85, 91, 197
multiculturalism, 266–67
multiverse, 130
Mumford, Lewis, as quoted, 341
Murray, W. H., as quoted, 181
music:
  appeal of praise worship style, 296–97
  emotional response to, 150, 294–95
  evolutionary understanding of, 238–39
  as human universal, 238
  in support of the Great Story, 296–97

Muslim, *see* Islam
mythopoeic drive, 34–35

naturalism vs. supernaturalism, 78
  see also REALizing
natural selection:
  genetic drives of, 164
  how it works, 32, 46
  role in happiness, 164, 166
  as unknown to ancient peoples, 143
nature:
  ancestral view of as divine, 105
  desacralization of, 111, 128, 130, 131
  divorce of God from, 108, 130–31
  as primary revelation of God, 127, 128, 131, 134, 328
  resacralization of, 127–28, 131
nature vs. nurture, 155
  see also Original Sin
nematode worms:
  characteristics shared with humans, 141
  as vulnerable to nicotine addiction, 148
neocortex, *see* neo-mammalian brain
neo-mammalian brain:
  description of, *149*, 150–52
  as rational part of our brain, 150
  scenario-building function of, 150
  see also Monkey Mind
neoteny, as characteristic of humans, 236
nested emergent:
  creativity of Universe, 84–85, 124, 129, 130, 316, 337
  quality of divine, 106, 119, 126, 134, 195, 247, 253, 316, 335, 337, 346
  quality of Universe, 103, 119, 244, 251–54, 262, 289
nested stories, 22
New Testament, as early Christian scriptures, 197, 204, 265
Newton, Isaac, 73, 80
Nicene Creed, as literalist commentary, 73
Niebuhr, Reinhold, as quoted, 303
night language:
  definition of, 113–14
  examples of, 117, 122, 133, 162, 167, 184, 193, 195, 199, 213, 214, 255, 260, 262, 267, 282, 292, 316, 339, 359, 365
  for rationally understanding miracle stories, 359–60
  see also day vs. night language
Noah's Flood:
  consequences of belief in, 66, 67
  as inflicted by a cruel God, 320
Nohria, Nitin, as quoted, 151, 230
noosphere, as term, 299

obesity, evolutionary understanding of, 175
O'Murchu, Diarmuid, role in evolutionary spirituality, 275
Oprah, television show, 309
original blessing, gifts of our animal body, 144, 174
Original Sin:
  understanding of via science, 144–47, 184, 187, 204
  vs. "blank slate" understanding of psychology, 144
  ways to make REAL, 140, 144–47, 158, 166–68
Osteen, Joel, as charismatic televangelist, 295, 309
overpopulation, 96, 300–301
Overton, Patrick, as quoted, 200
ovulation, concealed, 173
oxygen:
  early build-up in atmosphere, 54
  role in formation of iron deposits, 280
  as toxic gas, 280
Oxygen Crisis, 280
ozone shield, as unknown to ancient peoples, 143

paleo-mammalian brain:
  controversy over, 66
  description of, 148–50, *149*
  functions of, 144, 148–49
  history of, 144
  as seat of limbic system and emotions, 148–49
  vulnerability to addictions, 148, 150
  see also Furry Li'l Mammal
panentheism, 129, 131
Pangaea, break-up of supercontinent, 95
pantheism, 129–30, 131
Parable of the Pickle Jar, 163
parables, evolutionary forms of, 26, 101, 189
paradigm, scientific revolutions of, 80, 112, 128, 244
  see also worldview
passenger pigeon, extinction of, 257
Passino, Kevin, as contributor to ecological improvement, 312
Peak Oil, 301
Pearce, Joseph Chilton, role in popularizing brain science, 275
Pentecost, evolutionary view of, 214
Pentacostal worship, 3
perception, as influenced by language, 104–5
periodic table of elements, 86, 119
personification:
  playful examples of, 127–28
  practical value of, 126–27

photosynthesis:
  origin of, 280
  as unknown to ancient peoples, 142–43
Pinker, Steven, role in advancing evolutionary
    brain science, 275
planets:
  misunderstood as "wandering stars," 142
  origin of, 279
plants, evolutionary history of, 281
plate tectonics:
  as cause of tsunamis, 23, 305
  as crucial for cycling of atoms for life, 23, 279
  discovery of, 95
  religious interpretation of, 23
  as unknown to ancient peoples, 142
play, importance of for mammal, 236–37
Pleistocene Rewilding, 258–59
Pluto, as dwarf planet, 26, 27
pollution:
  on early Earth, 53–54
  prospects for reducing, 312–13
Popper Karl, as quoted, 151
Pratney, Winkie, as quoted, 226–27
prayer, as enriched by sacred evolutionary
    worldview, 124
predation, evolution of, 280–81
prefrontal cortex (of brain):
  delayed development of in teens, 174
  description of, 149, 152–53
  functions of, 152–53
  role in choice and decision-making, 152, 167
  as seat of self-awareness, 153
  see also Higher Porpoise
Primack, Joel R.:
  as quoted, 13, 19, 22, 85–87, 93, 107, 118, 126,
    134, 137, 277, 293, 307, 358
  role in evolutionary spirituality, 276
private revelation, see public revelation
process theology, 2, 130
programmed cell death, 94
progress in evolution, see directionality
prophetic inquiry, questions for religious
    reflection, 18, 140–41, 157, 169, 182, 195,
    200, 251, 256, 261, 266, 275, 328, 335, 357,
    362, 365
Protestant Reformation, 201
psychology:
  "blank slate" worldview, 144
  developmental stages of individuals and
    societies, 335
  see also evolutionary psychology
psychopharmacology, challenges of, 155
public revelation:
  correlation with day vs. night language,
    113–14

of cosmology, 277
definitions of, 14, 65
of evolutionary brain science, 163, 164
examples of, 73–75, 84, 142
history of relative importance of, 285–86
hopeful vision of, 316, 324
in science vs. religion, 65–72, 75, 244, 349
shifting importance of today, 330–32, 361
in truth-seeking, 197, 360
in understanding Jesus' teachings, 337
purpose, see human role; meaning-making

quadrune brain:
  affirmations for self-help support, 243
  appreciation of as salvific, 264
  blessings and burdens of, 289
  as correlated with evolutionary history, 147,
    163–64, 189
  description of, 148–54
  differing responses to music of, 237–38
  illustration of, 149
  ways of engaging for health, 157, 219–20
  see also neo-mammalian brain; paleo-
    mammalian brain; prefrontal cortex;
    reptilian brain
Qur'an, portrayal of God as sometimes cruel in,
    320–22

randomness:
  as distinct from spontaneity, 282
  distinction between fact and interpretation
    of, 110
  as faulty reason to oppose evolution, 31–32,
    122
  limited role in evolution, 37–38, 134
rape, prevalence in history of, 321
Reality (Ultimate):
  perception of as shaped by metaphors, 104–5
  as term for God or the Whole, 68, 135
REALize:
  definition of, 16, 140
  four criteria of, 16, 140, 363
  night vs. day language distinction as helpful
    for, 140
REALizing:
  the biblical story of the Fall, 141–46, 158–62,
    164, 166–68
  the body of Christ, 266, 346
  collective sin and salvation, 251–60
  core religious doctrines, 140–41
  God, 119–20, 203, 324
  God's will, 163, 184, 201, 205, 209, 229, 313, 324
  the Gospel, 200–205, 337, 344
  heaven, 123, 198, 202, 226, 262–66, 291,
    336–39, 344

hell, 185, 202, 345
holy scripture and divine revelation, 328–30, 360
Jesus as Lord, 263, 336–39
"judge not," 223–25
the Kingdom, 140, 226, 261–66, 274, 338
"love God and your neighbor," 226–28
"love your enemies," 217–21
miracles, 357–76
Original Sin, 144–47, 158, 166–68, 189
the Rapture, 205, 210, 260, 344, 360
religion, 205
"remove the plank," 221–23
the Resurrection and Ascension, 204, 260, 316, 364–67
salvation, 181–200, 210, 237, 344–46
Satan, 169–70, 176
"saving faith," 195–200
scriptural passages of God's cruelty, 322–24
speaking in tongues, 212–14
the Virgin birth, 362–64
worship, 246
recovery programs, *see* twelve-step recovery programs
Red Giant star, chemical creation within, 90, 91, 101
religion:
    conflict with science, *see* conflict religion vs. science
    as enriching science, 12–13, 347–48
    as evolving, 75–77, 310
    finding ways to evolve, 140–41, 315–16
    as human universal, 34
    as supporting systems of governance, 285
religious belief vs. knowledge, 66–67, 83, 106–7, 124–26, 129–30, 204, 210, 310, 324, 346, 350–56, 361
religious conflict, causes of, 135
religious conservatism vs. liberalism:
    as complementary, 273
    as transcended via evolutionary faith, 205, 338
religious diversity, ways to appreciate, 61, 205
religious intolerance, 135, 352, 355, 359
religious liberalism, decline of church membership in, 296
religious naturalism, 67, 130
religious nontheism, 129
reptilian brain:
    definition of, 144
    description of, 148, *149*
    evolution of, 144, 147, 148
    functions of, 144, 148
    instinctual drives of, 148, 151, 161
    role in addictions, 148

    role in sex scandals, 160–62
    see also Lizard Legacy
respect, human need for, 233–34
responsibility:
    as aspect of deep-integrity, 60, 186–89, 274, 309
    for one's behavior, 166
resurrection:
    evolutionary interpretation of, 100, 204
    see also REALizing the Resurrection and the Ascension
revelation:
    as insufficient for belief, 353–54
    as ongoing, 68–72, 87, 135, 197, 244–45, 260, 329, 341, 346
    special vs. natural, 329
    see also public revelation
Rewilding, 258–59
Rhodes, Tom, as quoted, 101
Ridley, Matt:
    as quoted, 261
    role in advancing evolutionary psychology, 275
ritual:
    as contributor to human health, 237–39
    evolutionary understanding of, 237–38
Robin, Vicki, as quoted, 239–40
Rohr, Richard, role in bringing evolution to Progressive Christians, 276
Rue, Loyal:
    as quoted, 67, 84
    role in evolutionary spirituality, 275
Rumi, as quoted, 343
Rupp, Joyce, as contributor to evolutionary spirituality, 275
Russell, Peter, role in evolutionary spirituality, 275

Sagan, Carl:
    and his *Cosmos* series, 89
    as quoted, 7, 89, 107
Sahtouris, Elisabet:
    as quoted, 248
    role in evolutionary spirituality, 275
salvation:
    through deep integrity, 338, 344–46
    as ecological sustainability, 135
    through evolutionary understanding of the Gospel, 198
    by grace through faith, 61
    interpreting the one-way doctrine, 210
    at the level of collective, 251–60
    new criteria for, 198
    process of in a megachurch, 294–95
    in this world, 73, 135, 263–64, 346
    through understanding of evolutionary brain science, 144–54, 163–95, 198, 237, 344–46

Sandburg, Carl, as quoted, 169, 171, 174, 176, 181
Satan, 1
  as name for powerful forbidden urges, 161, 182
  nonsupernatural usage of term, 169–70, 176, 260, 345
saved by grace through faith, 61
Schenk, Jim, role in evolutionary spirituality, 275
Schlesinger, William H., as quoted, 23
science:
  as based on evidence, 350
  as contributor to worldviews, 29, 69–74
  as enriching religion, 12–13, 65, 75–76, 87, 105, 316, 324, 347–48
  as Ever-Renewing Testament, 330–32
  as gift of God, 7
  as global and diverse effort, 68
  as ground for Big Picture story, 25, 288
  historical impetus for, 108, 110
  history of advance, 49–50, 65, 68, 69–72, 78–80, 110, 121–22
  importance of viewing as sacred effort, 324
  as interpreted to the glory of God, 68, 75–76
  meaningful interpretations of, 29, 68–69, 75–76, 89–102
  need for respect of traditional wisdom, 144, 189
  as providing a globally unifying sacred story, 288
  revolutions in, 80, 112, 128
  as self-correcting, 11, 26, 27, 28–29, 65, 68, 78, 105
  see also conflict religion vs. science; public revelation
scientific method:
  evolution of, 72
  soundness of, 330, 350–56
scriptural literalism:
  and conflicts over natural history sites, 66
  cultural origin of, 28
  cultural problems caused by, 67, 73, 97
  as disempowering future generations, 324
  ethical arguments against, 319–27
  ethical problems caused by, 76, 94, 97, 324
  examples of, 11, 73–74, 133
  as including horrific moral dictums, 318–27
  as limited in moral precepts, 326
  as making an idol of language, 331
  as obstacle to naturalizing death, 94, 96–97
  as obstacle to religious faith, 324, 328, 357
  role in preventing religions from evolving, 110, 328
  as trivializing God, 324
  as trivializing holy texts, 12, 166, 316, 331–32, 360

  as true to the beliefs of time and place, 106
  see also REALizing miracles
Selby, John, as teacher of meditation practices, 211
Self:
  expansion of via evolutionary worldview, 44, 86–87, 181, 187, 242, 248, 270–71, 290–93
  as expression of the Universe, 57, 58–59
  as identifying with Earth, *247,* 248
  as understood via evolutionary brain science, 181
self-acceptance affirmation for, 242
self-deception, evolutionary advantages of, 156–57, 322
self-help:
  evolutionary expansion of, 283–88, 292, 313
  via evolutionary affirmations, 241–43
  via evolutionary view, 147, 157–58, 162–65
  see also deep integrity
selfish gene theory, 172, 174, 261, 325
self-organization, 46, 121
Serenity Prayer, as quoted, 303
service:
  as aspect of deep integrity, 60, 178, 184
  as component of deep integrity, 186–89, 240, 274
  to larger holons of existence, 132, 240
  see also Higher Porpoise
sex, cellular origin of, 280
sex, nonprocreative acts of in nature, 161
sex addiction, role of reptilian brain in, 148, 150, 171, 172
sex education, evolutionary version of, 173
sex scandals, evolutionary causes of, 158–62, 195
sexual instinct:
  evolution of, 148, 170–74
  see also reptilian brain
sexuality:
  as complicated by language, 174
  future evolution of in human, 177
  when confused with need for intimacy, 231
sexual reproduction, evolution of, 69, 177
sexual selection, 46, 143
Shelley, Percy, as quoted, 288
Shermer, Michael:
  as quoted, 325, 333
  role in advancing evolutionary psychology, 275
  as skeptic of scriptural morality, 325
Shiva, Vandana, role in evolutionary spirituality, 275
sin:
  atoning for collective, 258
  of the collective, 251–60

corporate forms of, 254, 259–60
  as crimes against Creation, 254
  definition of in nested Universe, 251
  evolutionary perspective on, 166–68, 251–54,
    256
  institutional forms of, 259–60
  naturalizing the concept of Original Sin,
    144–47, 158, 166–68, 189
  our evolving understanding of, 252–54
  as portrayed through evolutionary worldview,
    158–62
  as stimulant for growth, 61
skeptics, prominent examples of, 326
social darwinism, 156, 318
sociality, as cause of new instinctual drives, 149
social systems, evolutionary understanding of,
  284–88
soil, death as generative for, 95
solar system, *see* Sun
Song, Choan-Seng, role in evolutionary
  spirituality, 275
songs:
  importance of for human health, 237–38
  lyrics illustrating evolution, 170, 367
soul, whether exists after death, 97
Southard, Mary, as quoted, 75
speaking in tongues, to quell Monkey Mind,
  152
Spencer, Herbert, 318
Spiral Dynamics, as model of development, 153,
  275
spiritual awakening, as happening today, 128
spirituality, as right relationship at all scales, 16
spiritual practices:
  core aspect of, 123
  evolutionary examples of, 210–30
  as evolving, 211
spiritual transformation, Great Story as catalyst
  for, 141
Spong, John Shelby:
  as quoted, 362, 365–66
  role in evolving Christian faith, 275, 325
Sproul, R. C., as Christian evangelist, 345
stardust:
  awareness of for awakening to Epic of
    Evolution, 89–90, 92, 93, 97–100, 288
  as celebrated in song *Woodstock*, 53, 288
  as composing our bodies, 53, 85–87, 89–93,
    101, 335
  for meaningful understanding of death, 95,
    97–98
  ritual for celebrating, 97–98
  understanding of as bridge to religion, 89–93,
    100, 347
stardust ritual, 97–98

stars:
  as comprehended by ancient peoples, 142
  ongoing birth and death of, 87–94
  as our ancestors, 85–86, 89–90, 91
  as source of complex atoms, 89–91, 133, 279
  their source of power, 53, *90*, 95
status, effect on testosterone levels, 161, 194
status-seeking:
  basic need for respect, 233
  evolutionary understanding of, 149, 151–52,
    157, 161–63, 165, 167, 169, 175, 190, 287
  as instinct of social mammals, 149, 151
stellar nucleosynthesis, creation of atoms, 101
stewardship, as limited understanding of
  human bond with Earth, 248
Stewart, John:
  his recommendation for vertical markets, 313
  as quoted, 10, 283
  role in understanding evolutionary
    directionality, 275, 283, 309
stories of awakening (to evolution), 11, 33, 35,
  44, 87, 89–90, 91, 93, 100, 163, 203, 229, 307
Story of the Universe, 2, 24–25, 50, 61, 69, 73,
  89–90, 136, 328
storytelling, evolutionary importance of, 34–35,
  72, 101, 105–6, 115, 298
stress, causes of in social mammals, 151–52
strife, as evolutionary driver, 40–41, 43, 44
subsidiarity, principle of, 287
suffering:
  biblical explanation of, 73, 96
  as caused by consumerist culture, 135
  examples of among animals, 172
  as felt by God, 106, 121
  transformed into growth, 198–99, 210, 366
Sun:
  birth of heliocentric theory, 78
  its future and lifespan, *90*, 91, 95
  its location in Universe, 87, 278
  as middle-aged average star, 91
  our intimacy with, 102
  source of its power, 91, 102
Sunderland, Jabez, as quoted, 132
supernatural, *see* naturalism vs.
  supernaturalism
supernova, as origin of heavy metals, 53, *90*, 93,
  101, 142, 279
supervolcano eruption, destructiveness of,
  303–4
Supreme Wholeness, as name for God, 123, 125,
  127, 131, 135, 163, 179, 260, 263, 282, 338
Surette, John, as quoted, 328
survival of the fittest, as term, 42
sustainable progress, how to measure, 271–72
Swaggart, Jimmy, sex scandal of, 159

Swimme, Brian:
  mention of, 3
  as quoted, 6, 21, 102, 124, 133, 288–89, 367
  role in evolutionary spirituality, 275
symbiosis:
  definition of, 42
  as evolutionary driver, 42–45, 280
  as unknown to ancient peoples, 143
synergy:
  in cultural contexts, 239–40, 312–14
  definition of, 43
Szathmáry, Eörs, as evolutionary theorist, 69

Tao, as "the Way," 263
Taoism, literalist interpretations of, 73
technology:
  cautionary approach toward, 271, 302
  for defending against asteroid impacts, 304
  effects on social evolution, 286, 287
  for increasing feedback, 310
  as increasing the damage of sinful actions,
    251, 253
  role in evolution, 70–71
  role in evolution revolution, 202
  role in paradigm shifts, 128
  role in solving environmental problems,
    312–14
  as seemingly miraculous, 367
Teilhard de Chardin, Pierre:
  as quoted, 77, 239
  role in evolutionary spirituality, 29, 275, 299
televangelists, attraction of, 294–95
Ten Commandments, as outdated basis for
  morality, 326
territorial aggression, role of reptilian brain in,
  149, 322
terrorism, official definition of, 320
testosterone, as increasing with high status,
  161–62
theism, origin of, 111, 129, 130, 131
theory, falsification of, 78
  see also fact vs. theory
time, as both cyclical and linear, 36–40
Todd, John, as contributor to ecological
  improvement, 312
Todd, Nancy Jack, as contributor to ecological
  improvement, 312
Toland, John, as originator of name for
  pantheism, 130
tongues, speaking in, 1, 212–14
touch, importance of for mammals, 231–33
Toulmin, Stephen, as quoted, 143
trees, evolution of, 36, 41, 42, 44
triune brain, *see* quadrune brain
trust, as aspect of deep integrity, 186–88, 274

Trusting the Universe:
  as accepting our own imperfections, 61–62
  advocacy of, 30, 49–57, 68
  affirmations to support, 241–43
  as aspect of enlightenment, 61, 185
  as being in action, 200
  as faith in God, 58, 124–25, 180, 185, 366
  pragmatic value of, 58–59, 61, 221, 263–64,
    304
  as surrender to what is so, 100, 124–25, 185
trusting time, 203
truth, new ways of understanding, 2, 198
  see also science, as self-correcting
tsunami, mega, 54, 304–5
tsunami of 2004, religious responses to, 22–23
Tucker, Mary Evelyn, role in evolutionary
  spirituality, 275
twelve-step recovery programs, 158, 309, 346
  helpful practices of, 221–22, 223–24

Ultimate Reality:
  God as name for, 60, 85, 86, 109, 119–23, 127,
    129, 131
  metaphors as shaping our relationship with,
    109–12, 123, 130
  value of personal metaphors for, 130
  see also Supreme Wholeness
Unitarian Universalism:
  as a liberal religious faith, 296
  as tradition that celebrates evolution, 88, 100,
    101, 123
United Church of Christ Statement of Faith, 116
Universalizing:
  core religious doctrines, 76, 141, 182, 265, 316,
    324, 335, 364
  God's will, 205
  the Kingdom of Heaven, 263
  spiritual practices, 211
  see also REALizing
Universe:
  ancient view of as divine, 109, 126
  as communion of subjects, 247
  desacralization of, 111
  history of, 24, 90, 133, 278–81, 288
  human relationship to, 89–90, 109–12, 133, 136
  increasing knowledge of, 87–88
  as like a clock, 109, 110–12, 121
  metaphors as shaping our relationship with,
    109–12, 126, *245, 246*
  as multiverse, 130
  nested emergent creativity of, 84–85, 124, 129,
    130, 316, 337
  origin of, 69, 89
  as primary revelation of God, 134
  size of, 87, 142, 251

Universe Story, 2, 24–25, 50, 61, 69, 73, 89–91, 136, 328

Vail, Tom, as author of creationist book, 66
values:
 Christ-like, 265, 274
 evolutionary parables as useful for teaching, 101, 189
 evolutionary worldview as superior for teaching, 326–27
 see also morality
Van der Ryn, Sim, as contributor to ecological improvement, 312
vision, evolution of, 38, 39
Volk, Tyler, as quoted, 27

Wallace, Emory, 116–17
war:
 mammalian examples of, 322
 religious causes of, 67, 352
Watts, Alan, as quoted, 290–91
Wessels, Cletus, role in evolutionary spirituality, 275
West, Paul, as quoted, 123, 200, 273
Westminster Catechism, 132
Whitehead, Alfred North, as quoted, 80, 105
Wiki, technology for collective intelligence, 274–75, 276
Wilber, Ken:
 his work in integral thinking, 275
 as quoted, 85, 86, 114
wild cards:
 asteroid impact, 304
 definition of, 299, 303
 disease epidemics, 306–7
 evidence of extraterrestrial life, 306
 extreme solar activity, 304
 Gulf Stream shutting down, 305–6
 mega tsunami, 304–5
 pole shift or magnetic reversal, 305
 supervolcano eruption, 303–4
Wilson, David Sloan:
 and convergent evolution, 38
 as quoted, 12, 13, 112–13, 141, 144, 162, 205, 209, 266, 287–88, 334
 role in evolutionary spirituality, 275

Wilson, Edward O.:
 and convergent evolution, 38
 as enjoying televangelists, 294–95
 as quoted, 24, 25, 34, 53, 294
 role in evolutionary spirituality, 275
Wink, Walter, as quoted, 259
witness capacity, as developed via evolutionary understanding, 164, 169, 171
World Trade Center, 3
worldview:
 as affecting environmental ethics, 127, 245
 blank slate belief about human psychology, 144
 deep-time, 75, 298–317
 evolutionary, 10, 24, 47, 63, 73–75, 84–100, 121–22, 127–32, 195, 198, 244, *245*, *246*, 290–93, 315–16
 evolution of, 283
 as expressed via metaphor, 244–47
 flat-earth, 11, 66, 73–75, 94, 142, 198, 244, *245*, *246*, 310, 349
 mechanistic, 10, 86, 110–12, 121–22, 130–31, 324
 metareligious form of, 26, 129
 nestedly creative, 84–86, 112–13, 121–22, 129, 131
 practical value of evolutionary form of, 209–10, 310, 319, 324, 337
 role in interpreting causes of evolution, 41
 as shaped by creation stories, 24
 sin as portrayed by flat-earth view, 166–67
 sin as portrayed via evolutionary view, 158–62, 251–60
Wright, Robert:
 his view of expanding cooperation, 308–9
 as quoted, 146, 158, 166, 185
 role in advancing evolutionary psychology, 163, 173, 275
writing:
 as locus of moral authority, 333
 role in cultural evolution, 285
 role in scriptural literalism, 28, 328

Yellowstone Park, as supervolcano site, 303–4
Yunus, Muhammad, as recipient of Nobel Peace Prize, 287

Grateful acknowledgment is made for permission to reprint the following copyrighted works:

Selection by Halsey Barlow. Used by permission of the author.

Excerpt from "I Don't Know That Guy" by Greg Brown. Used by permission of S.A.D. Management & Booking.

Excerpts from *Endless Forms Most Beautiful: The New Science of Evo Devo and the Making of the Animal Kingdom* by Sean B. Carroll. Copyright © 2005 by Sean B. Carroll. Used by permission of W. W. Norton & Company, Inc.

Excerpts from *The Making of the Fittest: DNA and the Ultimate Forensic Record of Evolution* by Sean B. Carroll. Copyright © 2006 by Sean B. Carroll. Used by permission of W. W. Norton & Company, Inc.

Excerpt from *Awakening Universe, Emerging Personhood: The Power of Contemplation in an Evolving Universe* by Mary Coelho. Used by permission of Wyndham Hall Press.

Selection by Edward Collins. Used by permission of the author.

"Good and Bad Reasons for Believing" from *A Devil's Chaplain* by Richard Dawkins (Houghton Mifflin, 2003). Used by permission of the author.

Excerpts from *The Symbiotic Man: A New Understanding of the Organization of Life and a Vision of the Future* by Joel de Rosnay. Copyright © Joel de Rosnay, 2000. Used by permission of The McGraw-Hill Companies.

Selections by Michael S. Earl. Used by permission of the author.

Excerpt from "Science as Epic: Can the Modern Evolutionary Cosmology Be a Mythic Story for Our Time?" by William Grassie, *Science & Spirit*, vol. 9, issue 1. Used by permission of *Science & Spirit*.

Quotation by Ken Ham from "Answers in Genesis" podcast. Used by permission of Answers in Genesis.

Selection from *Secrets from the Center of the World* by Joy Harjo (University of Arizona Press, 1989). Used by permission of the author.

Selections by Philip Hefner. Copyright © Philip Hefner, 2002. Used by permission of *Zygon: Journal of Religion and Science*.

Selections by Barbara Marx Hubbard. Used by permission of the author.

Excerpts from "Evolution of the Scientific Method" by Kevin Kelly. Used by permission of the author.

Embraced by Religious , Scientific, and Cultural Leaders

Complete versions can be found at ThankGodforEvolution.com

"Amidst the quarrels between scientists and fundamentalists, theists and atheists, liberals and conservatives, comes this wonderful book about our common journey and the stupendous story that we all share. Michael Dowd's worldview is desperately needed if we are to find a future together on this tiny little planet." —PETER MAYER, SINGER-SONGWRITER

"A wonderful tool for making meaning of our past, present, and future as a species."
—MICHAEL LINDFIELD, BUSINESS CONSULTANT; AUTHOR OF *THE DANCE OF CHANGE*

"Whatever our faith traditions and our ways of explaining the infinite majesty of the universe, this book will help us reach out to bridge some of America's most profound and most harmful divides."
—PAUL LOEB, AUTHOR OF *SOUL OF A CITIZEN*

"Michael Dowd is an indispensable new voice now brilliantly bridging the science and religion gap."
—DAVID LOYE, EVOLUTIONARY SYSTEMS SCIENTIST; AUTHOR OF *DARWIN'S LOST THEORY*

"An energizing, spiraling flow from stardust to human belief systems; transformation can happen!"
—HASITA NADAI, GEOLOGIST, CREATOR OF YOGAGAIA

"Michael Dowd is one of the most popular presenters we've had the pleasure to host at the First Unitarian Society of Madison. Now, with a book-length treatment of his 'gospel of evolution,' he provides us with a treasure trove of cogent information, practical application, and just plain inspiration. Dowd successfully meets the challenge of addressing seekers across the theological spectrum, and gives modern, secular readers a fresh perspective on traditional religious themes and usages. This is a book I would gladly (and securely) recommend, not only to members of my own humanist-inclined congregation but to my evangelical relatives as well!"
—REVEREND MICHAEL SCHULER, SENIOR MINISTER, FIRST UNITARIAN SOCIETY OF MADISON

"A bold and inspiring synthesis that can help us celebrate our differences and shape life-affirming futures, together. Yes!"
—JUANITA BROWN AND DAVID ISAACS, CO-ORIGINATORS, THE WORLD CAFÉ

"A treasure trove of insights fired by inspiration and expressed with a rare passion for truth and integrity. This is a book that could transform your life." —CHRISTIAN DE QUINCEY, PROFESSOR OF CONSCIOUSNESS STUDIES, JOHN F. KENNEDY UNIVERSITY; AUTHOR OF *RADICAL NATURE*

"*Thank God for Evolution* is a rare combination of virtues: it is passionately and deeply spiritual and reveals an amazingly good grasp of what the evolutionary sciences have discovered about our basic human nature." —KIMMO KETOLA, RESEARCHER, THE CHURCH RESEARCH INSTITUTE OF FINLAND

"Like two orphans reunited, spirituality and evolution come face-to-face in this daring new book by Michael Dowd. *Thank God for Evolution* invites us to explore new family traditions of knowledge informed by science and mystery inspired by God. As a pastor, I know the temptation may be to continue in a divided house—evolution versus religion. But Dowd skillfully redefines the terms of the relationship, creating hope and practical ways to awaken to the future of living as part of an integrated family that can tackle the challenges of today."
—SPENCER BURKE, THEOOZE.COM (EVANGELICAL), FORMER MEGACHURCH PASTOR

"Dynamite—almost literally; this deceptively low-key book will blast your mind open."
— **PAUL RAYNAULT**, FOUNDER, STUDENT WORLD ASSEMBLY

"Read this book! Then gather together a group of people from your church or place of worship and read it with them. Now is the time and here is the book to settle the evolution/religion misunderstandings."
— **LAWRENCE EDWARDS**, CHEMICAL PHYSICIST, COSMOLOGIST, EVOLUTIONARY EDUCATOR

"A timely and significant contribution to a new Living Cosmic Bible."
— **JIM CONLON**, AUTHOR OF *FROM THE STARS TO THE STREET*

"We can now realize and experience that God is here, now, that creation is ongoing, and that God is greater than the limited knowledge and expression of historic human thought and language."
— **MARK R. JONES**, CEO OF THE INTEGRAL WELLNESS GROUP AND THE SUNYATA GROUP

"Dowd shows that there are many ways to be a spiritual person and that all of them are enriched by an understanding of modern science, especially evolution. This is a creative, provocative book that sheds light on just about any spiritual path one might be on. Many will find their faith revolutionized."
— **EUGENIE C. SCOTT**, DIRECTOR OF THE NATIONAL CENTER FOR SCIENCE EDUCATION

"Dowd belongs to that tradition of religious thinkers, from Thomas Aquinas to Thomas Berry, that illuminates the Universe as revelation of the Divine. Aimed straight at the unnecessary split between Christianity and current cosmology, this significant work has revolutionary implications for our collective worldview. Michael Dowd is a voice the world needs."
— **DREW DELLINGER**, DIRECTOR OF SOCIAL ECOLOGY, JOHN F. KENNEDY UNIVERSITY; FOUNDER OF POETS FOR GLOBAL JUSTICE; AUTHOR OF *LOVE LETTER TO THE MILKY WAY*

"Bravo! For too long, science and religion have been missing what the other is truly seeking. Michael Dowd has been seeking what each of them alone has been missing. And he found it: the Great Story—a science-based creation story, a sacred epic with sufficient depth and meaning to quench our hot thirst for a deep remembering of who we truly are."
— **JIM MANGANIELLO**, FOUNDER OF MESICS TRAINING; AUTHOR OF *UNSHAKABLE CERTAINTY*

"*Thank God for Evolution* conveys the awesome grandeur of evolution and the Universe."
— **CHARLENE BROTMAN**, AUTHOR OF *THE KIDS' BOOK OF AWESOME STUFF*

"There is so much in this book to admire and to take inspiration from. Michael's gutsy reframing of Original Sin is especially welcome. The book is a timely bridge—a reconciliation not only of science and religion but of all that keeps us separate. It is above all a compassionate and masterful exploration, a prophetic vision of our evolving human nature. Michael delivers the magnificent story of the universe straight to the heart."
— **PAULINE LE BEL**, COSMOLOGICAL PLAYWRIGHT AND SINGER

"A beautiful book that embraces everyone, trumps outdated disagreements, and takes us all to the next level of hope and inspiration."
— **CINDY WIGGLESWORTH**, PRESIDENT OF CONSCIOUS PURSUITS, INC.

"A passionate, forceful, intelligent, and irenic voice in the all-too-often strident exchanges between creationists and evolutionists."
— **G. PETER SCHRECK**, PROFESSOR OF PASTORAL CARE, PALMER THEOLOGICAL SEMINARY

"Sometimes a book is more than a book. In this case it's also a virtual transpartisan table—an invitation to dialogue, with open seating, for those across a variety of divides: believers, semi-believers, and nonbelievers of all stripes and lineages, as well as aetheists, agnostics, a range of spiritual devotees and activists, and those in the worlds of science and other ways of knowing. *Thank God for Evolution* offers a historic opportunity for many Americans who rarely if ever talk to one another to do so. One can imagine Meet Ups of a whole range of folks having a new basis to exchange ideas, beliefs, passions, meanings of life in a way that has never happened before. I see this book making it possible to find new steps to fulfill America's promise."

—JOHN STEINER, CHAIR OF THE EXECUTIVE COMMITTEE OF REUNITING AMERICA

"The gospel (good news) of our times...a radically new relationship between science and spirit."

—DANA LYNNE ANDERSEN, ILLUSTRATOR OF *BORN WITH A BANG* TRILOGY

"Touches something that resonates deep within all of us."

—MILT MARKEWITZ, JEWISH RENEWAL MOVEMENT, PORTLAND, OREGON

"The great scourge of our time is not immorality, drugs, or greed. Rather, the greatest threat of our age is that so many of us believe in our hearts that life is without meaning or purpose, and thus our actions reflect this. Michael Dowd has methodically demonstrated in this book that a modern, rational, scientific worldview does not lead to nihilism. Rather, a full embrace of a scientifically based understanding of the world is key to a powerful religious faith that offers us meaning, wonder, joy, and tremendous hope for solving our most pressing global and human problems."

—SALLY BETH SHORE, UNITARIAN UNIVERSALIST CHURCH OF ASHEVILLE, NORTH CAROLINA

"With passion and wit, Michael Dowd does what very few before him have been able to accomplish—he demonstrates how it is possible to productively blend spirituality and science. In *Thank God for Evolution* Dowd draws from both science and religion and evokes a world in which evolution and creation are not at odds with each other—and he does so without compromising the basic principles of either." —MICHAEL ZIMMERMAN, BIOLOGIST, FOUNDER OF THE CLERGY LETTER PROJECT AND "EVOLUTION SUNDAY"

"This book enlightens with each turn of the page." —DANIEL DANCER, *ART FOR THE SKY*

"Few if any religious books quote Richard Dawkins in his own terms, or with approval. That this book does both hints at the breadth of audience it will intrigue. *Thank God for Evolution* has become my most quoted pastoral resource as a Christian minister. I cannot recommend it too highly."

—REVEREND JASON JOHN, SCOTS CHURCH ADELAIDE (PRESBYTERIAN), AUSTRALIA

"Michael Dowd's book, *Thank God for Evolution*, is a doorway into a new and glorious transformational moment for the history of religion as it enters into its cosmological phase."

—GAIL WORCELO, SGM, GREEN MOUNTAIN MONASTERY

"It is a bold adventure to take on evolution and religion in the same book, and Michael Dowd has done so beautifully. *Thank God for Evolution* gives us the tools to understand this great debate so that we might have an open and sane national conversation."

—DONNA ZAJONC, AUTHOR OF *THE POLITICS OF HOPE: REVIVING THE DREAM OF DEMOCRACY*

"If ever a book could successfully bridge opposing religious positions, this book can. It joyously spans not only 14 billion years, but history, science, theology, and self-help."

—TAHDI BLACKSTONE, INSTITUTE OF NOETIC SCIENCES WORLD COUNCIL

"*Thank God for Evolution* 'binds us back to origin,' which is the root meaning of the word religion." —**BERNADETTE BOSTWICK**, SGM, GREEN MOUNTAIN MONASTERY

"Nearly twenty years ago, Michael Dowd helped us awaken to the fact that we were Cosmic entities, part of a 14-billion-year process. It has never been the same since. In his new book, he integrates and weaves the best understandings and revelations of the divine in evolution and science into a deep spiritual experience, to be had by all who are seeking and awakening." —**JIM** AND **MARY JO BRAUNER**, NEW AWAKENINGS COMMUNITY

"A treasure for our time; a great resource and invitation to evolve." —**MARY SOUTHARD**, CSJ, ARTIST AND COFOUNDER OF SPIRITEARTH

"A publishing event of rare significance, this exhilarating book moves us at last beyond false separations of science and religion. With true evangelical fervor, it liberates both heart and mind in service to the great adventure of our time: the rediscovery of our mutual belonging and our kinship with all life. With breathtaking clarity, Michael Dowd not only treats us to a panoramic vision of our origins, but reveals how this Great Story can grace our lives right now with courage, creativity, and peace of mind. Such gifts are not only our birthright, they are essential now for our survival." —**JOANNA MACY**, AUTHOR OF *WORLD AS LOVER, WORLD AS SELF*

"A beacon of light destined to become a classic." —**PAUL** AND **LAYNE CUTRIGHT**, AUTHORS OF *STRAIGHT FROM THE HEART*

"This book is a gift that far exceeds the evolution vs. God debate. It is for anyone who wants to feel hopeful that we can and will successfully evolve as a human species." —**DAVID GERSHON**, FOUNDER AND CEO OF EMPOWERMENT INSTITUTE

"So many today are starving for a meaningful, timely spirituality. *Thank God for Evolution* will help you to celebrate that the Creator (however you understand that mystery) has used an evolutionary mode of creating this magnificent universe and invites each of us to be 'co-creators' at this crucial time in Earth's history." —**PAULA GONZALEZ**, SC, UNIVERSE STORY PRESENTER AND SOLAR ACTIVIST

"Opens the space for a true reconciliation; can help us evolve into a wisdom culture." —**ADIN ROGOVIN**, FACILITATOR, CO-INTELLIGENCE INSTITUTE

"*Thank God for Evolution* is a heroic accomplishment. Not only does it succeed in blazing new trails between the heart and the mind; it does so with the exciting, seemingly dangerous, soul force of an Old Testament prophet. No idle entertainment here—this book fills you with wonder and changes your life!" —**JESSICA RICE**, MUSICIAN/EDUCATOR

"A timely and healing gift to the world." —**LAUREN DE BOER**, PAST EDITOR OF *EARTHLIGHT MAGAZINE*

"That evolution is a divine force has not been well appreciated, or articulated, up till now. But through the efforts of this brave book and others inspired by it, this realization is certain to be more widely embraced in the coming decades." —**KEVIN KELLY**, FOUNDING EDITOR OF *WIRED MAGAZINE*; AUTHOR OF *OUT OF CONTROL*

"For Catholics who are worried that evolution has nothing to offer them but random Godlessness, this is the book to get. Mind you, it may sometimes be challenging. Traditional concepts, like 'original sin,' get re-expressed with new words, like 'our lizard legacy.' But the courageous will discover the same joy, energy, and spirituality that fill the author as he explores an evolutionary view of God and Creation. Take courage—and read it!"
—CHRIS CORBALLY, SJ, VICE DIRECTOR OF THE VATICAN OBSERVATORY

"A landmark contribution; a vision of the infinite calling each one of us to fulfill our destiny."
—RABBI MARC GAFNI, AUTHOR OF MYSTERY OF LOVE AND SOUL PRINTS

"Deftly transcends the raging battle between science and religion; an integral masterstroke!"
—SUSAN CANNON, INTEGRAL EDUCATOR AND FUTURIST

"As Dowd makes clear, even atheists are left in a glowing state of spirituality once the awe-inspiring wonder of the Great Story is understood. Thank evolution for Michael Dowd!"
—MICHAEL SCHACKER, AUTHOR OF 21ST CENTURY TRANSFORMATION

"With creativity and enthusiasm, Michael Dowd skillfully presents an ecological theology emerging out of the new cosmology. He makes evolution and complex scientific concepts more understandable, while making the central tenets of Christianity a little more acceptable and a lot more hopeful. This book is a significant contribution to revisioning Christian faith within the Universe story."
—FATHER JOSEPH MITCHELL, CP, DIRECTOR OF THE PASSIONIST EARTH & SPIRIT CENTER

"A joy and inspiration to read, Michael's visionary words lay to rest the tired, if not fraudulent, argument that there's a conflict between faith and science." —JIM SCOTT, SINGER-SONGWRITER

"Dowd shows us that the evolutionary perspective can be helpful to any human being seeking to live in integrity." —SUSAN CAMPBELL, AUTHOR OF GETTING REAL: 10 TRUTH SKILLS YOU NEED TO LIVE AN AUTHENTIC LIFE

"This is a book to read and reread, but also to share with others and discuss."
—JIM SCHENK, EDITOR OF WHAT DOES GOD LOOK LIKE IN AN EXPANDING UNIVERSE?

"Oh my God, Michael, you've done it! Your book accesses the entire universe of compassion through honoring evolution as divine creativity. By your magnificent portrayal of our Quadrune Brain (Chapter 9), you opened a door to meeting Jesus' mandate to 'love your neighbor as yourself.' This book 'cracks the code' of self-deception and honors all of who we are. Suddenly, the enigma of the human condition makes sense!" —KAREN KUDEBEH, FOUNDER OF TIMETRACE, INC.

"Michael Dowd's marvelous use of scientific information, religious metaphor, humor, and sheer delight in his subject(s) carries us to new places where evolution and the deep stories of religion converge. His book is a gift for our time."
—REVEREND LAUREL HALLMAN, SENIOR MINISTER, FIRST UNITARIAN CHURCH OF DALLAS